Pesticides and agriculture

Profit, politics and policy

It is widely recognised that agriculture is a significant contributor to global warming and climate change. Agriculture needs to reduce its environmental impact and adapt to current climate change whilst still feeding a growing population, i.e. become more 'climate-smart'. Burleigh Dodds Science Publishing is playing its part in achieving this by bringing together key research on making the production of the world's most important crops and livestock products more sustainable.

Based on extensive research, our publications specifically target the challenge of climate-smart agriculture. In this way we are using 'smart publishing' to help achieve climate-smart agriculture.

Burleigh Dodds Science Publishing is an independent and innovative publisher delivering high quality customer-focused agricultural science content in both print and online formats for the academic and research communities. Our aim is to build a foundation of knowledge on which researchers can build to meet the challenge of climate-smart agriculture.

For more information about Burleigh Dodds Science Publishing simply call us on +44 (0) 1223 839365, email info@bdspublishing.com or alternatively please visit our website at www.bdspublishing.com.

Related titles:

Critical issues in plant health: 50 years of research in African agriculture
Print (ISBN 978-1-78676-232-0); Online (ISBN 978-1-78676-234-4, 978-1-78676-235-1)

Integrated disease management of wheat and barley
Print (ISBN 978-1-78676-216-0); Online (ISBN 978-1-78676-218-4, 978-1-78676-219-1)

Integrated weed management for sustainable agriculture
Print (ISBN 978-1-78676-164-4); Online (ISBN 978-1-78676-166-8, 978-1-78676-167-5)

Chapters are available individually from our online bookshop:
https://shop.bdspublishing.com

climate•SMART•publishing
IN AGRICULTURAL SCIENCE

BURLEIGH DODDS SERIES IN AGRICULTURAL SCIENCE
NUMBER 67

Pesticides and agriculture

Profit, politics and policy

Dave Watson

burleigh dodds
SCIENCE PUBLISHING

Published by Burleigh Dodds Science Publishing Limited
82 High Street, Sawston, Cambridge CB22 3HJ, UK
www.bdspublishing.com

Burleigh Dodds Science Publishing, 1518 Walnut Street, Suite 900, Philadelphia, PA 19102-3406, USA

First published 2018 by Burleigh Dodds Science Publishing Limited

Library of Congress Control Number: 2018949494

British Library Cataloguing in Publication Data
A catalogue record for this book is available from the British Library

ISBN 978-1-78676-276-4 (Print)
ISBN 978-1-78676-279-5 (PDF)
ISBN 978-1-78676-278-8 (ePub)
ISSN 2059-6936 (print)
ISSN 2059-6944 (online)

DOI 10.19103/AS.2018.0050

Typeset by Deanta Global Publishing Services, Chennai, India
Printed by Lightning Source

Contents

Series list

Title	Series number
Achieving sustainable cultivation of maize - Vol 1 From improved varieties to local applications Edited by: Dr Dave Watson, CGIAR Maize Research Program Manager, CIMMYT, Mexico	001
Achieving sustainable cultivation of maize - Vol 2 Cultivation techniques, pest and disease control Edited by: Dr Dave Watson, CGIAR Maize Research Program Manager, CIMMYT, Mexico	002
Achieving sustainable cultivation of rice - Vol 1 Breeding for higher yield and quality Edited by: Prof. Takuji Sasaki, Tokyo University of Agriculture, Japan	003
Achieving sustainable cultivation of rice - Vol 2 Cultivation, pest and disease management Edited by: Prof Takuji Sasaki, Tokyo University of Agriculture, Japan	004
Achieving sustainable cultivation of wheat - Vol 1 Breeding, quality traits, pests and diseases Edited by: Prof. Peter Langridge, The University of Adelaide, Australia	005
Achieving sustainable cultivation of wheat - Vol 2 Cultivation techniques Edited by: Prof. Peter Langridge, The University of Adelaide, Australia	006
Achieving sustainable cultivation of tomatoes Edited by: Dr Autar Mattoo, USDA-ARS, USA & Prof. Avtar Handa, Purdue University, USA	007
Achieving sustainable production of milk - Vol 1 Milk composition, genetics and breeding Edited by: Dr Nico van Belzen, International Dairy Federation (IDF), Belgium	008
Achieving sustainable production of milk - Vol 2 Safety, quality and sustainability Edited by: Dr Nico van Belzen, International Dairy Federation (IDF), Belgium	009
Achieving sustainable production of milk - Vol 3 Dairy herd management and welfare Edited by: Prof. John Webster, University of Bristol, UK	010
Ensuring safety and quality in the production of beef - Vol 1 Safety Edited by: Prof. Gary Acuff, Texas A&M University, USA & Prof. James Dickson, Iowa State University, USA	011
Ensuring safety and quality in the production of beef - Vol 2 Quality Edited by: Prof. Michael Dikeman, Kansas State University, USA	012
Achieving sustainable production of poultry meat - Vol 1 Safety, quality and sustainability Edited by: Prof. Steven C. Ricke, University of Arkansas, USA	013
Achieving sustainable production of poultry meat - Vol 2 Breeding and nutrition Edited by: Prof. Todd Applegate, University of Georgia, USA	014

Achieving sustainable production of poultry meat - Vol 3 015
Health and welfare
Edited by: Prof. Todd Applegate, University of Georgia, USA

Achieving sustainable production of eggs - Vol 1 016
Safety and quality
Edited by: Prof. Julie Roberts, University of New England, Australia

Achieving sustainable production of eggs - Vol 2 017
Animal welfare and sustainability
Edited by: Prof. Julie Roberts, University of New England, Australia

Achieving sustainable cultivation of apples 018
Edited by: Dr Kate Evans, Washington State University, USA

Integrated disease management of wheat and barley 019
Edited by: Prof. Richard Oliver, Curtin University, Australia

Achieving sustainable cultivation of cassava - Vol 1 020
Cultivation techniques
Edited by: Dr Clair Hershey, formerly International Center for Tropical Agriculture (CIAT), Colombia

Achieving sustainable cultivation of cassava - Vol 2 021
Genetics, breeding, pests and diseases
Edited by: Dr Clair Hershey, formerly International Center for Tropical Agriculture (CIAT), Colombia

Achieving sustainable production of sheep 022
Edited by: Prof. Johan Greyling, University of the Free State, South Africa

Achieving sustainable production of pig meat - Vol 1 023
Safety, quality and sustainability
Edited by: Prof. Alan Mathew, Purdue University, USA

Achieving sustainable production of pig meat - Vol 2 024
Animal breeding and nutrition
Edited by: Prof. Julian Wiseman, University of Nottingham, UK

Achieving sustainable production of pig meat - Vol 3 025
Animal health and welfare
Edited by: Prof. Julian Wiseman, University of Nottingham, UK

Achieving sustainable cultivation of potatoes - Vol 1 026
Breeding improved varieties
Edited by: Prof. Gefu Wang-Pruski, Dalhousie University, Canada

Achieving sustainable cultivation of oil palm - Vol 1 027
Introduction, breeding and cultivation techniques
Edited by: Prof. Alain Rival, Center for International Cooperation in Agricultural Research for Development (CIRAD), France

Achieving sustainable cultivation of oil palm - Vol 2 028
Diseases, pests, quality and sustainability
Edited by: Prof. Alain Rival, Center for International Cooperation in Agricultural Research for Development (CIRAD), France

Achieving sustainable cultivation of soybeans - Vol 1 029
Breeding and cultivation techniques
Edited by: Prof. Henry T. Nguyen, University of Missouri, USA

Achieving sustainable cultivation of soybeans - Vol 2 030
Diseases, pests, food and non-food uses
Edited by: Prof. Henry T. Nguyen, University of Missouri, USA

Achieving sustainable cultivation of sorghum - Vol 1 031
Genetics, breeding and production techniques
Edited by: Prof. William Rooney, Texas A&M University, USA

Achieving sustainable cultivation of sorghum - Vol 2 032
Sorghum utilization around the world
Edited by: Prof. William Rooney, Texas A&M University, USA

Achieving sustainable cultivation of potatoes - Vol 2 033
Production, storage and crop protection
Edited by: Dr Stuart Wale, Potato Dynamics Ltd, UK

Achieving sustainable cultivation of mangoes 034
Edited by: Professor Víctor Galán Saúco, Instituto Canario de Investigaciones Agrarias (ICIA), Spain & Dr Ping Lu, Charles Darwin University, Australia

Achieving sustainable cultivation of grain legumes - Vol 1 035
Advances in breeding and cultivation techniques
Edited by: Dr Shoba Sivasankar et al., formerly International Crops Research Institute for the Semi-Arid Tropics (ICRISAT), India

Achieving sustainable cultivation of grain legumes - Vol 2 036
Improving cultivation of particular grain legumes
Edited by: Dr Shoba Sivasankar et al., formerly International Crops Research Institute for the Semi-Arid Tropics (ICRISAT), India

Achieving sustainable cultivation of sugarcane - Vol 1 037
Cultivation techniques, quality and sustainability
Edited by: Prof. Philippe Rott, University of Florida, USA

Achieving sustainable cultivation of sugarcane - Vol 2 038
Breeding, pests and diseases
Edited by: Prof. Philippe Rott, University of Florida, USA

Achieving sustainable cultivation of coffee 039
Edited by: Dr Philippe Lashermes, Institut de Recherche pour le Développement (IRD), France

Achieving sustainable cultivation of bananas - Vol 1 040
Cultivation techniques
Edited by: Prof. Gert Kema, Wageningen University, The Netherlands & Prof. André Drenth, University of Queensland, Australia

Global Tea Science 041
Current status and future needs
Edited by: Dr V. S. Sharma, formerly UPASI Tea Research Institute, India & Dr M. T. Kumudini Gunasekare, Coordinating Secretariat for Science Technology and Innovation (COSTI), Sri Lanka

Integrated weed management 042
Edited by: Emeritus Prof. Rob Zimdahl, Colorado State University, USA

Achieving sustainable cultivation of cocoa 043
Edited by: Prof. Pathmanathan Umaharan, Cocoa Research Centre – The University of the West Indies, Trinidad and Tobago

Robotics and automation for improving agriculture 044
Edited by: Prof. John Billingsley, University of Southern Queensland, Australia

Water management for sustainable agriculture 045
Edited by: Prof. Theib Oweis, ICARDA, Jordan

Improving organic animal farming 046
Edited by: Dr Mette Vaarst, Aarhus University, Denmark & Dr Stephen Roderick, Duchy College, Cornwall, UK

Introduction

This book aims to unreveal the history of both agricultural pesticides and the industry that evolved around them. The book is divided into chapters that review discreet periods in history within which pesticides, agriculture and human societies evolved.

Each chapter is structured in a similar way to both facilitate comparisons between chapters and provide a familiar format to aid digestibility. Chapters begin with an outline of the key characteristics of the farming systems, which sets the background against which pesticides evolved. This is followed by an overview of the key external drivers that catalysed the development of pesticides. Next the reader is provided with the key characteristics of pesticide development processes and the manufacturers who began producing pesticides. Detailed analysis is provided for the most important pesticides and pesticide manufacturers and the processes in which pesticides came into, and remained in, common usage. The concluding section of each chapter focuses on the key outcomes of pesticide development and application. This includes both intended outcomes such as increased agricultural production and unintended outcomes linked to environmental pollution and human health.

Chapter 1 discusses the evolution of agriculture and the early development of the pesticide industry through to the early twentieth century. The chapter describes the evolution of agriculture from traditional to revolutionary and international phases in the face of forces such as urbanisation and the increased demand for food, and the increase in the global food trade. It reviews the consequent intensification of agricultural production, in which the emerging pesticides industry played an increasingly important role.

Chapter 2 moves onto discuss developments in the 'Productivist Period' from the 1930s to the 1970s, influenced by Fordist models of mass production. Supported by governments concerned about food security, agriculture in developed countries became even more intensive, reliant on mechanisation and extensive use of chemical inputs such as fertilisers and pesticides to achieve dramatic increases in production. This period was also one of major expansion of pesticide industry that became increasingly embedded both in agriculture itself and the broader policy/regulatory framework around agriculture, building on the development and widespread use of synthetic organic pesticides such as organophosphates, carbamates and pyrethroids, as well as a new generation of herbicides and fungicides.

The next group of chapters are unusual in that they cover different aspects of the same period: 1974–2017. Chapter 4 sets the scene for this 'Divergent Period' which saw the collapse of the Productivist paradigm and what has been called the Second Food Regime. Chapters 4–7 look at different responses to this crisis, from Residual Productivism and Neo-Productivism to Post-Productivism and the emergence of a new 'Sustainability Paradigm' for agricultural production. In each case, the chapters chart the various ways the pesticide industry sought to manage and adapt to a more challenging environment, whether by developing new products in such areas as biotechnology. broadening their global operations, seeking to influence debate about the role of pesticides or repositioning themselves through horizontal and vertical mergers. The final chapter speculates as to how the global food system is likely to evolve during the next 10–15 years and the role that science and technology, especially pesticides and biotechnologies, may play.

Chapter 1

The evolution of the pesticide industry in the Pre-Productivist Period (2500 BC–1929 AD)

1.1 Introduction

The Pre-Productivist Period is described as the extensive phase of pesticide and pesticide industry development from the earliest recorded history of pesticide use through to 1929. The Period itself can be subdivided into three phases: Traditional (2500 BC–1749 AD); Revolutionary (1750–1846) and International (1847–1929). Each phase is demarcated by the changing social, political and economic context of pesticide development, as well as by the characteristics of crop protection agents and their manufacturers.

1.2 Traditional Phase (2500 BC–1749 AD)

1.2.1 Key characteristics of farming systems

To understand the evolution of pesticides in the 'Traditional Phase', it is important to understand their evolution in the broader context of farming systems and the even wider context of human economic, social and political development. At the beginning of this Phase, farming systems are characterised by the input extensive, subsistence and small-scale market-based production of a wide range of crop plants and animals. By this time, sedentary agriculture had become the favoured food security strategy for much of the world's population and food was principally consumed at or near its point of production (Atkins and Bowler 2001). Whilst there are no recorded examples of crop breeding, it is likely that farmers would have chosen the next season's seeds from either the healthiest or surviving crop plants and therefore inadvertently selected seed germplasm with at least some pest and disease resistance.

1.2.2 Key external drivers of pesticide evolution

1.2.2.1 Increasing food demand

Concomitant with population growth, increasing food demand was the principal driver associated with the development and application of crop protection agents during this Phase (see Fig. 1.1). Increased food demand, which was manifest through local and national and latterly from international markets, was underpinned by population growth, particularly in Europe. In 3000 BC, the global population was estimated to be between 14 000 000 (Census 2003) and 25 000 000 (ATT 2003), increasing to between 629 000 000 and 961 000 000 by 1750 (Census 2003).

1.2.2.2 Increasing knowledge of agricultural production and agroecosystems

Even though this Phase is characterised by a relatively limited understanding of the basic principles of agricultural production and agroecosystems, it was during this Phase that farmers slowly began to associate the presence of plant 'pests' (particularly weeds and insects) with the likelihood of lower yields. The capacity to associate plant pests with the likelihood of reduced yield gave farmers an incentive to use pesticides. Farmers used pesticides in the hope that they would reduce the negative impact of the pest and leave more agricultural produce. In turn, this would enhance the household food security of subsistence farmers and/or provide a larger surplus of agricultural produce for farmers to sell or barter in the market place (see Fig. 1.1).

1.2.2.3 Political and economic intervention by sovereign governments

The later part of this Phase witnessed the first sustained political intervention in agricultural markets. In 1689, British farming and political interests secured protection from cheap cereal imports through the imposition of import tariffs underwritten by a series of parliamentary Corn Laws (1689–1846). Ironically, the Corn Laws did not demonstrate a growing concern by the British Government for the welfare of its population but instead demonstrated a realisation of the growing capacity of food production to generate considerable wealth and power for those that controlled the key factors of production (land, labour and capital). The enclosed brackets fail to include an equally important facet of production, that of the power to manipulate the political and economic context of

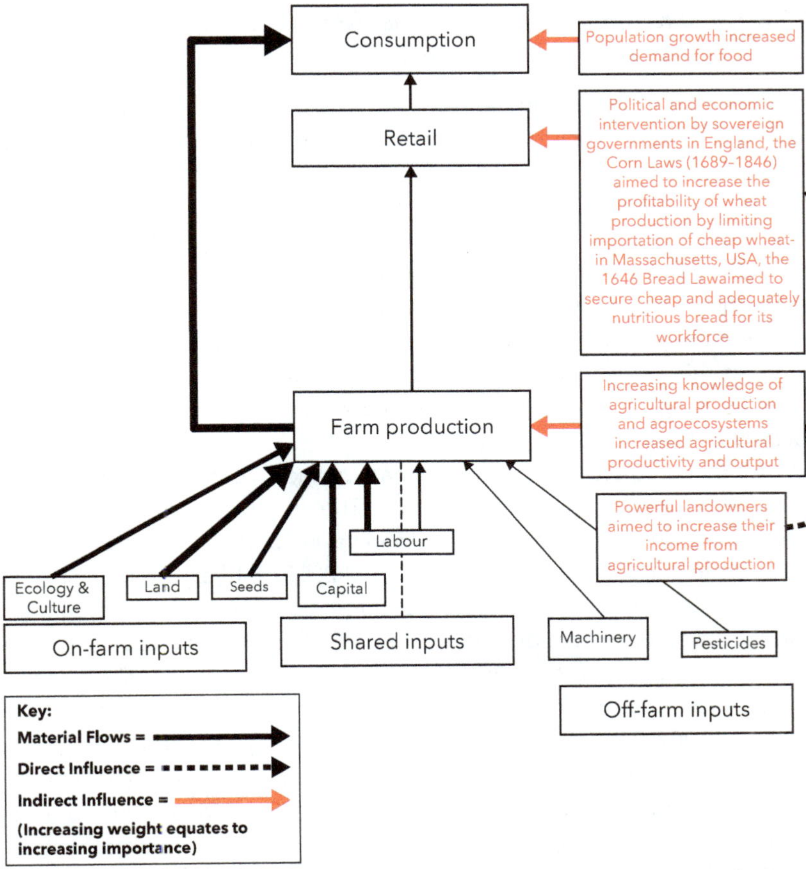

Figure 1.1 Food system dynamics during the Traditional Phase (2500 BC–1749 AD) of pesticide and pesticide industry development.

food production. Namely, when the Corn Laws were passed, and throughout the period that they remained operable, members of the British Government were landowners themselves or were in some manner connected to food production and therefore directly benefited from the considerable financial rewards that ensued from this legislation (see Fig. 1.1). Political concern for agriculture and food also manifested itself in the USA. In 1646, Massachusetts passed the Bread Law, which set a price for a loaf of bread at one penny and decreed how much it should weigh. However, while this legislation sought to limit the exploitation of the poorest in the American society by directly regulating the quality and price of bread, it is likely that the principal motive for this intervention was to secure the provision of an affordable and nutritious staple food for America's workforce (see Fig. 1.1).

1.2.3 Key characteristics of pesticide development processes and manufacturers

Plant protection strategies during the Traditional Phase of pesticide development were a mixture of acute observation (through trial and error) and folklore, superstition and religion. Indeed, during this Phase, the Romans were attributed with using magic rituals to protect vines from moth attack, including wiping the pruning knife with beaver skin prior to use (Ordish 1976). In France, it was common to excommunicate caterpillars, or to convict grasshoppers in court for the crime of attacking crops (Ordish 1976). However, while folklore, superstition and religion continued to be used as crop protection strategies throughout this Phase, latterly, the use of crop protection agents based on either basic inorganic elements and/or compounds or naturally occurring plant extracts (botanicals) became more popular.

Early inorganic crop protection agents include compounds of sulphur, mercury, lead, arsenic, sodium and fluorine. The first documented application of crop protection agents was by the Sumerians in 2500 BC, who used sulphur compounds to control insects and mites (CABI 2003). Biblical armies were also accredited with the use of non-selective herbicides when they used salt and ash on the fields of the conquered in 1200 BC. The Chinese also used compounds of mercury, lead, arsenic and fluoride as early as 1200 BC, to control body lice. In Europe, Homer refers to the use of sulphur to avert pests and diseases in 1000 BC (Hough 1998) and bitumen boiling was used to rid grapes leaves of pests in the Roman Empire (Boardman 1986). In AD 70, Pliny the Elder recommended the use of arsenic to kill insects in Southern Europe. By the 1500s, the Chinese were using arsenic sulphide to control insects (Pimentel et al. 1993). Later, in France (1600), mercury (mixed with meat) was used as a rat poison (Estienne and Liebault 1600), and, in England (1659), mercuric chloride was used for the poisoning of rooks, crows, jackdaws and magpies (Speed 1659). In 1660, brine was used as a treatment on wheat seeds to prevent mould (Pimentel and Pimentel 1979). By the late seventeenth century, the use of arsenical compounds as bird and ant baits was well established in Northern Europe (Pimentel and Pimentel 1979).

The earliest documented use of botanical crop protection agents relates to the use of hellebore by the Romans and Greeks to control rats, mice and insects. Rotenone, a botanical derived from the roots of the derris plant, was used to paralyse fish in South America from as early as 1649, and to control insect pests in Asia since at least the 1840s (Pesticide News 2001). Nicotine, derived from extracts of tobacco leaves, was first introduced into Europe during the seventeenth century, where it was used to control plum curculio and the lace bug. Pyrethrum was used in the seventeenth century against human pests (Boardman 1986).

1.2.4 Key outcomes of pesticide development and application

Throughout the Traditional Phase, an imprecise and often total lack of understanding of plant protection resulted in the application of non-selective, persistent, toxic, and, in many cases, ineffectual crop protection agents (Hough 1998). Occasionally, however, crop protection agents did confer an advantage to farmers by helping the farmer attain higher and stable crop yields and prevent post-harvest losses. In turn, the production of food surpluses on the farm helped to ensure farm household food security. Moreover, when surpluses were sold or traded on local markets, material well-being was improved. However, throughout this Phase, the use of crop protection agents remained the exception rather than the rule. Other facets of production remained far more important to the farmer than off-farm inputs (such as machinery and pesticides), namely on-farm characteristics such as soil fertility, pH, texture, structure, depth, landscape topography and rainfall (see Fig. 1.1).

1.3 Revolutionary Phase (1750-1846)

1.3.1 Key characteristics of farming systems and broader economic, social and political contexts

The Revolutionary Phase corresponds to the period of agricultural and industrial revolution in Europe. The start of the Revolutionary Phase differed from the beginning of the Traditional Phase in four important respects: (1) Increased population levels and concomitant level of food demand. (2) Increased importance of small- and medium-scale market-based agricultural production compared to subsistence agriculture. (3) Increased scientific understanding of the basic principles of agricultural production and agroecosystems. (4) Increased intensity of agricultural production, including the use of crop protection agents. The capacity of farmers and farming systems, in the latter part of the Traditional Phase, to provide marketable food surpluses enabled the balance between market-based agricultural production and subsistence production to shift, particularly in Europe. Agricultural surpluses enabled those previously involved in agriculture to undertake non-agricultural employment in the growing villages and towns across Europe and to purchase their food from the marketplace.

1.3.2 Key external drivers of pesticide evolution

Four important drivers promoted the development and application of crop protection agents throughout the Revolutionary Phase. (1) Population growth and the increased demand for food. (2) Rapid urbanisation. (3) Political and economic intervention by sovereign governments. (4) Public and private sponsorship of agricultural research, development and extension.

1.3.2.1 Population growth and the increased demand for food

According to some of the most conservative estimates of global population growth, the world population almost doubled between 1750 and 1846 (Census 2003). Continued population growth, particularly in Europe and its colonies, and the concomitant increased demand for food were principal drivers for pesticide evolution throughout this Phase (see Fig. 1.2). Increased food demand, which was increasingly manifested through buoyant local, national and international markets, provided farmers with the financial incentive to produce food surpluses.

1.3.2.2 Rapid urbanisation

Urbanisation, referring to the outflow of people from areas of low population densities to areas of high population densities, is a structural tendency in social and economic development that has been evident since the beginning of sedentary agriculture.

However, it was the Industrial Revolution which promoted unprecedented levels of urbanisation and employment opportunities outside the agricultural sector in Europe (see Fig. 1.2). Industrialisation is a process whereby industrial activity comes to play a dominant role in the economy of a nation or region. It refers to the augmentation or replacement of small-scale production for either personal use or a limited local market by a type of activity characterised by a much larger scale of productive unit and by mechanisation. Evolving industries were either located in existing urban areas to tap the growing abundance of labour or because of other structural factors such as transport infrastructure and natural resources.

Combined with rapid population growth and the physical demand for more food, the considerable exodus from rural to urban areas meant that, aside from urban gardens, urban populations were no longer involved in the production of their own food and became increasingly reliant on their domestic agricultural sector or on foreign imports to supply their food needs. High levels of population growth and urbanisation thus created a seemingly overwhelming demand for enhanced agricultural productivity and/or a significant expansion of farmed areas.

1.3.2.3 Political and economic intervention by sovereign governments

Political and economic intervention in agriculture, which began in the late part of the Traditional Phase, continued throughout the Revolutionary Phase (see Fig. 1.2). The rationale behind this intervention was to harness agriculture as an engine of economic growth and generator of foreign exchange that would

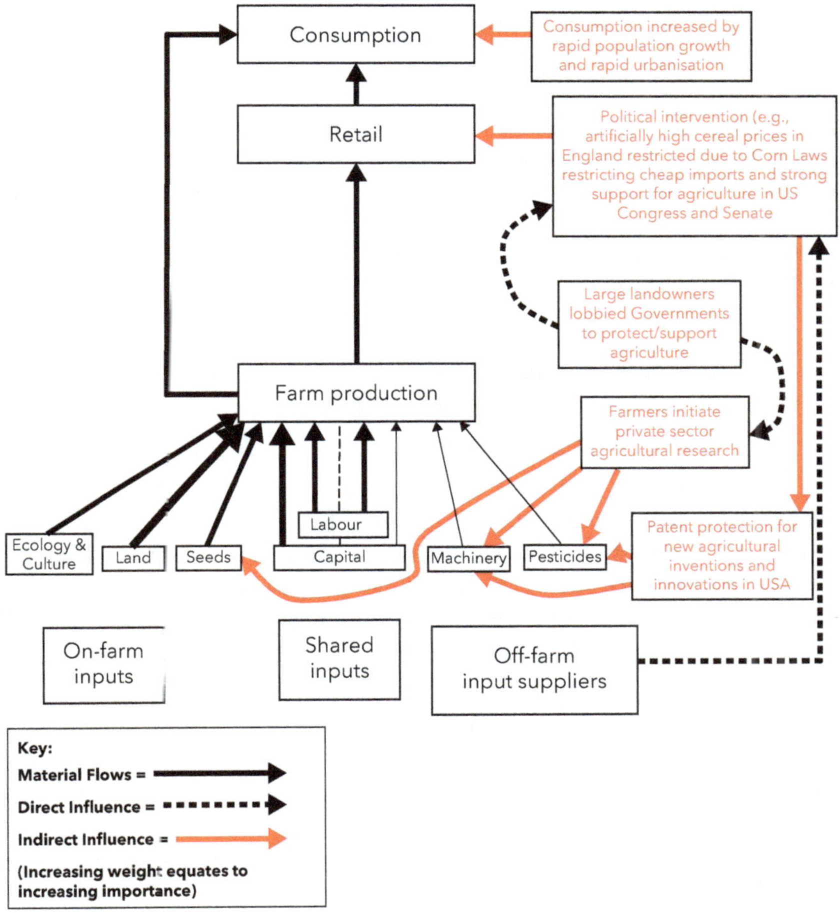

Figure 1.2 Food system dynamics during the Revolutionary Phase (1750–1846) of pesticide and pesticide industry development.

promote social stability at the same time as securing significant profits for vested interests. In Britain, the landowning and politically elite continued to ensure the profitability of agriculture through maintenance of the Corn Laws (1689–1846) (NUS 2003). In the USA, George Washington developed an interest in the political and economic importance of agriculture as early as 1766, when he suggested that Congress should establish a National Board of Agriculture (USDA 2003). However, it was not until 1819 that agriculture, as a political interest, was firmly entrenched (institutionalised) as a core, and durable, facet of American society through the formation of the New York State Board of Agriculture. Agriculture's growing importance to the national interest quickly assured its committee-level status in the House of Representatives in 1820 and in the US

Senate in 1825. In 1839, the US Patent Office was created in State Department in order to ensure potential income streams from new agricultural inventions and innovations. As early as 1785, land transfers, purchases and treaties added territory to the USA, and policies were formulated to encourage the movement of the population and industry to fill the space, creating numerous family farms. Land distribution policies culminated in the Homestead Act of 1862, which provided free distribution of land to anyone who would settle and farm it. In a sense, US policy developments in this period laid down the foundation for public policy towards the agricultural sector.

1.3.2.4 Public and private sector sponsorship of agricultural research, development and extension

This Phase witnessed a rapid increase in the scientific understanding of the basic principles of agricultural production and agroecosystems. The embrace of science evolved through the process of 'Enlightenment' at the end of the eighteenth century. The Enlightenment Period can be characterised by the power of reason over ignorance, the power of order over disorder and the power of science over superstition. Application of this knowledge and the ensuing improvement in farmer productivity became known as the Agricultural Revolution (see Fig. 1.2). During this Phase, cause and effect became increasingly apparent and the first books and papers devoted entirely to pest control, including cultural, biological and chemical methods, were produced (Dent 2003). Farmers began to develop a basic understanding of pest identification, biology and ecology, plant nutrition and crop rotation. However, even though plant protection grew increasingly scientific, trial and error remained the principal route of discovery and practice.

Initially, interest in agricultural research, development and extension came from the private sector. This was primarily in the form of experimentation by gentleman farmers and the establishment of numerous agricultural societies devoted to the betterment of agriculture and profit. In 1843, the world's first agricultural research station was established at Rothamsted in the UK (Rothamsted 2003). In the USA, an increasing number of progressive farmers, journalists, educators and producers of commercial farm inputs began to focus on the improvement of farm productivity and profitability. In the 1820s, farmers began to organise themselves into state and county agricultural societies and to promote the need for specialised training and scientific research to advance the productivity and professionalism of the industry (Agriculture Outlook 2000). Much of the support for research and training came from older farming regions of the South and East, which had begun to suffer from competition with newly opened lands on which staples like wheat, cotton and livestock could be produced more cheaply. This ultimately forced farmers in older, settled regions to evaluate their production methods. Indeed, years of cultivation without attention to preserving fertility of the soil had led to falling yields and even the abandonment of land, particularly in areas growing cotton and tobacco. Farmers increasingly looked to scientific research, development and extension for improved crop and soil management technologies. In Britain and the USA, a growing preponderance of horticultural and agricultural husbandry literature provided potential recipes for higher net yields, increased profit and, in turn, greater use of crop protection agents.

1.3.3 Key characteristics of pesticide development processes and manufacturers

Even though simple inorganic and botanical compounds continued to be used, this Phase witnessed the development of more complex pesticides, particularly insecticides. By the late 1700s, a combined wash of tobacco, sulphur and unslaked lime was recommended for the control of insects, as well as fungi, on a range of horticultural crops, and arsenic was used for killing flies and other insects (Duke 1990). By the early 1800s, pesticidal soap was recommended for the control of aphids, alum (hydrated potassium aluminium sulphate) dissolved in tobacco liquor was recommended for caterpillar control on gooseberry bushes and sulphuric acid was recommended for the control of slugs and other surface insects (Duke 1990). As crop protection agents were expensive to buy, they were dusted, sprinkled or syringed onto the plants. Because of their cost, they were generally only applied to high-value horticultural crops (Bailey 1909).

In addition to the development of insecticides, this Phase also witnessed the development of the first selective herbicides, copper sulphate and ferrous sulphate (Bailey 1909). Copper sulphate and ferrous sulphate were recommended for use as cereal seed steeps, used to inhibit the germination of weed seeds. This was done despite concerns that copper may also adversely affect wheat seed germination. In addition, both herbicides were applied as selective contact herbicides but were not widely used due to their poor efficacy and expense (Bailey 1909).

As in the case of insecticides, early fungicides were predominantly used on garden and field vegetables because of their expense and high labour intensity (Bailey 1909). One of the few exceptions was the use of inorganic fungicides for protection against bunt and smuts in cereal seed steeps, which became very popular in both Europe and America during this Phase. These steeps often included alum, salt and lime, saltpetre (Mortimer 1721), arsenic, copper sulphate (Holmes 2001), copper acetate, sulphuric acid and, increasing, formaldehyde (Bailey 1909). Several new chemicals (including mercuric chloride) also demonstrated fungicidal properties but were often either too toxic or too expensive for use on cereals (Holmes 2001).

1.3.4 Key outcomes of pesticide development and application

As with the Traditional Phase of pesticide evolution, the use of crop protection agents in the Revolutionary Phase, whilst still modest (see Fig. 1.2) compared to other off-farm inputs, contributed to the production of higher and more stable crop yields and reduced post-harvest losses, particularly in Western Europe and the USA. Increased farmer productivity, brought about, in part, by the growing use of crop protection agents, generated two key outcomes. First, increased levels of food surpluses compared to the Traditional Phase helped to ensure both farm household food security and material well-being. Second, and more importantly, crop protection agents enabled the generation of surplus agricultural production by fewer farmers. In turn, it not only did allow the growing numbers of industrial workforce to be fed, but also released farm labourers to join the newly emerging industrial labour markets in Western Europe and the USA.

1.4 International Phase (1847–1929)

1.4.1 Key characteristics of farming systems and broader economic, social and political contexts

Many of the key structural trends evident at the beginning of the Revolutionary Phase were still evident at the start of the International Phase. Population levels continued to rise throughout the Revolutionary Phase, reaching a high point of between 1 128 000 000 and 1 402 000 000 by the mid-nineteenth century (Census 2003). By the start of the International Phase, approximately 30% of Western Europe's population was urbanised and severely limited with respect to food production and/or collection opportunities. Conversely, in the same vein, an increased capacity of farmers and farming systems in the Revolutionary Phase to provide greater volumes of marketable food surpluses enabled the balance between market-based agricultural production and subsistence production to shift, particularly in Western Europe and the USA. Agricultural surpluses continued to facilitate the release of agricultural labour to supply the growing demand for industrial workers, predominantly in urban areas, and supply adequate quantities of reasonable quality food to this growing non-agricultural workforce. By the mid-nineteenth century, market-based agriculture had become the dominant model for agriculture over much of the developed world. Additionally, the appliance of scientific principles of agricultural production and management was much more widespread than at the beginning of the Revolutionary Phase, as were the relatively higher levels of agricultural production intensity, which incorporated the use of crop protection agents.

However, there were two key differences between the beginning of the Revolutionary Phase and the start of the International Phase. First, there had been a significant shift in political and economic power within sovereign states from the landowning elites to a new breed of industrialist entrepreneurs, which had accumulated vast profits and the concomitant political power on the back of industrialisation. Second, Western European countries had amassed significant overseas territories through which trade in agricultural products and manufactured goods had grown.

1.4.2 Key external drivers of pesticide evolution

The development and application of crop protection agents throughout the International Phase was promoted by two key drivers: (1) Increasing food demand, due to escalating population growth in Western Europe and the colonies, and continued urbanisation, and; (2) Deepening industrialisation of capitalist production and consumption relations and emergence of the so-called First Food Regime.

1.4.2.1 Deepening capitalist production and consumption relations and emergence of the First Food Regime

In contrast to the Revolutionary Phase, when politically powerful landowners and farmers were key actors in promoting the profitable development of agriculture, the International Phase witnessed the growing political and economic power of industrialist entrepreneurs that had different priorities from those of the powerful landowners and gentleman farmers. Specifically, power and wealth generating industrial activities also relied heavily on food production. However, as they were not either directly or indirectly involved in agricultural production, they did not benefit from the increased volumes of highly priced agricultural commodities supported, for example, by the Corn Laws in Britain. Indeed, even though industrialists benefited from reliable supplies of nutritionally adequate food, they were dependent on the provision of this food at low cost. The prosperity of industrialists was heavily dependent on a highly productive yet cheap workforce. In order to be productive, the workforce had to be healthy and well fed. In Western Europe and North America, industrialists secured a healthy workforce through the provision of basic health care and disease prevention strategies. However, the prevention and cure of disease was only part of the labour productivity equation, as a healthy workforce also required regular supplies of nutritionally adequate food. Even though industrialists paid pitifully low wages to their often-oppressed workforces, their employees still required a living wage for themselves and their families. During this Phase, employees spent the lion's share of their wages on food. Therefore, the cost of basic staple foods was the major determinant of base-line industrial wages.

A regular supply of cheap, yet nutritionally adequate, foods was therefore imperative for food security and social stability of an increasingly urban-based industrial workforce and the profitability of labour-intensive industrial activities. During this Phase, industrialists became concerned that as populations grew and urbanisation continued apace, the ratio of rural food producers to urban-industrial food consumers would reach unsustainable proportions, and that increasing productivity would no longer be able to meet the growing demand for food. If this occurred, and regular food supplies could not be assured or that restrictions in supply resulted in a price increase for staple foods, profitable industrial activity and indeed social stability could not be assured.

In order to assure regular supplies of cheap, yet nutritionally adequate, food, maintain social stability and maintain high and stable profits from industrial activities, an ambitious strategy was pursued. It was during this Phase that, industrialists, in cahoots with national governments, used their political and economic power to reorganise food production, distribution and consumption to serve their needs. Indeed, industrialists were the principal architects of what was latterly termed as the First Food Regime.

1.4.2.1.1 Emergence of the First Food Regime

Key components of the First Food Regime were. (1) A sharp increase in international food trade. (2) The reorganisation and industrialisation of food production, processing, distribution and consumption in nationally organised economies. (3) Reorganisation of agricultural production. (4) An increase in the intensity of agricultural production on existing cultivated areas.

1.4.2.1.1.1 Increased international food trade

Order in world politics is typically created by a single dominant power. Since regimes constitute elements of an international order, this implies that the formation of regimes normally depends on hegemony (Keohane 1984). The First Food Regime developed during the era of British imperialism and was constituted by the role of Britain as 'workshop of the world', and by the related politics of building and maintaining a global food system consistent with this role. Essentially, the First Food Regime was underpinned by two key trade-related components. First, powerful industrialists in Britain and Western Europe secured cheap imports of unprocessed and semi-processed foods and materials, principally wheat and meat (wage-foods), from white settler countries (see Fig. 1.3). In effect, international trade facilitated the relocation of commercial farm production from Western Europe to the extensive settler frontiers in Africa, South America and Australasia and to the new agricultural areas in North America. The introduction of refrigerated ships in the 1880s increased both the

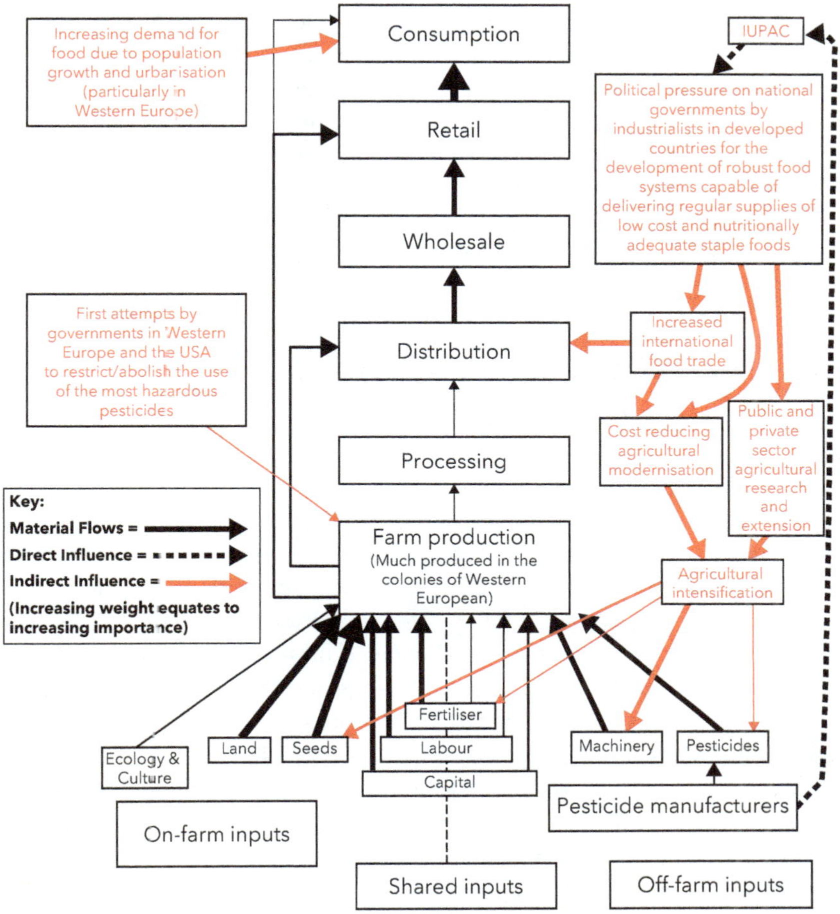

Figure 1.3 Food system dynamics during the International Phase (1847–1929) of pesticide and pesticide industry development.

range of produce that could be supplied by distant colonies and the distance over which perishables such as butter, meat and tropical crops could be transported to the metropolitan economies (Peet 1969). Importantly, in Britain, this demanded repel of the Corn Laws that had restricted the British corn market for nearly 200 years.

Second, the First Food Regime instigated a new division of industrial and agricultural labour across space and social units; that is imported foods were exchanged for exported European manufactured goods, labour and capital. Eventually, complementary product exchange gave way to competitive product trade according to comparative economic advantage and trade became increasingly multilateral. This caused the breakdown of prior trading monopolies of the European colonial system, led to a reconstitution of the world economy as an international economy and established the domination of dynamic world market prices over food production. As a recently independent colony of Britain, the USA was an anomaly in this new world order, in as much as they had both the military and economic power to rival the Europeans. In addition, the USA also possessed a comparative economic advantage with respect to the production of a range of staple foods (including the all-important wheat and meat). This meant that the USA could potentially be a major food importer as well as a competitive food exporter.

1.4.2.1.1.2 Reorganisation of food systems

Institutionalisation of liberal trade in agricultural commodities, and the associated heightened importance of competitive agricultural production, implied a substantial reorganisation/reconstruction of food production, processing, distribution and consumption into urban-centred industrial agro-food complexes within nationally organised economies. Use of the term 'urban-centred' not only alludes to the fact that the world's population was increasingly urbanised, but more importantly, refers to the transfer of power over food production, distribution and consumption from rural elites to urban-centred elites. The term agro-food complexes is used to acknowledge the fact that food production and consumption is a complex system comprised of distinct yet interactive components that are reliant on each other (see Fig. 1.3).

It was during this Phase, and the promise of substantial international revenues from agricultural exports, that the American Government took over the mantle as protector of the agricultural sector from the British. The unique situation of the USA meant that, as a relatively prosperous and independent nation, it had the capacity to mobilise resources to ensure domestic supplies of cheap yet adequate food to its workforce. At the same time, the USA was able to create an export-driven agricultural sector second to none that could generate foreign exchange and act as an engine for growth within the American economy. Throughout this Phase, US Government policy focused on the promotion and commercialisation of farming by investing substantial federal funds in a series of measures. Alongside the establishment of the United States Department of Agriculture (USDA) in 1862, the Federal Government passed a series of Acts. First, the Morrill Act, passed in 1862, created a system of state-level (Land Grant) agricultural colleges and experiment stations for the conduct of agricultural research (Sheingate 2001). Second, the Hatch Act, passed in 1887, established an adult education system, and the USDA's Cooperation Extension Service, established in 1914. In 1862, the Homestead Act encouraged the westward expansion of agriculture, and the Pacific Railroad Act, also passed in 1862, connected developing agricultural production areas on the frontier with the urban markets of the East. In addition, the Federal Government enacted pro-agricultural market regulation and provided credit (Agriculture Outlook 2000). Between the mid-nineteenth century and World War I, farm acreage and production steadily increased and the farm production component of the food system prospered.

In Japan, the Meiji era (1868–1912) also signalled the beginning of the modernisation of Japanese agriculture. Initially, Japan aimed to accelerate agricultural development by the direct transposition of British and American technologies. However, in 1890, after discovering that American machinery was not suited to Japanese conditions, Japan reorganised its agricultural development strategy, seeking agricultural improvements through agricultural research and education. Following the establishment of the National Agricultural Experiment Station network in 1893, the national government provided subsidies to aid the development of prefectural experiment stations (Barker et al. 1985).

The expansion of lucrative world trade also initiated the establishment of international industry organisations. Founded in 1919, the International Union of Pure and Applied Chemistry is an international organisation dedicated to the advancement and application of chemical science and technology (Hough 1998). It includes crop protection chemical manufacturers among its range of interests (Hough 1998) (see Fig. 1.3).

This Phase also witnessed the first concerted attempts by national governments to restrict the use of the most hazardous crop protection products, generally in response to popular appeal. By 1900, State legislature was introduced in the USA to control the use of 'Paris Green' (copper arsenite), discussed later. In the early 1900s, the US Federal Government passed two key acts: The Federal Food and Drugs Act (1906), creating the Food and Drug Administration (FDA), and The Federal Insecticide Act (1910). Ironically, pesticides were believed to be safe when correctly formulated and applied. As a result, instead of being designed to protect consumers, much of the early pesticide legislation focused on the protection of farmers from fraudulent sales of misbranded or adulterated pesticide products (Burgett 2003). However, by the late 1910s, the extensive application of arsenical insecticides in the USA began to cause widespread public concern as fruits, vegetables and cotton were increasingly shown to contain poisonous residues. This stimulated two things: (1) the establishment of arsenic tolerance levels on apples by the FDA in the 1920s and (2) the search for less toxic crop protection agents.

1.4.2.1.1.3 Reorganisation of agricultural production

This Phase witnessed a reorganisation of farm-level production in response to the increased commercialisation of the evolving international food system. Reorganisation generally responded to two key structural dynamics of the international food system. First, phases of interrupted food supplies and concomitant hikes in agricultural prices provided the more commercially oriented family farmers with economic incentives to increase profits. The growing importance of the profit motive catalysed a more business-like approach to farming and forced an evaluation of the combination and organisation of the factors of production. Improvements in the organisation of labour included: specialised divisions of labour and new work practices. Second, the rigours of the growing capitalist market place increasingly demanded cost reductions in order to remain competitive (see Fig. 1.3). Cost reductions were often achieved through increasing economies of scale and through more efficient uses of natural and capital inputs into the production process. These inputs included the use of machinery, fertilisers and crop protection agents (see Fig. 1.3).

1.4.2.1.1.4 Intensification of agricultural production

Throughout the International Phase, the expansion of agriculture to new territories was matched by new levels of production intensity (see Fig. 1.3). Although agricultural profitability fluctuated between sectors and over time during this Phase, international trade generally had the effect of encouraging farmers to lower production costs in order to remain competitive (see Fig. 1.3). Within an increasingly commercialised environment, farmers were continuously faced by limiting factors of production, irrespective of whether or not they required higher yields to take advantage of high food prices or in order to reduce unit costs of production to remain profitable. This generally meant an increased use of crop protection products. Ironically, many of the industrialists that orchestrated the development of the First Food Regime, in order to capture the indirect benefits of low food costs, also profited through the production and supply of a growing range of machinery, industrial fertilisers and crop protection products. When combined, these drivers created an international market infrastructure containing significant economic incentives for the development and commercialisation of new pesticide technologies.

1.4.3 Key characteristics of pesticide development processes and manufacturers

The development and use of insecticides during the International Phase is characterised by two parallel activities: (1) the use of toxic substances for insect control and (b) the search for less toxic insecticides. The first activity involved the production and use of toxic chemicals, often generated as by-products from industrial processes. One of the most notable, and commercially important, insecticides was the toxic dye copper acetoarsenite (Paris Green). In 1867, the combined negative economic impacts of Colorado beetles on potato crops and gypsy moths on fruit tree crops in the USA prompted entomologist Charles Riley to pioneer the use of Paris Green and London Purple (an arsenical dye residue) as insecticide sprays (Boardman 1986; Hough 1998). Even though this pesticide was incredibly toxic, its inherent efficacy and cost effectiveness enabled it to quickly dominate the insecticide market in both Europe and the USA (Holmes 2001).

Conversely, Paris Green's unenviable record with regard to human health was to stimulate research towards the development of less toxic pesticides and the succeeding years witnessed the successful introduction of new materials containing arsenic, copper, mercury and sulphur. Within 25 years, lead arsenate was released as a less toxic alternative to Paris Green, which was, in turn, joined by zinc and calcium arsenate in the early 1900s (Bailey 1909). By 1917, calcium arsenate and lead arsenate were the most widely used insecticides. Lead arsenate contained less than half the amount of arsenic as Paris Green (Bailey 1909).

At this time, there was an ever-growing list of commercially available insecticides including botanicals and new chemicals derived, often as by-products, from the emerging coal-gas production, petrochemical, synthetic dyes and pharmaceutical industries, including products such as mercuric chloride, kerosene, hydrogen cyanide, tar oils, selenium compounds and cryolite (Bailey 1909). Mercuric chloride solutions were used to control soil borne pests during the 1860s and Kerosene emulsion was used as an insecticide on dormant deciduous fruit trees during the 1870s. Hydrogen cyanide, first used as a fumigant in museums during the 1870s, was used against scale insects on citrus trees in California during the 1880s. Phenyl emulsions were used against scale and mealy bug infestations (Boardman 1986). Tar oils, first used to control aphid eggs on dormant trees, were eventually replaced by dinitro-o-cresol, which was subsequently patented as a selective herbicide against weeds in cereal crops. Arsenicals, lime-sulphur, petroleum oils and nicotine were the principal insecticides pre-World War I (Pimentel and Pimentel 1979). Early organics insecticides included nitrophenols, chlorophenols, creosote, naphthalene and petroleum oils. Fluorine compounds, thiocyanates and dinitro compounds, also appeared between 1914 and 1939 (Pimentel and Pimentel 1979). However, most of the products used were non-selective and only marginally less toxic than the pesticides they were to replace. Even though botanicals such as nicotine and pyrethrum had been used in Europe and the USA since the mid-1800s, and often offered more specific pest control, they were not stable enough for use in agriculture given their rapid breakdown in the environment. An exception to this rule, at least for a short while, was the use of rotenone; its production and use rapidly expanded during the 1920s after being patented in 1912.

While insecticides received the greatest attention during the International Phase, several important herbicides and fungicides were also developed. In 1908, Bolley reported successful weed control in wheat using salt, iron sulphate, copper sulphate and sodium arsenate (Pimentel and Pimentel 1979). Ammonium sulphate was also used as a selective herbicide during the International Phase.

During this Phase, organomercurials were first used as fungicidal seed dressings against cereal smut and bunt diseases. Other fungicides include lime sulphur for use against apple scab and San Jose Scale in California. However, the most important fungicide discovered during this Phase was Bordeaux mixture (copper sulphate and lime). Bordeaux mixture (discussed in detail below) was discovered and developed by a French professor of botanics at the Faculty of Science, Bordeaux University, in 1882. Interestingly, it was also credited with being the first foliar-applied liquid fungicide and the catalyst that led to the development of the first knapsack sprayer. Paris Green was also used with Bordeaux mixture against Grape Phylloxera in French vineyards and in 1910 (Boardman 1986).

The International Phase also witnessed the emergence of the first documented manufacturers and suppliers of agricultural pesticides. Whilst the USA witnessed the creation of small pesticide manufacturing businesses during the mid-nineteenth century, the industry remained relatively small scale and limited to the production of a narrow range of pesticide products (Boardman 1986). This Phase was also characterised by a significant amount of fraud, especially through mail order sales (Boardman 1986). However, as attention began to turn towards the potential profits associated with the newly emerging synthetic organic compounds, larger specialist pesticide producers or evolving diversified corporations with a pesticide division, which could afford to employ research staff and exploit economies of scale, were endowed with significant advantages compared to smaller producers (Boardman 1986). This was particularly the case where diversified corporations produced a wide range of industrial chemicals, and intermediates, with potential pesticidal activity. Ultimately, these corporations were limited to a few countries in the developed world, namely the USA and Germany. In the USA, companies included Dow, founded in 1897 as a bleach manufacturer (Dow 2003a). Dow began producing chemical products for fruits and flowers in 1906, giving rise to the company's agricultural chemicals division. In Germany, Rohm and Haas, founded in 1907 by two young German entrepreneurs (chemist Otto Rohm and businessman Otto Haas), manufactured and sold the leather bate, Oropon, to tanners (Rohmhaas 2003). By 1917, the company was well established in Europe and the USA. Between 1921 and 1930, Rohm and Haas expanded its franchise into speciality chemicals and produced its first pesticide lethane, a synthetic organic insecticide used for killing houseflies (Rohmhaas 2003). German chemical companies also produced the first organomercury seed dressing at the start of World War I (Boardman 1986).

1.4.4 Key impacts of pesticide development and application

As with both the Traditional and Revolutionary Phases of pesticide evolution, an increased use of crop protection agents in the International Phase contributed to the production of higher and more stable crop yields and reduced post-harvest losses. However, while the use of inorganic pesticides during the International Phase was not yet common practice, their use was much more widespread than in either of the previous phases (Hough 1998). In turn, the application of inorganic pesticides contributed to increased world agricultural output, which, in turn, contributed to food security, economic development and the formation and stability of independent nation states. Increased farmer productivity, brought about in part by the growing use of crop protection agents, generated three key outcomes: (1) increased food supplies; (2) improved profitability and household food security of farm-families and (3) negative environmental and human health externalities.

1.4.4.1 Increased food surpluses

As with the Traditional and Revolutionary Phases, the principal impact of pesticide use during the International Phase was to contribute to the production of higher net agricultural yields and world food supplies. Two of the most important pesticides with regard to increasing net food supplies were the arsenic-based insecticides and the fungicide Bordeaux mixture. Paris Green (lead arsenite) was first successfully used in the State of Mississippi, USA, in 1867, against Colorado beetle, which had been introduced into the USA in 1860 and was quickly spreading across the country causing considerable damage. It was also used against gypsy moth, cotton worm, canker worms on apple trees, caterpillars and grasshoppers, elm leaf beetle, plum curculio, rose chafer, grape rootworm and the economically important codling moth (Bailey 1909). During the latter part of the nineteenth century, more than 2 000 tons of Paris Green was applied annually in the USA (Bailey 1909).

Bordeaux mixture quickly became an important fungicide because of its effectiveness against late blight of potatoes, which was almost single handily responsible for the great potato famine in Ireland in 1845/47 that claimed 1 500 000 Irish lives. At this time, almost 90% of the Irish population was wholly dependent on potatoes for their food security. The widespread use of a modern potato variety 'Lumper' (Duke 1990), which was high yielding, yet highly susceptible to late blight of potatoes, enabled the disease to claim approximately 40% of the potato harvest in 1845 and 100% of the harvest in 1846. Even though Ireland was free of the disease by 1847, a shortage of potato seeds underpinned another poor potato harvest and the start of mass emigration; over 1 000 000 people emigrated from Ireland during this time. Bordeaux mixture made a significant contribution in the battle to control this devastating disease. It should be noted that, according to Agrios (1988), the potato famine 'led to the development of the science of plant pathology'.

1.4.4.2 Improved farm-family household food security and profit

By 1910, lead arsenate and Paris Green were the most widely used insecticides, sold on a commercial basis, with sales reaching £10 000 000 each year. By this time, aggregated estimates of economic losses to insects in the USA amounted to US$700 000 000, more than the entire expenditure of the US Government (Bailey 1909). As well as the capacity to control late blight of potatoes, Bordeaux mixture was especially effective in controlling downy mildew of grapes. Grapes, and ultimately wine, was vital to the French economy. During the 5 years since its introduction, downy mildew decimated French grape harvests and cost the French economy nearly $50 billion. The negative impact on wine revenues created considerable financial opportunities for the manufacturers and suppliers of Bordeaux mixture. The new economic context also created considerable financial incentives for the development of new pesticide application technologies that could increase the timeliness, and reduce the costs, of pesticide application. Until this time, pesticides were applied by hand, which was incredibly labour intensive and time consuming. Consequently, developed in conjunction with liquid Bordeaux mixture, in 1883, John Bean Spray Company (predecessor of FMC Corporation) was the first company to patent the first knapsack sprayer (FMC 2003). Agricultural machinery developed rapidly, and before the end of the 1920s, aeroplanes were being used to apply insecticide dusts in Ohio. Interestingly, the search for progressively less toxic pesticides initiated an increasingly competitive pesticide product treadmill, providing the financial incentives and rewards for inventive and innovative pesticide manufacturers. Recently established patent protection in the USA allowed the newly emerging pesticide manufacturers short-run monopolies on their products, further enhancing the profitability of pesticide research and development (R&D).

1.4.4.3 Increasing negative environmental and human health externalities

Traditionally, pests, not pesticides, were seen as a serious threat to human health and well-being (Meiners and Morriss 2001). Indeed, the first pesticide laws in the USA were state laws that tried to redress the problem that not enough pesticides were being used. For example, Washington State passed a law that encouraged free-ridding farmers to spray arsenic onto their apple orchards (Whorton 1974). In a similar vein, the secondary concern was that many of the pesticides on the market were of poor quality, ineffective, and often adulterated with a range of poisonous agents (Anderson et al. 1999). In

1910, the US Government introduced the Insecticide Act, the USA's first federal pesticide regulation, which, together with the Pure Food and Drugs Act (1906), aimed to reduce economic exploitation of farmers by pesticide manufacturers and distributors (Schierow 2004), and to assure public health, by removing misbranded, adulterated, ineffective and poisonous products from the market (Morriss 1997). The Insecticide Act resulted from pressures from farming organisations, sympathetic Congressmen and USDA officials. One of its key achievements was the specification of active ingredient (AI) concentration for Paris Green and lead arsenate (Boardman 1986).

In addition to public concerns, growers also worried about negative publicity associated with food contaminants. In 1891, the *New York Times*, and other broad circulation papers, reported that a fungicide residue had been detected on grapes. In response to public uproar, state health authorities began destroying contaminated grapes. Subsequently, analysis by the USDA showed that 'an adult would have to eat 300 pounds of grapes a day, including heavily coated stems, to get a harmful dose' (Dunlap 1981). Eventually, after a grape boycott, and subsequent fall in sales, the scare subsided. Concern over the use of pesticides on high-value food crops also spread to England. In 1925, and at considerable expense, growers hired independent chemists to inspect fruit imported from the USA in order to assure the public that their products were safe (Dunlap 1981). Indeed, in extraordinary cases, several pesticides were banned because of their public health risks. For example, arsenic was already banned in France following deaths from high levels of arsenic in bread, even though it continued to be used in England for several decades (Holmes 2001).

1.5 Summary

Reorganisation of food production, distribution and consumption within the framework of the First Food Regime substantially increased the capacity of farmers and farming systems in both the old and the new world to provide greater volumes of marketable food commodities. Growing agricultural surpluses continued to facilitate the release of agricultural labour to supply the growing demand for industrial workers, predominantly in urban areas, and supply adequate quantities of reasonable quality food to this growing non-agricultural workforce. However, throughout the International Phase, the application of chemical products continued to be haphazard and very imprecise as the technical material was often ineffective. The use of simple inorganic compounds often resulted in unsatisfactory outcomes because of the primitive chemistry and application methods.

Hence, at this time, although chemical pesticides continued to grow in importance, they did not yet dominate agricultural production, and biological control, resistance breeding and sound agronomy were still seen as the predominant tools in the control of crop pests and diseases. Even as late as 1935, frequent cultivations were still acknowledged as the most biologically and economically effective control of weeds (Anonymous 1935). Ultimately, however, the case of Paris Green and other arsenic based insecticides illustrated the early institutionalisation of the use of crop protection agents in intensive agricultural systems. That is, the process of institutionalisation is clearly demonstrated by the attitudes of, and subsequent actions taken by, governments in Western Europe and the USA with respect to growing concerns over the negative health effects of pesticide residues on their citizens. In general, their response was not to question the continued use of modern crop protection agents but to search for less toxic replacements, regulate the formulation of pesticides in order to eliminate toxic impurities or to regulate pesticide use in order to eliminate their misuse. Their attitudes and actions served to provide a green light for pesticide manufacturers to produce new pesticide products and to attribute pesticides as a key and unquestionable component of modern agriculture.

Chapter 2

The evolution of the pesticide industry in the Productivist Period (1930–1973)

2.1 Key characteristics of farming systems and broader economic, social and political contexts

Many of the key structural trends evident at the beginning of the Pre-Productivist Period were still evident at the start of the Productivist Period. Population continued to rise throughout the Pre-Productivist Period and had reached a high point of 2 000 000 000 by 1930. By the beginning of the Productivist Period, between 15% and 29% of the world's population was urbanised, and generally limited with respect to food production and/or food collection opportunities (University of North Colorado 2003). Market-based agriculture had become institutionalised as 'the' model for agriculture over much of the globe, and industrial principles of agricultural production and management were beginning to be widely applied. In general, this led to the continued intensification of agricultural production and the increased use of, and reliance on, crop protection agents. More importantly, however, the beginning of the Productivist Period is marked by a virtual collapse of the First Food Regime.

2.1.1 Collapse of the First Food Regime

To understand the development of pesticides and the pesticide industry during the Productivist Period, it is important to understand the context in which this development took place and, in turn, the key drivers that created this context. The overriding context in which pesticides and the pesticide industry evolved was that provided by a virtual collapse of the First Food Regime. This virtual collapse occurred after the protracted international economic depression of the 1920s and the 1930s and the extensive political and economic disruptions caused by the first and second world wars. That aside, several components of the First Food Regime managed to survive. These were food trade in dairy produce, meat and cereals originating in the extensively farmed rangelands of the Americas and Australasia, and the production of sugar, tropical tree crops (cocoa, coconut, rubber, palm oil, bananas) and beverages (tea, coffee) through quasi-plantation systems of production (Atkins and Bowler 2001).

2.2 Key external drivers of pesticide evolution

This section aims to explain the unprecedented and generally unquestioned rate of growth in the invention, innovation, development, production, distribution and use of an expanding range of specialised synthetic organic pesticides, which were manufactured and supplied by emerging corporate agribusinesses. In order to do this, we must first set the evolution of modern pesticides and the pesticide industry into the much broader context of a series of nested drivers. The evolution of pesticides and the pesticide industry needs to be contextualised within the parallel processes of the commoditisation, industrialisation and intensification of agriculture and its continued subsumption within urban-centred industrial agro-food complexes. In turn, these immediate drivers need to be couched within a much broader context of continued population growth (particularly in developing countries), increased urbanisation and industrialisation, rapid scientific and technological advances, growing concerns over food and national security, and the establishment of Fordism and the Second Food Regime (Perkins 1997).

2.2.1 Population growth, urbanisation, industrialisation and scientific and technological progress

Even though population growth was not a new phenomenon, there was a significant expansion in world population during the Productivist Period, particularly in developing countries. Between 1930 and 1973, the world's population nearly doubled from just over 2 000 000 000 to just less than 4 000 000 000 (UN 2003a). However, population growth rates in developed and developing countries varied considerably. Between 1950 and 1975, the average annual population growth rate in developed countries was 1.13%. This was in stark contrast to 2.33% in developing countries (Sarre and Blunden 1995). High population growth rates were particularly evident in Asia and Africa. Between 1950 and 1975, the population of Asia and Africa grew from less than 600 million and

170 million to 1.2 billion and 320 million, respectively (Sarre and Blunden 1995). As the world's population grew, so did the demand for goods, services and food.

During this time, the process of urbanisation also continued apace. This was mainly due to enhanced employment opportunities and higher pay in urban areas. However, as with population growth rates, the degree of urbanisation between developed and developing countries varied considerably. In 1950, 54.9% of the population in developed countries was urbanised. This was in stark comparison to only 17.8% in developing countries. By 1975, 70% of the population in developed countries was urbanised, compared to only 26.8% in developing countries (UN 1999). The growth of urbanisation was closely associated with the accelerated pace of industrialisation across the globe.

When combined, the parallel processes of population growth, urbanisation and industrialisation created an ideal environment for equally rapid scientific and technological progress, and, in most developed countries, a rapid growth in national economic output and an associated increase in living standards (see Fig. 2.2). During this Period, scientific and technological progress was underpinned by significant financial investments in research originating from both public and private sources and education underpinned by public finances. Significant scientific advancements were made in plant breeding, plant production and protection, and agricultural mechanisation (see Fig. 2.2). Scientific and technological innovations in non-agricultural sectors also generated important agricultural spillovers (see Fig. 2.2). These include spillovers from light and heavy engineering (for agricultural machinery), manufacturing of explosives (Haber process for nitrogen fertiliser) and developments in the petrochemical, synthetic chemicals and pharmaceuticals industries, particularly in relation to pesticide development. Some of the more important spillovers will be discussed later in this chapter. The combined effects of population growth, urbanisation, industrialisation and scientific and technological process set the scene for the dramatic restructuring of agro-food systems across much of the world. First, rapid population growth created the physical demand for more food. Second, the considerable exodus from rural to urban areas meant that, aside from the urban gardens, urban populations (whose numbers were increasingly swelled due to rapid population growth) were no longer involved in the production of their own food and became increasingly reliant on their domestic agricultural sectors or on foreign imports to supply their food needs. High levels of population growth and urbanisation thus created a seemingly overwhelming demand for enhanced agricultural productivity and/or a significant expansion of farmed areas. Third, it was the rapid pace of industrialisation, and parallel developments in science and technology that provided the potential means to supply food for this growing and increasingly urbanised population (see Fig. 2.2).

2.2.2 Food security and national security

The stage was set for the development of a modern agro-industrial food system that could deliver adequate supplies of low-cost quality foods in the post-war period. However, there was no guarantee that the components of this new system would miraculously come together or that the particular parts of new agro-food systems, namely farmers, had the capacity or desire to restructure production. The concerns outlined above were to preoccupy leaders of developed nations in the immediate post-war period and provoke detailed consideration of world food supplies, food security, economic growth and prosperity and, based on these concerns, national security (see Fig. 2.2). These concerns were particularly pertinent in the period directly after the Second World War, as the food production capacity of Europe had been decimated. Indeed, many European countries maintained wartime food rationing policies for several years after the war ended.

It was, however, to be the USA that took the lead role in the reconstruction of global food systems. The USA emerged from the war as not only the strongest military force in the world but also the most politically and economically powerful nation. It was thus, during the latter part of the war and post-war period that the USA asserted its newly forged hegemony by persuading governments of developed nations to underwrite the establishment of a range of key policies and institutions, which aimed to stabilise social, economic and political relations on an international basis. The key development was the declaration by 26 'United Nations' in 1942 'to continue their united fight against the Axis Powers'. This declaration was, in many ways, similar to the League of Nations, an organisation conceived under similar circumstances during the First World War, and established in 1919 under the Treaty of Versailles 'to promote international cooperation and to achieve peace and security'. The United Nations Declaration was quickly followed by the United Nations Monetary and Financial Conference at Bretton Woods in 1944. The final Bretton Woods agreement established international exchange rates based on the US$ and proposed establishment of the World Bank (International Bank for Reconstruction and Development) and the International Monetary Fund (IMF), which were both founded in 1945, along with the formal establishment of the United Nations (UN). Four years later in 1949, the North Atlantic Treaty Organisation (NATO) added a further tier of international military stability.

One of the main concerns of the new US-dominated world order was the reconstruction of war-torn Europe. America felt bound to find a solution to the widespread hunger, unemployment and housing shortages that faced Europeans in the aftermath of World War II. The US administration's concern culminated in U.S. Secretary of State George Catlett Marshall's Harvard University address on 5 June 1947. The address emphasised that the physical destruction of the war and the general economic dislocation threatened a breakdown of moral, social and commercial life. Raw materials and food were in short supply, and war-damaged industries needed machinery and capital before production could be resumed. Marshall devised a plan of action for Europe that, over a period of 4 years, aimed to redress these issues. Implementation of the Marshall Plan was always going to be a costly operation, possibly more than the average American taxpayer was willing to pay. However, the political will was mustered to deliver financial aid based on the American Administration's desire to stabilise the international order in a way favourable to the development of political democracy and free market economies and, more importantly, to prevent the spread of communism in Western Europe. Costing US$13.3 billion, the reconstruction of Europe was administered through the European Recovery Programme and newly introduced international and European institutions. One of the most active international institutions was the World Bank, making its first loan to France in 1947.

As the population of Europe faced acute food restrictions and a severely decimated agricultural sector, agricultural restructuring through a process of industrialisation and modernisation was one of the key facets of the Marshall Plan. However, a significant part of the Marshall Plan consisted of a process of self-help within Europe. Hence, within Europe, countries quickly developed comprehensive agricultural support mechanisms through which both American and domestic financial resources could be channelled. Established on 16 April 1948, the Organisation for European Economic Co-operation was one of the key European institutions through which the USA channelled financial resources.

2.2.3 Establishment of Fordism and the Second Food Regime

It was obvious to the architects of post-war reconstruction that lasting peace, social stability and economic growth required the establishment of an elaborate, yet durable, social, political and economic infrastructure akin to the First Food Regime (Atkins and Bowler 2001). With the demise of British hegemony after the two world wars, there followed an aid-based food regime. The post-World War II food regime was based on the politics of food self-sufficiency (disposal of any surplus production in the USA and other Organisation for Economic Cooperation and Development (OECD) countries by way of foreign food aid), and on the diffusion of American agricultural institutions, technologies and foods to the South. The Second Food Regime, which comprised part of the wider Fordist Regime of accumulation and period of intensive or organised capitalism, encompassed the long post-war boom between 1945 and the early 1970s. Accordingly, the Fordist Regime embodied a range of key social, political, economic, organisational and technological characteristics. (1) Continued urbanisation. (2) The concentration of wealth (capital). (3) The separation of business ownership and control. (4) The dominance of manufacturing industries concentrating on the mass production of standardised products. (5) Evolution of a mass consumption culture. (6) Mass advertising. (7) The 'scientific management' of labour and assembly line production. (8) The growth of the professional, managerial and administrative 'service class'. (9) The development of the nation state. (10) The regulation of national economies by the state in cooperation with business and the trade unions.

In essence, Fordism was based on the establishment of a durable macroeconomic interdependence between the mass production of standardised goods on the one hand, and the mass consumption of such goods on the other. The balance between the phases of production and those of consumption was ensured by high wages and the concomitant extension of the purchasing power of the population in exchange for improved productivity and higher profits in domestically oriented manufacturing industries in moderately closed economies. The balance between different classes was reinforced by the development of Keynesian welfare policies that guaranteed both a minimum level of subsistence for the working classes and subsidies to industrial and productive sectors. Indeed, in most advanced economies, the Fordist Regime elevated the nation state to the position of principal coordinator of economic and social development and broker of economic and political power. Nation states attempted to balance the interests of oligopolistic corporations and labour unions through market intervention and demand management, using corporate and other taxes to underwrite human and physical investment and to support demand in key sectors.

As with the extensive period of capitalism, cheap and reliable supplies of staple foods were a prerequisite of the Fordist Regime. However, after the food shortages experienced in many developed countries during the world economic depression of the 1920s and the 1930s, and throughout the two world wars, food security, national security and social, economic and political stability became the overriding objectives of nation states during the Second Food Regime. In contrast to the First Food Regime, the Second Food Regime replaced the free market philosophy of international trade in food commodities (based on comparative economic advantage) with market regulation based on the drive towards food self-sufficiency. Embedded in the overarching Fordist framework, the Second Food Regime developed under two key international agreements: the 1945 Bretton Woods Agreement governing the stability of exchange rates between national currencies (based on the dollar/gold standard); and the 1947 General Agreement on Tariffs and Trade (GATT) rules on international trade. The former underpinned the international diffusion of the national model of economic growth and the latter excluded agriculture from more liberal trade practices (Atkins and Bowler 2001). This set the scene for strong state protection of agriculture and organised the world economy under US geopolitical hegemony (Atkins and Bowler 2001). Much of the post-war international economic and military order was moulded to suit US interests, just as the previous pre-war regime had been moulded by the UK.

Militarily, the US structured NATO, the Southeast Asia Treaty Organisation, and the ANZUS pact between the USA, Australia and New Zealand (Hough 1998). Economically, the USA dominated international trade and economic development through the Bretton Woods system (Hough 1998).

Successful establishment of the Second Food Regime required an unprecedented restructuring of the world's principal farming systems, firstly in developed countries during the 1950s and 1960s and latterly in developing countries during the late 1960s and the early 1970s. The five key components of the Second Food Regime were: (1) national food self-sufficiency; (2) increased commoditisation of production; (3) increased industrialisation of production; (4) increased subsumption of farming systems within deepening urban-centred agro-food complexes and (5) evolution of productivist high-input/high-output farming systems (see Figs. 2.1 and 2.2). Even though the restructuring process embodied essentially the same characteristics of the First Food Regime, reorientation of agricultural production, distribution and consumption based on the goal of national food self-sufficiency produced totally new and distinctive agro-food complexes.

2.2.3.1 National food self-sufficiency

The UK was one of the first European countries to develop comprehensive agricultural support structures based on the goal of national food self-sufficiency. This was primarily accomplished through the Agriculture Act of 1947, which sought to increase domestic food production by providing guaranteed returns to farmers, in order to reduce reliance on imported food, and to foster rural economies. At the same time, the Government extended provision of free advisory services to farmers, funded agricultural R&D and provided grants to encourage investment in seemingly more effective and efficient production systems. Statutory bodies also promoted the marketing and distribution of certain agricultural products. A decade later, France, Belgium, the Netherlands, Luxembourg, Italy and Germany signed the 1957 Treaty of Rome and created the European Economic Community (EEC). In 1962, the EEC inaugurated its Common Agricultural Policy (CAP). In line with Article 39 of the Treaty of Rome, the aims of the CAP were to: (1) Increase agricultural productivity by promoting technical progress and by ensuring the rational development of agricultural

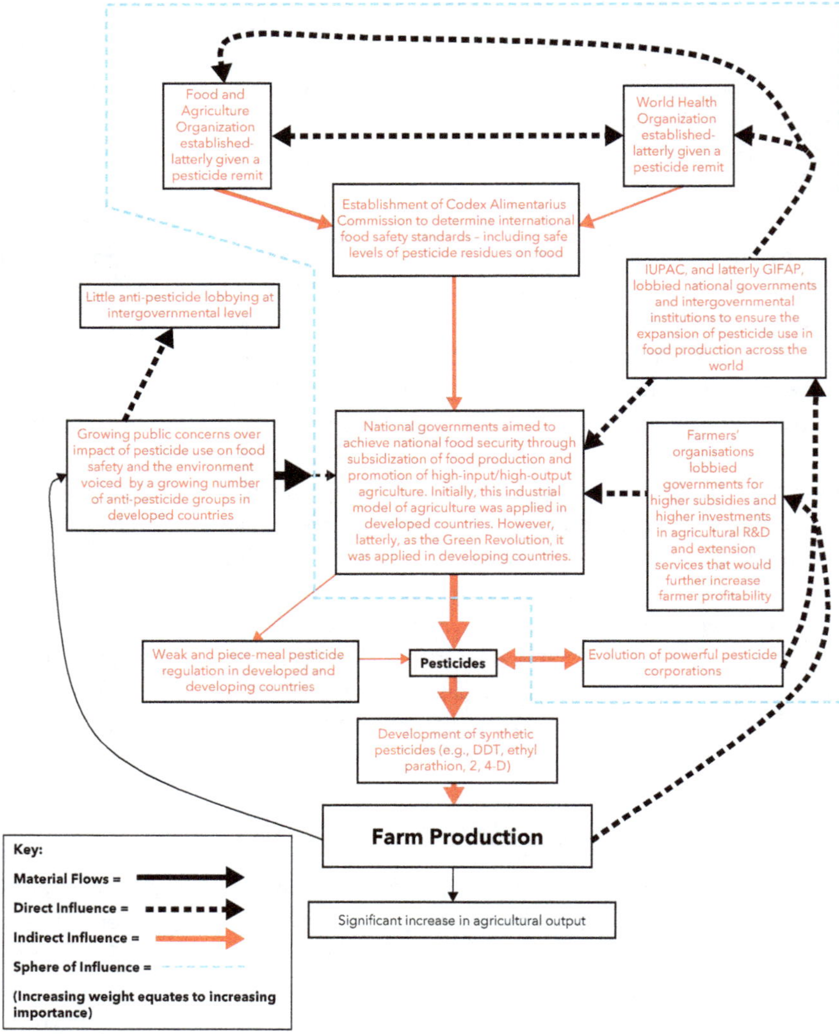

Figure 2.1 Pesticide manufacturers and the techno-policy model of agriculture.

production and the optimum utilisation of all factors of production, particularly labour. (2) Ensure a fair standard of living for the agricultural community, by increasing the individual earnings of persons engaged in agriculture. (3) Stabilise markets. (4) Guarantee supplies. (5) Ensure reasonable prices for consumers (Wrathall 1988). At the same time, the USA also focused attention on its own agriculture sector, which had survived the depression years of the 1920s and the 1930s, through direct government intervention in the form of farm income support (Agriculture Outlook 2000).

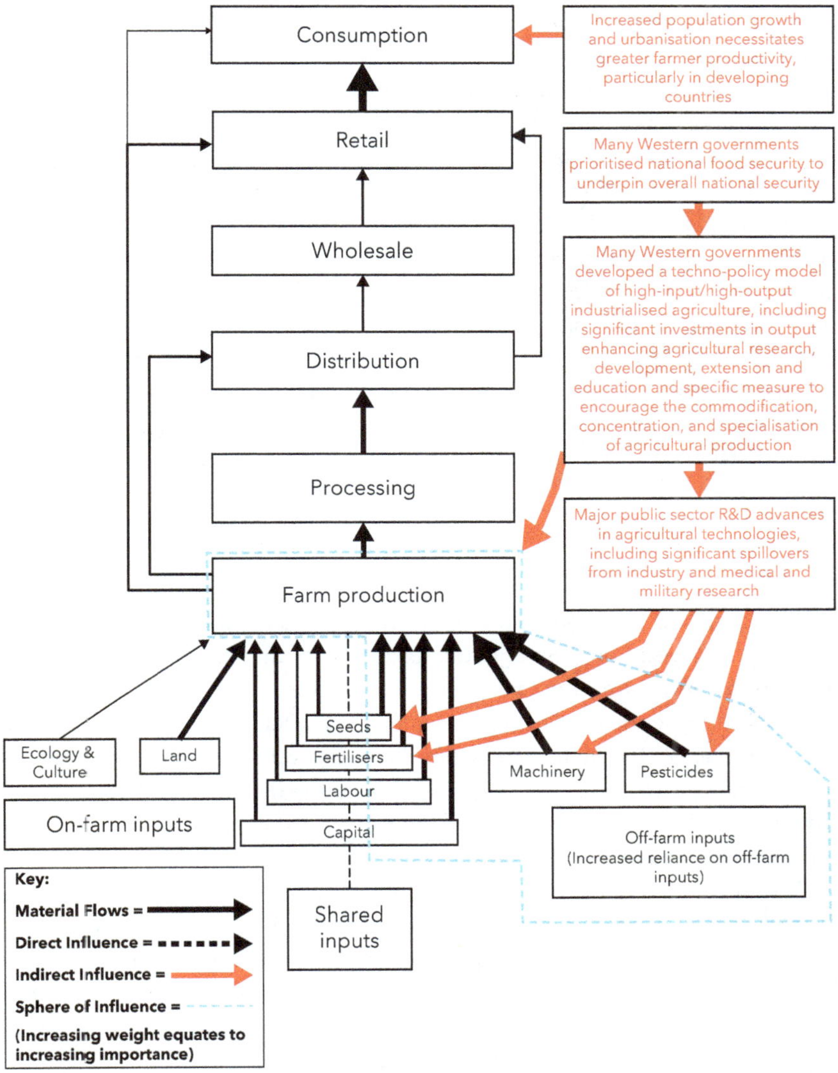

Figure 2.2 Food system dynamics during the Productivist Period (1930–73) of pesticide and pesticide industry development.

2.2.3.2 Agricultural commoditisation

The term 'agricultural commoditisation' describes the process of agrarian transformation from non-market (i.e. subsistence) to market forms of production (i.e. the inputs and outputs of agricultural systems increasingly become goods for market exchange). Indeed, the term 'commodity' applies to objects that are produced for the purpose of being exchanged, as opposed to subsistence societies where objects are produced simply so that people can carry out their everyday lives. The market is the central mechanism, which profitably brings together the producers and consumers of commodities. Commodity-based subsidies and other forms of government support acted in two ways. First, support encouraged farmers to generate greater amounts of marketable food surpluses. These surpluses were increasingly created by the use of purchased off-farm inputs. Second, evolving agribusinesses sought to capitalise on agricultural support by encouraging commoditisation of agricultural production (Bowler 1992). Agribusiness strategies focused on the extraction of surplus value through the provision of off-farm seeds, fertilisers, machinery and pesticides (Bowler 1992). Indeed, it was not until the transition to Second Food Regime that the term 'agribusiness' was coined. The term referred to vertically integrated businesses involved in the supply of agri-inputs (e.g. pesticides), the processing of agri-outputs and the distribution of food products (Wallace 1985).

2.2.3.3 Agricultural industrialisation

The industrialisation of agriculture refers primarily to the application of innovations that emerged from non-agricultural sectors of the economy and was part of the industrialisation of the whole agro-food system (discussed below). Key industrial innovations applied to agriculture include. (1) Adoption of economies of scale. (2) Production intensification. (3) Implementation of organisational features associated with the concept of the firm (specialisation of labour and mechanisation). (4) Production specialisation. (5) Production concentration (Symes and Marsden 1985).

The term 'economies of scale' refers to the concept that the unit cost of production declines as the volume of production increases. This concept assumes that the average costs of producing a tonne of wheat, for example, will diminish as each additional tonne is produced, since the fixed costs are spread over more grain. Widespread belief in the concept of economies of scale by both farmers and agricultural policymakers provided an impetus for the rapid expansion of farm size across much of the developed world.

Production intensification is defined as increased agricultural output per area of land and increased agricultural productivity. Increased agricultural output was achieved by utilising increasing amounts of industrial off-farm inputs. Off-farm inputs were usually new crop varieties and improved animal breeds, agricultural machinery, inorganic fertilisers, fossil fuels, credit and inorganic pesticides. Efficient combinations of these predominantly industrial inputs significantly enhanced the productivity of land, labour and capital, particularly when capital was cost-effectively substituted for labour and land.

Pesticides played a significant role in both enhancing crop production and increasing farmer, and pesticide producer, profits. In response to demand, the production and use of pesticides rapidly expanded. For example, in 1962, synthetic pesticide production in the USA was 730 million pounds. By 1976, this increased to 1630 million pounds, an increase of 124%. Herbicide sales grew by 487% from 95 million pounds in 1962 to 558 million pounds in 1976 (Pimentel and Pimentel 1979).

Production specialisation refers to concentration of farm production on a narrower range of crop and livestock commodities. This usually went hand in hand with the search for economies of scale on the same production unit or through farm expansion. Production concentration is comprised of the spatial and size concentration of farm businesses. Spatial concentration refers to process by which production switches from marginal production areas to areas with a comparative advantage for producing a particular agricultural commodity. For example, by the late twentieth century, approximately 80% of Europe's agricultural production came from 20% of agribusinesses located in agri-industrial hotspots: Paris Basin (wheat), East Anglia (wheat and barley), Emilia-Romagna (milk) and the Netherlands (hydroponic crops) (Byé and Font 1991). Size concentration refers to the process by which marginal farmers go out of business or become absorbed by larger production units. Industrialised agriculture was also the linchpin to urbanisation; as labour productivity increased with the introduction of machinery, higher-yielding varieties, fertilisers and pesticides, it freed up rural labour to fuel urbanisation and provided cheap labour for the growing urban-centred industries.

2.2.3.4 Agriculture's increasing subsumption within urban-centred industrial agro-food complexes

Although urban-centred industrial agro-food complexes were present during the First Food Regime, it was not until the Second Food Regime that food production and processing, distribution and retail sectors received substantial and prolonged interest from non-agriculture-based private companies. This interest materialised in the form of increasing subsumption and appropriation. Direct or real subsumption describes the process by which traditionally non-farm businesses take direct control of the means of agricultural production. This generally entails the purchase of agricultural holdings by non-farm businesses and the vertical integration of production/processing and, in some instances, the wholesaling and/or retailing of food products. CWS Agriculture, which is a direct subsidiary of the UK-based Co-operative Wholesale Society, is a good example of direct subsumption (Sauer 1990; Bowler 1990). On the other hand, indirect or informal subsumption describes the processes by which non-farm businesses take greater control of, and extract a higher proportion of value added from, the agricultural production process, without directly owning, or controlling, the means of production. Obvious examples would be the production of agricultural machinery, pesticide and fertiliser inputs or the provision of credit for agricultural investments (Sauer 1990; Bowler 1990).

Non-farm businesses were also keen to appropriate components of agricultural production. That is, elements of farm production were replaced by industrial inputs. An example of this would be the replacement of plant nutrients (of organic origin) by industrially processed inorganic plant nutrients, supplied by fertiliser manufacturers (Cloke and Goodwin 1992), or the replacement of sound agronomy (including crop rotations, manipulation of planting dates and the use of resistant varieties) by pesticides. This definition also encompasses the substitution of off-farm credit in place of farmer profits, for agricultural investments, or adding value through processing, packaging and sale of foods (Cloke and Goodwin 1992). Non-farm businesses also increased their proportion of value added through the process of substitutionism. This is the process by which agricultural outputs in the food chain are increasingly substituted by non-agricultural substitutes.

Throughout the Second Food Regime, agrarian economies were increasingly drawn into, and subsumed within, broader structures of the capitalist world, to such an extent that industrial conditions of production prevailed in most agricultural sectors in the developed world and increasingly so in developing countries. Throughout this Period, nationally oriented urban-centred industrial agro-food systems increasingly replaced subsistence and regionally focused commercial agricultures.

2.2.3.5 Productivist agriculture

Productivist agriculture refers to the development, and institutionalisation, of high-input/high-output industrialised agricultural systems underpinned by science and technology. Throughout the Productivist Period, governments of most developed countries cultured broad-based consensus and support from agricultural scientists, farmers, the public and emerging agribusinesses for the so-called techno-policy model of productivist agriculture and the associated processes of agricultural modernisation and commercialisation. This was achieved through the development of new and/or improved agro-industrial institutions, specifically the emerging agricultural policy communities (Perkins 1997; Agriculture Outlook 2000) that were created shortly after the Second World War.

Until the late 1800s, gains in crop yields resulted from area expansion and experimentation by amateur gentleman farmers. Even though scientific organisations, such as the American Association of Economic Entomology (founded in 1889) (Dent 2000), enhanced the pace of scientific discovery, it was the political, economic and institutional commitments of OECD governments (Perkins 1997) that made the rapid appliance of science and technology to agricultural intensification possible. This enthusiastic endorsement of the techno-policy model of agricultural development was also bolstered by the growing general acceptance of scientific innovation in society. For scientists weaned off the religious paradigm onto the scientific and public policy paradigm, positivist science held that science was value free. Hence, there was no need for ethical evaluation of scientific research or its products. Naïve economic utilitarianism re-enforced this assumption by providing reason to think that technologies adopted by farmers were inherently acceptable and desirable in ethical terms, and that a good scientist increased food production. Belief in the productivist paradigm of agricultural development was further reinforced through the recruitment of agricultural scientists from the ranks of farming, who also uncritically accepted productivism. Many agricultural scientists were, and in many respects, still are, the second sons or daughters of 'a farm too small' (Perkins 1997). By 1970, scientists working for corporate agribusinesses, public sector R&D institutes were highly trained in educational programmes dedicated to the production of agricultural scientists. In addition, they had the means of communicating their work that included both scientific and popular outlets, and they had substantial results to convey to farmers and the public (Perkins 1997).

In many respects, support of farmers mirrored that of the scientists, with the exception of a few so-called laggards. Indeed, many farmers across the developed world still adhered to the Protestant work ethic. The fruits of their labours and skill could be seen in the autumn in a crop of standing corn or fat cattle. Due to the conservatism and traditionalism of rural people, continuance of this work ethic, institutionalised through years of religious doctrine, was virtually assured. Here, the pursuit of abundance was a sign of grace (god's reward) on earth (Thompson 1971); indeed, the phrase 'this is god's country' encompasses grace but also a well-managed Garden of Eden, tended by god's faithful stewards (Thompson 1971). Farmers' perceived nature as something to be managed by human beings for productive purposes, and that setting land aside or reverting it back to nature were perceived as chaos, disorder and perverse (Thompson 1971).

In addition, the change from subsistence to commercial agriculture also had a major impact on farmers' attitudes to pest control, largely because commercialisation introduced the need to borrow or make cash investments, which had not been necessary in subsistence agriculture. Insect pest problems caused yield loss in subsistence agriculture but unless they were catastrophic they did not risk the farmer's livelihood, whereas in commercial agriculture, insect problems threatened the safety of other cash investments and thereby threatened the farmer's continued ability to stay in business (Dent 2000). This meant that the standards for acceptable levels of insect control increased from those that had been acceptable in the past (Dent 2000). The need to reduce risk by ensuring effective pest control became a prime consideration among commercial farmers.

Because public acceptance of, or support for, change needs to be understood as the aggregate attitude of individuals towards an innovation, it is often difficult to predict. In general, an individual's attitude towards an innovation is dependent on his/her perception of the risks and benefits associated with change, which, in turn, is underpinned by socially communicated values and the perceived trust in institutions. However, unquestioned and unconditional support of productivist agriculture was virtually assured from the citizens of OECD countries for two key reasons. (1) Citizens of these countries had experienced hunger and malnutrition during the global economic recessions of the 1920s and 1930s and during the Second World War. (2) This model of agricultural development promised to deliver a regular supply of a wide range of quality foods at reasonable prices as well as contributing to world peace. Likewise, support from agribusinesses (both upstream and downstream of the farmers) was also assured. Agribusinesses such as plant and animal breeders, and machinery, fertiliser and pesticide manufacturers and suppliers all stood to profit from a rapid expansion of input markets, and to a more limited extent, the sale of advice (Dent 2000). There were numerous opportunities for commercial companies that could provide efficient and effective low cost pesticides and application technology. Food processors, wholesalers and retailers also stood to profit from the larger volumes of a more diverse range of food products.

Although political and corporate advocates of agricultural science came closer to an explicit endorsement of the techno-policy model of productivist agriculture than farmers, productionism would not have succeeded if farmers had not been willing conspirators (or convinced of the merits of modern agriculture) (Thompson 1971). The conservative nature of farmers made change a slow process even though greater productive use could be made of their gardens.

Throughout the halcyon days of pesticides during the Productivist Period, stakeholder dynamics within evolving agro-industrial policy community institutions changed dramatically, and with them, so did the outcomes. Between 1950 and the early 1970s, farmers and agribusinesses built substantial power bases within agro-industrial policy communities and, thus, were able to significantly modify the direction of agricultural policy and more importantly the distribution of the associated costs and benefits of its implementation. Both farmers and the corporate sector quickly formed into lobbying organisations so that individuals could legitimately and authoritatively speak for a significant number of actors in the process of agricultural modernisation and commercialisation. At this stage, there was little involvement of social organisations, neither consumers nor taxpayers. Throughout this Period, IUPAC developed a close relationship with two key intergovernmental organisations: the United Nations' World Health Organization (WHO) and the United Nations' Food and Agriculture Organization (FAO) (Hough 1998) (see Fig. 2.1). These organisations regularly obtained technical advice on pesticides from IUPAC, and, subsequently, IUPAC has enjoyed 'specialized consultative' status with FAO since 1955 (Hough 1998). Under this arrangement, representative of FAO attends all IUPAC meetings, whilst IUPAC is allowed to send representatives to FAO Conference and Council sessions (Hough 1998). Thus, IUPAC carries a significant level of respect as a source of authoritative knowledge on the issues of increasing crop yields using pesticides (Hough 1998). As such, the IUPAC was/is ideally situated to influence key developments in international agriculture.

Aside from the direct financial benefits secured by the farming lobby and corporate agribusiness interests, there were also indirect benefits ensuing from state-sponsored agricultural research, development and extension, particularly when agribusiness interests were able to help determine the direction of R&D. The power of agribusiness lobbies enabled them to guide the work of technologists and scientists towards solutions that made sense for them (Perkins 1997). Once adopted, these new technologies increased both the wealth and power of their advocates, thus increasing their ability to influence the next round of scientific and technological investments (Perkins 1997). For example, the research and teaching agenda of Land Grant Universities focused much of their research time on blind tests of efficacy of new pesticides paid for through graduate assistantships by chemical companies (Thompson 1971). Professional science required: (1) the desire to develop better agricultural science; (2) the ability of society to support a cadre of scientists and (3) the power to allocate resources towards the research enterprise (Perkins 1997). These powerful corporate agribusiness interests underpinned the development and institutionalisation of the techno-policy model of intensive agricultural development.

2.2.3.5.1 Applying the urban-centred industrial agro-food system model to developing countries

After being demonstrably successful in stabilising the social, economic and political infrastructure of post-war Europe and the USA, and promoting rapid economic growth and prosperity for Europeans, Americans and their trading partners, the modern agro-food system was perceived as the rational mechanism with which to achieve food security and economic prosperity (see Fig. 2.1). The USA was keen to extend this model to the developing world (Bowler 1992), particularly where communist insurgence was apparent. In the words of President H. Truman, 'more than half the people of the world are living in conditions approaching misery. Their food is inadequate; their poverty is a handicap and a threat both to them and to more prosperous areas. The US is pre-eminent among nations in the development of industrial and scientific techniques. Our imponderable resources in technological knowledge are constantly growing and are inexhaustible. I believe that we should make available to peace-loving peoples the benefits of our store of technical knowledge in order to help them realize their aspirations for a better life' (Perkins 1997). Believed to have been underpinned by both food security and national security objectives, this statement institutionalised political and economic commitment from the US Government, and major US philanthropic foundations (e.g. Rockefeller and Ford), to the modernisation of agricultural systems in developing countries through a process that has since been termed as the 'Green Revolution' (Perkins 1997).

International support for agriculture manifest itself through establishment of the FAO established in 1945, and the Consultative Group on International Agricultural Research established in 1972 (Perkins 1997). Political and economic support for the modernisation of agricultural systems in developing countries was soon to follow from most of the remaining OECD nation states. Even though multinational corporations (MNCs) certainly found it in their interests to sell their products in the Third World, evidence indicates that the Green Revolution was not of their initiative. However, when MNCs realised the commercial opportunities that the Green Revolution offered, they quickly entered into the race to industrialise agriculture in developing countries (Perkins 1997). MNCs also incorporated former colonies into the global agro-food system. The extension of food trade between metropolitan nation states and newly independent post-colonial states was led by nationally based agribusinesses from North America and Europe (e.g. Coca-Cola, Del Monte, Heinz, Kellogg's, Nabisco, Pepsi and Unilever) as they sourced their raw materials through production contracts on a global basis (Atkins and Bowler 2001). Indeed, facets of the Green Revolution can be interpreted as appropriation, that is replacement of native seeds with commercially grown miracle strains of hybrid wheat and maize under the control of non-farm businesses in developed countries (Atkins and Bowler 2001).

However, in contrast to many developed countries, agriculture in developing countries was principally seen as an engine of economic growth, rather than as a means to achieve national food security or economic and social stability in rural areas. Instead of pump-priming agricultural modernisation, most developing country governments tended to tax their agricultural sectors to generate the financial resources to underpin urban-based public spending and industrial development. Indeed, throughout most of the 1960s, 1970s and 1980s, agriculture in developing countries was severely disadvantaged through the application of a range of macroeconomic policies and interventions. A World Bank study concluded that agricultural producers in 18 developing countries were taxed at a rate of 30% on average during 1960–84 because of both direct and indirect policies (E-council 2003). For example, overvalued real exchange rates stifled agriculture by raising domestic prices relative to world prices and by reducing the purchasing power of farm households; such interventions accounted for 75% of the tax on agriculture (E-council 2003).

To compensate farmers for these losses, governments in developing countries generally turned to subsidising credit and agricultural inputs such as irrigation, fertilisers and pesticides. In some cases, consumers have been subsidised to keep food prices low (E-council 2003). Pesticide policies include price and non-price subsidies, varying from free grants and preferential tax treatment for pesticide research to the excessive promotion of pesticide use.

Recognising the importance of agriculture to national food security and the economy, many developing countries constructed national agricultural development plans. Brazil's First National Development Plan was adopted in 1972. This plan laid down the policy of increasing agricultural productivity, and reducing pesticide imports by developing the Brazilian pesticide industry. Pesticide promotion intensified and thousands of sales representatives were employed. This encouraged agricultural concentration on soy, wheat and cotton crops, which require larger quantities of pesticides (Knight 1997). The Bank of Brazil began to make rural credit dependent on farmers applying 15% of the sum granted to buy pesticides.

2.3 Key characteristics of pesticide development processes and manufacturers

Establishment of a Second Food Regime, and the restructuring of nationally based agricultures, set the scene for a period of unprecedented, and generally unquestioned, rate of growth in the invention, innovation, development, production, distribution and use of an expanding range of specialised synthetic organic pesticides that were manufactured and supplied by emerging corporate agribusinesses. The production of pesticides became highly industrialised, even more so than agriculture, and increasingly science-based. Indeed, most pesticides during this Period were synthetic organic pesticides. Synthetic organic pesticides are those that do not occur naturally in the environment, but are synthesised by man, and are classified as organic because their structure is based on carbon and hydrogen atoms.

The Productivist Period also witnessed the development of large-scale research based on mass screening and testing of chemicals and high rates of increase in pesticide production and consumption (Boardman 1986). This Period corresponds to an era of rapid scientific and technological discovery in the USA and Europe and can be considered as the halcyon days of pesticides. Pesticide development during the Productivist Period was directly affected by two parallel developments: (1) medical and military research; (2) spillovers form industrial research.

2.3.1 Medical and military research

2.3.1.1 DDT development and the pesticide industry

Many synthetic organic pesticides emerged as direct spillovers from both medical and military research (see Fig. 2.2). The two most important and noteworthy examples are the development of organochlorine and organophosphate (OP) pesticides. Although first prepared in 1874, it was not until 1938 (Hoechst 2003a) when Dr. Paul Muller, an entomologist working for the Geigy Chemical Company, discovered that dichlorodiphenyltrichloroethane (DDT) was highly effective in destroying the mosquito species that transmitted malaria and yellow fever to humans as well as in killing many other insect-vectors for other diseases such as typhus and plague (Dunlap 1981). The Geigy Chemical Company was a relative newcomer to the pesticide sector, having only begun to produce insecticides in 1935 (Hoechst 2003a). In parallel to their investments in the evolving pesticide market, the Geigy Chemical Company also invested in pharmaceuticals, establishing a pharmaceutical department in 1938 (Hoechst 2003a).

It was DDT's ability to control malaria that attracted the attention of the US military. Work on the large-scale production of DDT began in 1943 in order to control malaria among US troops. By 1945, production of DDT had reached 140 000 tonnes/year. When hostilities ceased in 1945, and Allied troops withdrew from mosquito-ridden battlefields, the US military had amassed significant stockpiles of DDT. This prompted the US Government to search for non-military uses for the insecticide. Field trials of DDT in Switzerland against Colorado potato beetle in 1941 had already demonstrated DDT's pesticidal qualities, namely low cost, broad-spectrum insecticidal activity with low phytotoxicity and low acute mammalian toxicity.

It was therefore just a few years before stockpiled supplies of DDT were switched from military uses to both agricultural and health care sectors. By 1946, DDT had become a widely used agricultural insecticide. The discovery of DDT was acclaimed worldwide and, in 1948, Dr. Muller was awarded the Nobel Prize in Medicine and Physiology. DDT was the standard-bearer of modern pesticides and acknowledged as yet another major technological breakthrough and important public good. Since the 1940s, DDT has been used throughout the world. It is estimated that, during the 30 years prior to its eventual cancellation, a total of approximately 1 350 000 pounds of DDT was used in the USA, and that more than 4 billion pounds of DDT have been made and used worldwide (EPA 2003a). DDT was one of the most significant events in history of insect control (Boardman 1986). Indeed, it has been accredited with saving more lives than penicillin (Boardman 1986).

In turn, the success of DDT stimulated the search for similar chemicals and the subsequent development of other chlorinated hydrocarbons. Most of the organochlorines were released by either US or British companies. In 1945, Velsicol (USA) released chlordane, an insecticide used for the control of insects on maize, citrus, deciduous fruits and nuts and vegetables (EPA 2003b). In 1946, Allied Chemical Co. (USA) released mirex and heptachlor, insecticides for use on maize, wheat, sorghum, pineapple and other fruit crops (BCERF 1998). In 1947, Hercules Powder Co. (USA) released toxaphene, an insecticide for use on cotton, vegetables, livestock and poultry, soybeans, alfalfa, wheat and sorghum (EPA (2003c). In 1948, Shell (UK) released the insecticides aldrin, dieldrin and HCH (Lindane), followed by endrin, an insecticide for use on cotton, maize, sugarcane, rice, cereals, ornamentals and other crops, in 1951 (Greenpeace 2002; EPA 2003b). Finally, endosulfan was released in 1954 by Hoechst to control agricultural insects and mite pests on a variety of field, fruit and vegetable crops (EPA 2003e).

2.3.1.2 Organophosphate development and the pesticide industry

The story of OP pesticides is, in many ways, similar to the story of DDT. However, the principal difference is that instead of being appropriated by the military, OPs were explicitly developed for military purposes. Major research on OPs, by Dr Gerhard Schrader and co-workers at IG Farben in Germany, began in the 1930s and continued throughout the Second World War.

OP compounds demonstrated high volatility, rapid skin penetration and potent effects on the central nervous system. They work by interfering with the activity of an enzyme, cholinesterase, which is necessary for proper nerve function. Without this enzyme, impulses continue to pass down the nerve fibre disrupting the nervous system and, ultimately, result in death by respiratory failure. These unenviable qualities, and vis-à-vis potential uses, were quickly recognised. In turn, OPs were developed by the Germans as nerve agents for use against Allied troops. Nerve agents developed during and after the Second World War include Tabun, Sarin and Soman and the so-called V agents, such as VX. These are less volatile but more toxic than the earlier nerve agents. Thankfully, discovery of these nerve agents came relatively late during the war and OPs found no significant use as chemical warfare agents (Berenbaum 1995).

As with DDT, by the end of the war IG Farben was endowed with both a scientific and technological understanding of OPs and was keen to investigate and exploit their non-military applications. The anti-cholinesterase properties of OPs made them potential candidates for agricultural insecticides. At this time, the global insecticide market was growing and thus offered one of the most promising non-military applications, with the potential for high financial returns. The first OP insecticides to be tested included the powerful insecticide schradan (octamethylpyrophosphoramide), which acts as a systemic insecticide against aphids and red spider, and the contact insecticide parathion (O,O-diethyl-p-nitrophenyl phosphorothionate), which is remarkably effective against aphids, red spider and nematodes. However, as with the nerve agents of the Second World War, these pesticides were also highly toxic to humans and animal livestock and were therefore of little practical application to agriculture.

The first successful agricultural OP insecticides were tetraethyl pyrophosphate (also known as TEPP) released in 1946, and ethyl parathion released in 1947. The introduction of OPs brought a new class of insecticides with reduced persistence and lower risks to both users and the environment. However, it must be noted that, excessive exposure to these early OP insecticides still presented a high risk of irreversible OP poisoning, especially to agricultural workers applying OP pesticides and to non-target animals.

2.3.1.3 Phenoxyacetic acid development and the pesticide industry

The 1930s and the 1940s also witnessed the emergence of the modern era of synthetic organic herbicides. In 1943, Templeman and Secton, working for Imperial Chemical Industries (ICI) in England, discovered the herbicidal activity of the phenoxyacetic acids. These chemicals are based on the plant hormone indoleacetic acid (IAA or auxin), discovered by Fritz Went in 1929. Phenoxyacetic acid herbicides possess the capacity to kill broadleaved plants but not grasses/cereals. They were the first successful selective herbicides to be developed. The most notable phenoxyacetic acid herbicides were: 2,4-dichloro phenoxyacetic acid (2,4-D), released as 2,4-D Weedone® by Amchem (USA) in 1942; 2,4,5-T Weedone® released by Amchem (USA) in 1944; 2-methyl-4-chloro-phenoxyacetic acid (MCPA) Agroxone® released by ICI (UK) in 1945; and Butyrol® and Butoxone® (both 2,4-DB) released by Amchem (USA) in 1947 (Agranova 2001).

The efficacy and selectivity of phenoxy herbicides, particularly 2, 4-D and MCPA, compared favourably to the best alternatives at the time, allowing these herbicides to dominate the herbicide market for several decades (Duke 1990). However, as with organochlorine and OP insecticides, the release of phenoxy herbicides for agricultural purposes was greatly accelerated by the significant military R&D efforts at Camp Detrick, Maryland, USA, both during and after the Second World War (Detrick 2003). The defoliation properties of phenoxyacetic acids were highly valued by the military, particularly for use in military campaigns in areas of dense vegetation. The USA began work to test the defoliation properties of 2,4,5-T and 2,4-D in 1941, in preparation for use against the Japanese, but were never used (Hough 1998). Indeed, even though the phenoxies were not used in action during the Second World War, the British were the first to use them during the Malayan emergency; 2,4,5-T and 2,4-D were used to clear communication lines and destroy food crops in an attempt to halt the communist uprising. ICI provided technical assistance to the UK and Malayan Governments (Hough 1998). However, spraying stopped when studies suggested that it was more economical and effective to remove vegetation by hand (Hough 1998). By far the most significant use of phenoxy herbicides under wartime conditions was undertaken by the USA in the Vietnam War. Between 1962 and 1971, the US Air Force dropped a variety of phenoxy mixtures, including the infamous Agent Orange (Hough 1998). US scientists estimated that approximately 10% of Vietnam's inland forests, 36% of mangrove forests and 3% of cultivated land were affected by the programme code named Operation Ranch Hand (Hay 1982).

Other important herbicide discoveries include the first triazine-based herbicides Simazine and Atrazine (Hoechst 2003a) developed by Geigy in 1956. Simazine was introduced as a pre-emergence herbicide for the control of broadleaved and grassy weeds in a variety of deep rooted crops such as artichokes, asparagus, berry crops, broad beans and citrus and stone fruit orchards (EPA 2003g), and Atrazine as a grassy weed herbicide in maize, sorghum, soybean rangeland, sugarcane and other minor crops (EPA 2003h). Paraquat was developed in 1955, by ICI, as a non-selective herbicide for use in orchards and plantation crops, and non-cropped land, and as a defoliant on certain crops before harvesting but was not commercialised until 1961 (Ekboir 2002; PAN-UK 2003b.

2.3.2 Spillovers from industrial research

In 1951, Kittleson, working for the Standard Oil Company in America, discovered the fungicidal properties of N-trichloromethyl-thiotehydrophthalimide. Marketed as Captan, this fungicide had outstanding properties as a protectant against a wide spectrum of pathogenic fungi on fruit and vegetable crops. Subsequently, a number of other N-trichloromethylthio compounds have been marketed as foliar fungicides. Based on the work of Chain and Flory in 1940, useful antibiotics such as Aureomycin and streptomycin were soon discovered to demonstrate activity against certain fungal pathogens and bacterial diseases of plants, leading to the commercial fungicidal use of streptomycin by 1952.

2.3.3 Carbamate development and the pesticide industry

Carbamates were first discovered by researchers working for the Geigy Company in 1947, and were developed as an important class of agricultural insecticides from the late 1950s. Like OPs, carbamates are cholinesterase inhibitors. However, their inhibition of this enzyme is reversible. Therefore, compared to OPs, people excessively exposed to carbamates have a greatly reduced likelihood of acute nerve poisoning and a greatly increased recovery rate. Carbamates, like OPs, do not accumulate in the environment or fatty tissues of mammals. Both carbamates and OPs act as contact insecticides and have some stomach poisoning activity. Examples of carbamate insecticides include: Carbaryl released in 1959 by Union Carbide (USA) (EPA 2003j); Methomyl released in 1968 by DuPont (USA) (EPA 2003k); Propoxur released in 1963 by Chemagro (USA); and Aldicarb (sold as Temik) released by Union Carbide (USA) in 1970). In addition to their use as insecticides, a number of carbamates are also used as herbicides and fungicides.

2.3.4 Synthetic pyrethroid development and the pesticide industry

In 1973, Elliott and his team at Rothamsted Experimental Station in England demonstrated that it was possible to prepare synthetic analogues of permethrin that were much more potent insecticides than natural pyrethrum and were photochemically stable. These chemicals possessed many of the same characteristics as pyrethrum, but had greater killing power (almost ×10 of carbamate and OP insecticides), and were somewhat less irritating to the eyes and skin. In 1975, permethrin and cypermethrin, the first synthetic pyrethroid insecticides were released by ICI (UK) (Dent 2003). Within just a few years, permethrin had captured 60–75% of key herbicide markets in the USA (Boardman 1986).

2.3.5 Systemic fungicide development and the pesticide industry

This Period also witnessed a growing range of new fungicides, including: sulphanilamides to protect wheat seedlings from rust; the first dithiocarbamate fungicides Zineb and Copper Sandoz, introduced in 1943 (Hoechst 2003a); and Maneb, introduced in 1950. Copper Sandoz was Sandoz chemical company's first pesticide, having only entered the field of agrochemicals in 1939 (Novartis 2003a).

However, aside from notable examples such as the aforementioned wide-spectrum fungicide Captan (or N-trichloromethyl-thiotehydrophthalimide) released in 1951, it was not until the 1960s that effective systemic fungicides appeared on the market. Their development represented an important breakthrough in the field of plant protection. Having been forced to sever ties with its German parent company, the US division of Rohm and Haas chemical company introduced the first synthetic organic fungicide EBDC (ethylene bisdithiocarbamate) under the name of Dithane in the early 1960s (Rohmhaas 2003). In 1962, the company introduced a zinc and manganese combination product Dithane M-45, which rapidly became its bestselling product (Rohmhaas 2003). This was quickly followed by the release of Asulam (Asulox) by May and Baker (USA); Dicamba (Barvel®) by Velsicol (USA) in 1965; Carboxin, Ethirimol, cyanazine (Bladex®) by Shell (UK); and Benomyl (Benlate®) in 1972 by DuPont (USA).

Monsanto, another US diversified company, was also a late entrant into the pesticide market with its introduction of alachlor in 1966.

The Productivist Period witnessed the emergence of a considerable range of pesticides. Indeed, this Period also witnessed the discovery and/or development of new inorganic and botanical insecticides. These included the following: (1) The discovery of systemic insecticidal properties of inorganic selenium and the 'botanical' anabasine ((S)-3-(2-piperidinyl) pyridine) (Donga 2003). (2) Ryanodine (patented in 1945). (3) Development of the first dependable nicotine dust (nicotine-bentonite) combination (Donga 2003). (4) Introduction of the soil insecticide and nematocides chloropicrin and ethylene dibromide (EDB) in 1948 (IFAS 2003; EPA 2003i). (5) Commercialisation of *Bacillus thuringiensis*. (6) The soil fumigant methyl bromide in 1932. (7) The introduction of DNOC and DNC, and the first dinitrophenol herbicide, in the USA in 1938 (Agranova 2001).

2.3.6 Establishment of pesticide corporations

Throughout the Productivist Period, the industrialisation of agro-food systems and intensification of agricultural production provided ideal conditions for the rapid growth of specialist pesticide corporations or pesticide divisions of diversified corporations.

Geigy (Switzerland) founded in 1763 to trade materials, chemicals, dyes and drugs (Novartis 2003a), and incorporated in 1914 as J.R. Geigy Ltd (Novartis 2003b), was one of the most successful and influential pesticide corporations of the Period. Its commercialisation of DDT and the triazine-based herbicides, Simazine and Atrazine, were the basis for much of this success (Boardman 1986).

Hoechst (Germany), founded in 1863 to produce fuchsia dyes, began manufacturing and marketing pharmaceuticals in 1883 (Hoechst 2003a). One of its first major products was Antipyrin, an early analgesic and antipyretic. During World War I, Hoechst produced mustard gas and explosives (Hoechst 2003a).

Friedrich Bayer and Johann Friedrich Weskott founded Bayer (Germany) in 1863. Its purpose was to manufacture and market synthetic dyestuffs (Bayer 2003a). In 1865, Bayer acquired an interest in its first coal tar dyes factory in Albany, New York. Bayer's workforce expanded from 3 in 1863 to 300 in 1881 (Bayer 2003a). Between 1881 and 1914, Bayer developed into an international chemical company. Although dyestuffs remained the core business, Bayer diversified into new areas such as pharmaceuticals, launching Aspirin in 1899 (Bayer 2003b). By 1913, Bayer employed 10 000 staff, 9000 of which were based in Russia, France, Belgium, the UK and the USA (Bayer 2003b).

In the 1920s, Bayer and BASF (two important German dye-makers) pushed for a merger to create a single company. During these expansionary days,

German manufacturers dominated the international dye market, claiming 90% of the world market (Hoechst 2003a). Over 88% of Hoechst sales were international. In 1925, Hoechst, Bayer and BASF merged to form IG Farben (Interest Community of the Dye Industry, Inc.), which was the largest company in Europe and the fourth largest in the world (Gupta 2003). Subsequently, IG Farben developed the first OPs in the 1930s. Bayer began research on rubber and modern polymer chemistry in the early 1930s. In 1937, Otto Bayer invented polyurethanes (Bayer 2003c) and Gerhard Domagk received the Nobel Prize in 1939 for his discovery of sulfonamides, a key breakthrough in the chemotherapy of infectious diseases (Bayer 2003c).

After the war, several IG Farben executives were convicted of war crimes at the Nuremberg Trials. In 1952, after 6 years of US military control (Hoechst 2003a), IG Farben was divided into several independent firms, including BASF, Bayer and Hoechst (Gupta 2003). In 1957, Bayer entered the petrochemicals sector. By 1963, Bayer employed nearly 80 000 people (Bayer 2003d). BASF, founded by Friedrich Engelhorn to produce coal tar dyes (BASF 2003), moved into production of crop protection agents and fertilisers in the mid-1960s (BASF 1998).

Imperial Chemical Industries (UK), A.K.A. (ICI), was formed by merger of the UK's four principal chemical companies (United Alkali Co., Ltd., Nobel Industries, Brunner Mond Ltd. and British Dyestuffs Coronation Ltd.) in 1926 (Syngenta 2003). 2,4-D, MCPA, paraquat and synthetic pyrethroids have been of central importance to ICI's success in the pesticide marketplace.

'Kern and Sandoz' (Switzerland) was founded in 1886 by Dr Alfred Kern and Edouard Sandoz to manufacture and market dyes (Novartis 2003b). In 1895, the company produced its first pharmaceutical antipyrine, a fever-controlling agent, establishing a pharmaceuticals department in 1917 (Novartis 2003b).

DuPont (USA) entered the pesticide market in the 1930s as one of the world's largest diversified companies (Groundup 2003a). DuPont was initially founded in 1802 by E.I. DuPont to manufacture gunpowder (Genewatch 2003b). By 1905, it held 75% of the US powder market. In the early twentieth century, DuPont diversified its business into textiles, particularly artificial fibres, invented cellophane and acquired General Motors (Corporate Watch UK 2003b). During the 1940s, DuPont profited from the production and sale of military explosives, including plastic explosives, rocket propellants and chemical warfare. Indeed, DuPont supplied 40% of the explosives used by Allied Forces during World War I (Corporate Watch UK 2003b). Between 1941 and 1945, DuPont contributed to the Manhattan Project and the production of plutonium. However, unlike many other corporations, DuPont resisted the trend to diversify its business during the Productivist Period, and focused on its key areas and competencies (Corporate Watch UK 2003b).

John Francis Queeny founded Monsanto in 1901. One of its original products was the artificial sweetener saccharin. Aspirin production began in 1917 (Monsanto 2003a). In the 1920s, Monsanto was a leading manufacturer of basic industrial chemicals, such as sulphuric acid. Monsanto soon became the world's leading manufacturer of polychlorinated biphenyls (PCBs) (Genewatch 2003a). In the 1930s, Monsanto began a rapid expansion programme with the acquisition of companies dealing in key chemicals for rubber, textiles, paper and leather (Monsanto 2003a). Monsanto expanded its operations into Australia and Canada during the early 1930s. In the late 1930s, Monsanto also became involved in the Manhattan Project and the search for the world's first nuclear bomb (Monsanto 2003a). In 1945, Monsanto began producing insecticides and herbicides. In 1947, Monsanto opened an office in India. In 1960, Monsanto establishes it agricultural division, and, in 1963, opened its Brazilian office (Monsanto 2003a).

In 1942, Dow expanded its operations into Canada (Dow 2003a). In 1948, plastics accounted for 20% of sales (Dow 2003b). In 1957, nearly 8% of sales were overseas; chemicals accounted for 52% of sales and plastics 35% (Dow 2003b).

By the mid-1970s, pesticide corporations had developed several hundred AIs and were marketing 50 000–70 000 commercial formulations in the USA (Boardman 1986). In 1974, sales of 643 000 tonnes of synthetic organic pesticides in the USA totalled US$1.732 billion. In the UK, sales totalled £159 million in 1975 (Boardman 1986).

2.4 Key outcomes of pesticide development and application

As with the preceding Period of pesticide evolution, increased use of crop protection agents in the Productivist Period contributed to the production of higher and more stable crop yields and reduced post-harvest losses. In turn, this contributed to increased world agricultural output, which, in turn, contributed to food security, economic development and the formation and stability of independent nation states. Specifically, the growing use of crop protection agents continued to contribute to increased food supplies and food security. Indeed, by the end of the Productivist Period, the Fordist Food Regime had institutionalised regular aggregate global food surpluses and assured national food security in most countries, even though access to regular and sufficient quantities of nutritionally adequate food was not always assured in some developing countries. The use of crop protection agents also improved farmer profitability and household food security in many parts of the world. However, during the Productivist Period, structural tendencies embedded in the urban-centred industrial agro-food system created conditions under which, in order to remain profitable, farmers were obliged to 'jump' onto what came to be known as the technological and pesticide treadmills. Concerns also began to coalesce around growing negative impacts of pesticides on the environment, food quality and safety and human health.

2.4.1 Global food supplies, national food security and pesticide usage

Throughout the Productivist Period, state-subsidised industrialisation of farming systems succeeded in significantly increasing global food supplies and improving national food security in both developed and developing economies (see Fig. 2.1). Increased food supplies could be attributed to the adoption of high-yielding modern crop varieties, fertilisers and pesticides, particularly herbicides and insecticides (Hayes 1991). For example, before 1939, average cereal yields in the UK were approximately 2 tonnes/ha, demonstrating little improvement since records began in 1880 (Britton 1990). However, between 1952 and 1969, annual increases in oat yields averaged 2.3% (Britton 1990), matched closely by wheat (2.2%), barley (1.9%), potatoes (1.7%) and sugar beet (1.6%) (Britton 1990). During this time of continuous modernisation in agriculture, farmers' expenditures, chemical application and crop yields continued to increase. Farmers' expenditures on machinery and equipment increased 181% between 1950 and 1969. Farmers quadrupled chemical application from 8.5 million tons in 1940 to 37 million tons in 1968. World production of formulated pesticides increased from approximately 100 000 tonnes in 1944, to 450 000 tonnes in 1955, 1 000 000 tonnes in 1965 and 1 500 000 tonnes in 1970 (WHO 1990).

Due to the industrialisation of agriculture, crop yields for the major grain crops increased steadily between 1946 and 1969. For example, maize yields increased over 150% during this Period. In the 1950s and 1960s, pesticide use became routine in many developed countries. At this time, pesticide use in the USA averaged 1.5 kg/ha, 1.9 kg/ha in Europe and 10.8 kg/ha in Japan. Production of pesticide in Japan doubled between 1965 and 1970 and again between 1972 and 1974 (Boardman 1986). In the USA, pesticide use dramatically accelerated during the 1960s and early 1970s, and use spread to all major crops. By the 1950s, insecticides were being used widely on a variety of high-value crops including cotton, tobacco, peanuts, potatoes, fruits and vegetables. Insecticide use on major field crops accelerated during the 1960s and 1970s. For example, during the mid-1950s, less than 10% of the maize area was treated with insecticides. This expanded to 35–45% by 1975 (Pimentel et al. 1993). Herbicide use accelerated in the USA during the 1950s. Tables 2.1–2.5 demonstrate the growth in pesticide use in the USA during the 1960s and 1970s.

However, whilst pesticide usage grew rapidly in OECD countries (such as the USA), it remained relatively low in developing countries, Africa 0.13 kg/ha,

Table 2.1 Insecticide use on major crops in the USA (Pimentel et al. 1993)

Year	Cotton	Maize	Soybeans
	Million pounds of active ingredient		
1964	78	15.7	5.0
1966	64.9	23.6	3.2
1971	73.4	25.5	5.6

Table 2.2 Shares of insecticide classes on major crops in the USA (Pimentel et al. 1993)

Year	Organochlorines	Organophosphates	Carbamates
	Million pounds of active ingredient		
1964	70	20	8
1966	70	22	7
1971	45	39	14

Table 2.3 Herbicide use on major crops in the USA (Pimentel et al. 1993)

Year	Cotton	Maize	Soybeans	Wheat and small grains
	Million pounds of active ingredient			
1964	4.6	25.5	4.2	18.3
1966	6.5	46	10.4	13.2
1971	19.6	101.1	36.5	17

Table 2.4 Shares of herbicide classes on major crops in the USA (Pimentel et al. 1993)

Year	Phenoxies	Amides	Triazines	Nitro-phenols	Carbamates	Anilines
	Million pounds of active ingredient					
1964	46	6	15	1	7	1
1966	38	5	21	2	7	4
1971	16	22	28	21	5	6

Table 2.5 Pesticide use on major crops in the USA (Pimentel et al. 1993)

Year	Herbicides	Insecticides	Fungicides
	Million pounds of active ingredient		
1964	70.5	116.7	5.8
1966	101.2	108.3	6.0
1971	213.1	127.9	6.4

Table 2.6 Number of active ingredients registered for the first time as a pesticide under the Federal Insecticide, Fungicide, and Rodenticide Act (FIFRA)

Year	Insecticide	Herbicide	Fungicide
1967	4	2	2
1968	6	2	5
1969	7	4	0
1970	1	2	2
1971	0	1	1
1972	4	5	6
1973	5	3	4

Oceania 0.20 kg/ha and 0.22 kg/ha in Latin America (Hough 1998). The average compound rate of growth for pesticide use in developing countries during 1960–78 was 7.6%. However, there were a few positive deviants, such as Taiwan and the Philippines, which, between the early 1960s and the early 1970s, witnessed a significant increase in pesticide use on rice (Barker et al. 1985).

Pesticide manufacturers responded to the market opportunities by developing and releasing a wide range of pesticide products. By the mid-1940s, there were approximately 50 basic formulations of chemicals in the USA (Boardman 1986). Table 2.6 illustrates the steady stream of AIs registered for the first time as a pesticide under the FIFRA, which will be discussed later (EPA 1997).

2.4.2 Improved farmer profitability and household food security

Global food surpluses and national food security, at least for the majority, was primarily achieved through increased farmer productivity, both in the developed and developing world. For example, in the 1940s, US farmers could produce enough food for themselves and nine others. By the 1970s, this ratio had increased to themselves and 31 others (Green 1976). Significant increases in farmer productivity were closely associated with the use of fertiliser and insecticides (Hayes 1991). Pesticides provided farmers with a low cost form of insurance against yield fluctuations (Pimentel and Pimentel 1979). Indeed, as the relative price of pesticides decreased, many farmers began to use pesticides prophylactically, leading to their economically inefficient overuse, the destruction of beneficial organisms and natural enemies, acceleration of pesticide resistance development, and monoculture production (Pimentel et al. 1993).

During this Period, many cost-benefit analyses were undertaken to assess the rationale of pesticide use. In the 1960s, the benefit-cost ratio for the use of pesticides was between 4:1 (Pimentel and Pimentel 1979) and 6:1 (Headley

1968), and between 2:1 and 13:1 in the 1970s (Campbell 1978). However, it must be noted that the conclusions of cost-benefit exercises were often contested as the economics of pesticide use varied considerably between products, regions, local ecologies, crops, climate and pest and disease pressures (Barker et al. 1985). Rates of return on pesticide use were also affected by the growing problem of pesticide resistance, particularly in relation to insecticides in selected areas (Pimentel and Pimentel 1979). However, during this Period, failing pesticides were often replaced with new and equally effective pesticides.

2.4.3 The technological treadmill

The treadmill model is seen as a symbiosis between the interests of the state, agro-industrial corporations and agricultural science (Goodman and Redclift 1991), and is envisaged as an assemblage of structural conditions determined by international political and economic processes. During the Productivist Period, these conditions were embodied in agro-food policies of the developed capitalist world and encouraged farmers, with the promise of higher profits, to substitute industrial inputs for traditional inputs and labour (Ward 1993). Indeed, the arrival of pesticides shortly after the Second World War was greeted with euphoria since they allowed far more control of the agricultural system and assisted the process of agricultural industrialisation (Dent 2000). Initially, early adopters of output and productivity enhancing technologies reaped the benefits of lower unit costs of production, set against the backdrop of relatively stable commodity prices. However, as more farmers adopted new technologies and food supply outstripped food consumption, all farmers faced static or falling profits and were once again obliged to find new technologies that increased output and productivity and reduced the unit cost of production. As the supply of food commodities continued to outstrip demand during the Productivist Period, farmers found themselves on a continuous technological treadmill, forever searching for technologies that had an economic advantage in the face of stagnant commodity markets (see Fig. 2.2). Pesticides were seen as key candidate technologies that could potentially endow farmers with an economic advantage, thus creating a pesticide treadmill. However, only early adopters win because of the technological treadmill effect (Pimentel and Pimentel 1979).

2.4.4 The pesticide treadmill

Envisaged as a component of the technological treadmill, the pesticide treadmill is comprised of three core tenets. First, as with the general principles of technological treadmill, farmers find themselves continually searching for pest and disease control solutions that contribute to reduced unit costs of production at the same time as maintaining quality standards, which are often crucial to the profitability production of a particular crop commodity. Second, once a farmer is committed to, and reliant on, the use of pesticides inputs, it is difficult to explore alternative methods of pest and disease control and remain in viable business (Clunies-Ross and Hildyard 1992). Farmers therefore find themselves locked into the use of pesticides. Third, the pesticide treadmill concept also encompasses the knock-on effect of pesticide use on agroecosystems. This is particularly pertinent with the use of insecticides, as chemical pest control strategies generally ignore ecological factors that control insects. This often leads to problems with pest resurgence and the initiation, or acceleration, of an insecticide treadmill (Thompson 1971). For example, traditional cotton production in Central America focused on picking off insects by hand, and low levels of pesticide use such as tobacco infusions. There were three pests, red boll weevil, leaf-worm and desert locust. In the early 1950s, new chemical insecticides were applied, generally eight applications per season, and yields increased. By the mid-1950s, three new pests had emerged including bollworm. In the 1960s, insecticide use expanded rapidly; as many as 50 alternative pesticides were available for one pest. Two species disappeared but four new ones appeared, including armyworm and whitefly. Pesticide use increased to 50% of production costs, and pesticides were applied 28 times per season. By the early 1970s, there were eight important pests and, in some areas, 40 pesticide applications per season (Bull 1982). Examples can also be found in the developed world. In Texas (USA), the pink bollworm developed resistance to a range of cotton boll insecticides (Thompson 1971). Other examples include. (1) Cotton production in the Sudan where production costs increased four fold yet yields have dropped by 40% (Bull 1982). (2) India, where the application of DDT increased from 3 million kg in 1970 to 6 million kg in 1976 to maintain static levels of cotton production at just over 5 million bales of fibre (Norris 1982). (3) Nicaragua, where cotton yields fell 30% between 1965 and 1969 despite increased pesticide applications (Swezey et al. 1986). (4) Brazil (1964–79), where pesticide use increased by 421%, yet agricultural productivity increased by only 4.9% (Environmental Liaison Centre 1987). Finally, while insecticide application increased by a 1000 fold in the USA, crop losses remained relatively high at 12% (Pimentel et al. 1993).

Even given the disempowerment associated with pesticide use, the use of pesticide inputs became firmly institutionalised during the early part of the Productivist Period. The term 'institutionalised' alludes to the fact that pesticide use became a normal routine or convention and farmers implicitly expected to budget for them (Dent 2000). Indeed, pesticide use could be deemed as unquestionably institutionalised when the introduction of new pesticides seemed inevitable, natural (Burkhardt 1992) and logical (Ward 1993). In addition, the faith in, and addiction to, chemical technologies by farmers was assured through the continued development and availability of new pesticide

AIs coming onto the market. The pesticide treadmill was also perpetuated as farmers were persuaded to use more pesticides by company salespersons and advertisements and the lack, or perceived lack, of economically viable alternatives to the use of pesticides. This situation was exacerbated by the fact that the lion's share of private and public sector agricultural research was focused on scientific research that underpinned intensive industrial-based agriculture. Government policies, including tax incentives, made the production and use of pesticides increasingly attractive to agribusiness and farmers. Pesticide development, production and use became so institutionalised that farmers became increasingly dependent upon them (Dent 2000). In some respects, farmers have become increasingly like pesticide companies in as much as pesticide companies are constantly on a product treadmill, searching for new off-farm products that have a profitable market for a given number of years to produce a positive and often significant return on investment (Dent 2000). There is no sustainable product. As long as there is a good return on investment, companies remain on the product treadmill (Dent 2000). However, the subtle difference is that pesticide companies have substantially more control, than farmers do, over the prices that they receive for their products and therefore the returns on investments.

2.4.5 DDT and the control of pest-transmitted diseases

Malaria control is one of the true success stories of DDT. In 1955, in an attempt to reduce the 2.5 million lives per year claimed by malaria (Pampana and Russell 1955), the WHO launched a global malaria eradication programme (Hough 1998). DDT was the principal insecticide used to control mosquitoes, the insect vector of malaria, and virtually eliminated the disease wherever it was used. The most remarkable example is the use of DDT in Sardinia. In 1942, prior to the use of DDT, there were 78 000 cases of malaria per annum. In 1951, after several years of DDT treatment, only nine cases were reported (McEwen and Stephenson 1979). In Sri Lanka, malaria cases fell from 2.8 million and 7300 deaths per year before DDT spraying to 17 cases and no deaths after the introduction of DDT (Hicks 1992).

2.4.6 Pesticides and the environment

It is now widely acknowledged that, as part of the process of industrialisation, the application of the techno-policy model of productivist agriculture led, and continues to lead, to many negative environmental impacts (MacEachern 1992; Gasson and Potter 1988). These include problems such as: (1) Nitrate and pesticide pollution in watercourses. (2) Long-term environmental effects on soil structure (caused by the excessive use of nitrogen fertilisers, pesticides and intensive soil management techniques). (3) Concerns over animal welfare. (4) Degradation of valued landscapes and wildlife habitats in both the developed and developing world (Goodman and Watts 1997; Gasson and Potter 1988; Wrathall 1988; Ward 1993; Pimentel et al. 1995; Watson 1998; Ward et al. 1990). Pingali et al. (1997) provide an illuminating insight into the environmental impacts of the Green Revolution in Southeast Asia.

2.4.6.1 DDT

By the early 1960s, there was growing evidence that DDT, when sprayed over large areas, killed fish and birds, as well as the intended insect pest species, and, if consumed by cows, would contaminate milk. The widespread use of DDT also caused concern in the organic farming community, as, by their own definition, DDT contaminated their crops and fields so that they could no longer sell their produce as organic (Anderson et al. 1999). By the 1960s, scientists who studied birds were becoming increasingly convinced that DDT was decimating bird populations where it was sprayed widely. Studies of peregrine falcon and other birds of prey found evidence of a thinning of eggshells caused by DDT (Dunlap 1981). Conclusive or not, studies by qualified scientists pointed to wildlife problems stemming from extensive government use of DDT. The DDT metabolite residue DDE was linked to a decline in peregrine falcon, osprey and brown pelicans (Pimentel and Pimentel 1979).

2.4.6.2 Changing attitudes towards pesticide use in developed countries

Mounting evidence on the effects of particular pesticides on the environment and human health generated a growing chorus of concern from scientists and public interest groups (see Fig. 2.1). Established in 1948, the International Union for Conservation (IUCN) was among the first organisations to raise international concern over the use of pesticides (IUCN 1957). In 1961, IUCN's Commission on Ecology established a special group mandated to investigate the ecological effects of chemical methods of pest control (Boardman 1986). However, Rachel Carson made the most dramatic statement of these concerns in her classic work, Silent Spring, in 1962. It is often suggested that this publication not only did question the widely held perception of the environmentally benign nature of agriculture but also rekindled concerns over the impacts of pesticides in the human food chain, and on those involved in the manufacture and application of pesticides.

While pesticides were indeed boosting yields and bringing various pests under control, she warned that they were also accumulating in the environment and threatening the very existence of various forms of wildlife, and potentially of humans as well.

Whilst pesticide use continued to command significant support for their contribution to improving crop harvests and fighting diseases like malaria, this support now needed to be justified, and reason used to convince others that the benefits of pesticide use outweighed the costs. At the same time, environmentalists and sceptics had to mobilise opinion behind their new values. In this way, the arena of political contention was founded (Hough 1998). In 1966, IUCN prepared a draft policy statement. It recognised that pesticide use was still required for food production and vector control, but it expressed concerns over the impact on non-target species and ecosystems, monoculture farming and the need for biological control. It also stressed the need for minimum rates of pesticide use and increased pesticide specificity, and suggested that pesticides varied in their hazards depending on contexts (Boardman 1986). However, IUCN advisory experts were critical of the anti-pesticide line, which became almost fundamentalist in nature as more environmental groups evolved (Boardman 1986). In that same year, IUCN's General Assembly called for greater control over the trade in pesticides (IUCN 1967).

Public authorities and industry often expressed frustration with the public's attitudes towards food risks. They believed that the public worried about the wrong things, and that the public's 'misperception' skewed the agenda, which resulted in misallocation of efforts (Macfarlane 2002). The initial industry response to challenges to their new technology was swift and direct as it first attacked the messengers before addressing the message they carried. Rachel Carson was branded as hysterical, in league with communists, and unscientific by the pesticide industry (Vorley and Keeney 1998). Indeed, the pesticide industry equated Carson's position to a sort of latter day Luddite vision associated with the end of all human progress, reversion to a passive social state devoid of technology, scientific medicine, agriculture, sanitation or education. The pesticide industry characterised this world as one with disease, epidemics, starvation, misery and suffering incomparable and intolerable to modern man (Murray and Taylor 2000). As scientific research and popular media reports gave further credence to Carson's charges, the pesticide industry held steadfast to a strategy of denial and counterattack on those who would challenge the appropriateness of their technology. During the Productivist Period, it can be assumed that the majority of the population in most developing countries were badly informed about the issue of pesticides and completely absorbed with problems of everyday life.

2.4.6.2.1 UK

By the late 1960s, growing public concerns over the impact of industrial agriculture on the environment prompted the emergence of industry efforts to reconcile the interests of farmers and conservationists (Winter 1996). In the

UK, agro-food system interests established the Farming and Wildlife Advisory Group (FWAG) in 1970. FWAG comprised farmer members, agrochemical industry representatives, food retailers, consumers and environmentalists.

2.4.6.2.2 USA

In the USA, public concern manifested itself through the establishment and activities of a number of environmental organisations, including the National Audubon Society, Izaak Walton League, National Wildlife Federation and Environmental Defense Fund (EDF) (EPA 2003a). Following the environmental emphasis of Carson, both the Sierra Club and the National Wildlife Federation (founded in 1936) focused on the environmental impacts of pesticide use (NWS 2003). Anti-pesticide non-governmental organisations (NGOs) were given further impetus by the rhetoric of the Herbicide Assessment Commission, an offshoot of the American Association for the Advancement of Science, whose work led to the abandonment of Operation Ranch Hand in 1971 (Hough 1998). The EDF was established in 1967 by a group of scientists, lawyers and citizens disaffected with modern society's devastating impact on the environment. The EDF evolved from the Brookhaven Town Natural Resources Committee. This was a citizens' group, which, through legal action in 1966, had successfully stopped the use of the pesticide DDT by the Suffolk County (Long Island, New York) Mosquito Control Commission (EDF 1999). The EDF's principal aim was to litigate against injudicious treatment of the environment and increase public awareness to what members perceived as the most serious environmental problems (EDF 1999). Ironically, in 1968, the EDF received financial support for high-profile litigation against state and federal use of hazardous pesticides, namely DDT and dieldrin, from the same philanthropic foundations (particularly the Ford Foundation) that were funding Green Revolution research and encouraging the intensive use of pesticides in the Far East (EDF 1999; EPA 2003a; Ford Foundation 2003). In addition to DDT, dieldrin was added to the hit list after a veterinarian in Georgia reported that more than 100 cattle died after one USDA aerial bombardment of the pesticide. Even farmers in the area refused to contribute to the cost of aerial spray programmes (Anderson et al. 1999). Threats to wild birds from chemicals were one of the main concerns in the USA during the 1960s (Boardman 1986). Publicity resulting from these cases, together with advertising and direct mail campaigns, contributed to the growth in environmental NGOs (Anderson et al. 1999). Between 1971 and 1975, EDF membership grew from 20 000 to over 56 000 (EDF 1999).

During the 1950s, the promotion of DDT, and other pesticides, was justified on the same grounds as most other government programmes. Economists taught that government had an important role in subsidising 'public goods'. These are goods that will be undersupplied by the marketplace because no one person can capture all the economic benefits of providing them (Anderson et al. 1999). For example, if farmer A sprays his/her fields to control a pest, farmer B next door does not have to spray as much because farmer A inadvertently protects him/her from at least some of the pests that would have migrated to farmer B's farm in search of food. Thus farmer A is providing a 'public good' to farmer B, while farmer B is a 'free-rider' who benefits without paying farmer A. As a result, many economists believe that there will be too many pests and too little pesticide use. To get to the efficient level of spraying, farmer B must be forced to spray (the approach taken by Washington State with apple growers in the early 1900s) or all farmers must be subsidised via general treasury funds. Subsidies are the politically more attractive option since taxpayers are unlikely to be aware of how much of their taxes are going into subsidies or even which programmes their taxes support. In contrast, farmers who get the benefit of a subsidy know about it and appreciate their representatives in Congress who deliver it (Anderson et al. 1999).

After the Second World War, the USDA promoted widespread spraying of DDT, as well as several other pesticides, a programme that expanded in the 1950s with bipartisan support. As a member of the House Agriculture Committee said, 'We expect the Department of Agriculture, in cooperation with the land grant colleges and experiment stations and . . . with insecticide producers and the chemical industry, to develop pesticides that will control what is left of the immune insects that attack what we produce and present to the American consumer' (Bosso 1987). The Senate was equally enthusiastic. In 1953, Senator Allen Ellender said that pesticide use would help agriculture, thereby improving national security and helping defeat communism (Bosso 1987). In fact, DDT was so popular that it was even thrown instead of rice at some weddings (Whorton 1974). In 1957, Congress approved the Fire Ant Eradication Act, which aimed to spray 20 million acres (Anderson et al. 1999). The Department of Agriculture benefited as an agency from this expansion into active pest management, which traditionally had been left to individual farmers or to states. Members of Congress who sat on the agriculture committees could report to their constituents that they were generating benefits for them – the government would spray for pests and, at most, the beneficiaries would only have to pay a fraction of the bill. Chemical companies that manufactured DDT and other pesticides no doubt lobbied for expanded spray programmes, since it meant more sales of their products (Anderson et al. 1999).

However, by the end of the 1950s, government spraying programmes began to provoke public opposition. The USDA knew little about the effects of widespread dispersion of many of the chemicals it promoted and subsidised. One reason was that Congress directed funds for services such as spraying, not for background research needed to understand the environmental impacts of the programmes. In addition, because federal regulation was aimed at pre-empting state regulation, the federal government discouraged states and localities from controlling pesticide use. Farmers began to balk at paying for even a share of the spraying of chemicals that sometimes had unfortunate side effects. The department's solution was to offer the chemicals for free to those who would take them. Even in Alabama, where fire ants had first appeared via the port of Mobile and where people recognised the problems the ants posed, the legislature withdrew funding for spraying in 1959 after state game officials estimated that the spraying could eliminate up to 75% of the state's wildlife. The US Fire Ant Eradication Programmes regularly used aldrin, dieldrin or heptachlor, which also killed many birds (Pimentel and Pimentel 1979). Despite declining support, the USDA continued to spray without state or farmer input. Ultimately, opposition took its toll and the federal spray programmes gradually wound down (Anderson et al. 1999).

While pesticides such as dieldrin and DDT worked as advertised on pests, they also affected non-target species. When spraying harmed wildlife and domestic animals, the USDA did not have to compensate the owners and so failed to consider the full costs of the spraying (Anderson et al. 1999). Eventually, the pesticide spraying programmes ended when the political costs began to outweigh the political benefits. Politicians in Congress and the managers at the USDA had an incentive to maximise the net political benefits of pesticide application. The first spray programmes were conducted where the perceived benefits of spraying were largest, such as controlling malaria. As the programmes expanded, spraying was conducted where the benefits were smaller or localised, such as for gypsy moth control. Similarly, the spraying programmes began in places where the political costs of spraying were lowest (over swamps) and expanded into areas where the political costs increased (inhabited areas). Since Congress could mandate the use of tax dollars, spraying was extended beyond levels that individual decision makers would have reached. When public concern arose, however, especially after the publication of Silent Spring, the USDA and Congress were suddenly faced with a sharp increase in the political costs of such programmes (Anderson et al. 1999).

Although concerns were raised regarding DDT in the mid-1940s (EPA 2003a), it was not until the late 1960s that, DDT, once heralded as the world's no. 1 pesticide, became public enemy no. 1. Carson's book, which was serialised in part in the New Yorker before its publication, was well written, knowledgeably discussed scientific research and painted a horrifying picture of the world that was coming if DDT use was not restricted. Carson told her readers of a silent spring, when no birds would sing because they would all be dead from DDT exposure. Carson's assertions about the hazards of DDT were exaggerated, but there was enough to the story she told to fan public fears (Anderson et al. 1999) and catapult the concerns over the negative impacts of pesticides into the international arena (Hough 1998). Indeed, the President's Science Advisory

Committee (PSAC)'s 'Use of Pesticides' publication in 1963 recommended that 'the accretion of residues in the environment be controlled by orderly reduction in the use of persistent pesticides' (Pimentel and Pimentel 1979).

2.4.6.2.3 Japan

In a similar vein to the USA, Japan introduced legislation in 1948 to prevent the movement of inferior or illegal chemicals (Boardman 1986). However, it was not until the 1970s, after a number of animal poisonings and growing environmental concern, that a change began (Boardman 1986). This prompted the establishment of the Japanese Environment Agency, which came to share pesticide regulation functions with the Ministry of Agriculture and Forestry. This signalled stronger protection for human health and the environment (Boardman 1986).

2.4.7 Pesticides, food quality and food safety

Although recognition of the impact of pesticide residues on human health was not a new phenomenon - remember Paris Green and other earlier toxic inorganic pesticides - concerns over the potential negative human health impacts of modern pesticides increased throughout the 1950s, 1960s and early 1970s in Western Europe, the USA and Japan. Based on the work undertaken by its Expert Committee on Insecticides (Hough 1998), the World Health Assembly noted in 1953 that the increasing use of various chemicals in the food industry had, in recent decades, created a new public health problem. In response to this, the WHO, in conjunction with the FAO, initiated two series of annual meetings on food additives (Joint FAO/WHO Expert Committee on Food Additives) and on pesticide residues. The first meeting on food additives was held in 1956 and on pesticide residues in 1963 (IPCS 2003) (see Fig. 2.1).

The concept of the Joint Meeting FAO/WHO on Pesticide Residues (JMPR) was first proposed in 1959, when an FAO Panel of Experts on the Use of Pesticides in Agriculture recommended that FAO and WHO should jointly study a number of key indicators. (1) The hazard to consumers arising from pesticide residues in and on food and feedstuffs. (2) The establishment of principles governing the setting up of pesticide tolerances. (3) The feasibility of preparing an international code for toxicological and residue data required in achieving the safe use of a pesticide.

Consequently, in 1961, a Joint Meeting of the FAO Panel of Experts on the Use of Pesticides in Agriculture and the WHO Expert Committee on Pesticide Residues was convened. The report of the 1961 Meeting recommended the evaluation of toxicological and other pertinent data on those pesticides known to leave residues in food when used according to good agricultural practice (GAP) to the Directors-General of FAO and of WHO. The evaluations would include the estimate of an acceptable daily intake (ADI) and an explanation of its derivation. To implement this recommendation, the first Joint Meeting on Pesticide Residues was convened by FAO and WHO in September 1963. This meeting brought together the FAO Panel of Experts on Pesticide Residues in Food and the Environment and the WHO Core Assessment Group. During the meetings, the FAO Panel of Experts was responsible for reviewing residue and analytical aspects of the pesticides under consideration, including data on their metabolism, fate in the environment and use patterns, and for estimating the maximum residue levels (MRLs) that might occur as a result of the use of the pesticides according to GAPs. The WHO Core Assessment Group was responsible for reviewing toxicological and related data and for estimating, where possible, ADIs for humans of the pesticides under consideration (WHO 2003c). The Meeting adopted the concept of the ADI, which was based on a number of key facets. (1) The chemical nature of the residue. (2) The toxicity of the chemical based on data from acute, short- and long-term studies, and knowledge of metabolism, mechanism of action, and possible carcinogenicity of residue chemicals when consumed (usually determined in animals). (3) Knowledge of the effects of these chemicals on humans.

Separate meetings of an FAO Working Party on Pesticide Residues examined the issue of tolerances, later to become MRLs. In 1966, the remit of the JMPR incorporated both ADIs and MRLs. JMPR duly served as a scientific advisory body to FAO and WHO, to FAO and WHO member governments, and to the Codex Alimentarius Commission. Advice to the Codex Alimentarius Commission on pesticides is provided via the Codex Committee on Pesticide Residues (CCPR) (WHO 2003c).

The Codex Alimentarius Commission (which means 'food code' in Latin) was created in 1963 by FAO and WHO to develop universal food standards, guidelines and related texts such as codes of practice under the Joint FAO/WHO Food Standards Programme. The main purposes of the Programme are protecting health of the consumers and ensuring fair trade practices in the food trade, and promoting coordination of all food standards work undertaken by international governmental and non-governmental organisations (CODEX 2003).

Pesticide residues in food are dealt with by the CCPR, which bases its work on scientific approvals made by the independent expert panel, the Joint FAO/WHO Meeting on Pesticides Residues (see Fig. 2.1). One objective of Codex MRLs was to facilitate international trade: standards are intended to ensure that importing countries will not prevent import of food commodities because of residues that occur up to the level of the MRL. There is often a variation between national standards and international recommendations. Residue limits are set relying on a number of related concepts. (1) MRLs – the legal limit of pesticide allowed (expressed in mg/kg of produce) in food or animal feeds. (2)

ADI - the amount of chemical that can be consumed (in mg/kg bodyweight) every day for an individual's entire lifetime in the practical certainty, on the basis of all known facts, that no harm will result. (3) GAP - how a pesticide is used by the farmer. GAP represents nationally authorised safe pesticide use (amount, method, time of application, etc.) under actual conditions, so as to leave the smallest amount of residue that is toxicologically acceptable. Generally, the use of a pesticide will lead to residues in food and provided the pesticide is used according to the manufacturer's label directions and GAP, regulators calculate the residue should be within the MRL and ADI. Although most countries aim to observe Codex MRLs so that they can export produce, this does not necessarily mean that they have in place an effective system of national MRLs for domestically produced and consumed food.

2.4.7.1 USA

In the USA, federal pesticide regulation evolved in response to two unrelated drivers (see Fig. 2.1). The first driver was the public's growing concern for both human health and the environment, particularly in relation to the new synthetic pesticides (Anderson et al. 1999). The public called for legislation that would ensure safe food and reduce environmental damage. Perhaps more unexpectedly, the second driver was the pesticide industry's concern over the costs associated with inconsistent regulation within and between states. Industry called for national consistency in pesticide regulation (Morriss 1997). New regulation materialised in the form of the FIFRA 1947, which replaced the Insecticide Act of 1910 (USDA-ARS 2003).

FIFRA (1947) introduced three new mandatory regulations: (1) prohibition of the sale, or distribution of, 'economic poisons'; (2) pesticide registration; and (3) penalties for the violation of the Act (FWS 2003). However, in a similar vein to the 1910 Insecticide Act, and the Food, Drug, and Cosmetics Act (1938), FIFRA (1947) continued to assume that when unadulterated, pesticides should be efficacious, and that any public health or environmental problems were caused by misuse, which proper labelling would prevent (Anderson et al. 1999; WSU 2003a). In effect, it was not the validity of pesticide use that was questioned, but the misuse or adulteration of pesticides.

Under FIFRA, the USDA became the executive agency for administering the pesticide statutes during this Period (Schierow 2004). As part of its new remit, USDA was required to register all pesticides. This, in turn, ensured that labelling requirements (explaining how to use the compound effectively and safely) were the chief restriction to market entry (Anderson et al. 1999). However, pesticides to which the USDA objected could still be registered under protest. Once a pesticide was registered, it could be marketed unless the registration was suspended under an interim administrative procedure, pending the initiation of lengthy proceedings to cancel the registration (Anderson et al. 1999). FIFRA (1947) was an extension of the classic consumer protection objectives. However, there was a distinct lack of any review procedure and pesticide corporations could still market compounds, even when they were being challenged on safety grounds by regulatory bodies (Boardman 1986).

In addition, FIFRA (1947) failed to endow the USDA with any significant regulatory powers. For example, although the statute gave the USDA weak powers to require proof that the pesticide was effective, and required manufacturers to register their products with the agency, it denied the department the ability to refuse to register a product (Morriss 1997). Manufacturers thus had the best of all possible worlds. They had: no significant limitation on their ability to manufacture and sell effective pesticides; limits on competitors' ability to sell ineffective products; pre-emption of state regulation in many areas; and federal regulatory authority located in a sympathetic agency. While this was not the military-industrial complex that President Eisenhower warned of, it was a parallel set of interests – the private, congressional and bureaucratic interests that helped form the enduring 'iron triangle' that determines many policies in Washington. Members of Congress, overseeing numerous diligent employees at the USDA, were happy to spend taxpayers' money to subsidise agricultural production, all the while being patted on the back by the chemical producers. Powerful political forces were at play. However, while we can look back sceptically upon their behaviour now, no doubt the sense of mission was genuine and bolstered the special interests at work (Anderson et al. 1999).

In 1950, the Delaney Committee began a congressional investigation into the safety of chemicals in foods and cosmetics, which, in turn, laid the foundations of the Miller Pesticide Amendment in 1954 (FDA 2003a). The Act achieved two key things. First, it endowed the Federal Drugs Agency with powers to regulate pesticide residues in the food chain through residue testing (WSU 2003a). It spelt out procedures for setting safety limits for pesticide residues on raw agricultural commodities (FDA 2003a), and determined that food sold with residues above tolerance levels would be considered as adulterated (WSU 2003a). Second, it established the need for pesticide manufacturers to establish the safety of their products (FDA 2003b). Manufacturers were also required to obtain a certificate of usefulness from the USDA or an exemption (WSU 2003a).

A provision of the Food Additives Amendment in 1958, the Delaney Clause, banned any carcinogenic additive (FDA 2003b). The Delaney Amendment introduced zero tolerance for animal carcinogens in processed food but allowed carcinogens to be present in raw commodities. Industry wanted the Delaney Clause repealed, but anti-pesticide groups wanted children protected based on cancer risk (WSU 2003a).

Subsequently, in November 1959, the secretary of Health, Education, and Welfare announced that cranberries were contaminated with aminotriazole, a

USDA-registered herbicide, which the FDA said caused cancer in laboratory rats (Bosso 1987). This prompted another major public food scare about pesticides and food residues. Whilst the USDA insisted that cranberries posed no significant threat, and presidential candidates Nixon and Kennedy ate cranberries to demonstrate that they were safe, sales collapsed to one-third their normal level. The FDA recalled the US cranberry crop three weeks before Thanksgiving in order to test for contamination with aminotriazole. Cranberry sales dropped just before Thanksgiving, the critical market phase, and sales were halted in some states. Cleared berries were allowed a label stating that they had been tested and had passed FDA inspection, the only such endorsement ever allowed by FDA on a food product (FDA 2003a). However, even though the damage to the cranberry market eventually passed, the scare heightened US citizens' awareness of the potential dangers of pesticides (Anderson et al. 1999).

In May 1963, the PSAC released an influential report entitled the 'Use of Pesticides', addressing a range of costs and benefits associated with pesticide use (EPA 2003a). The following year, the US Government amended FIFRA (1947) through the federal Insecticide, Fungicide, and Rodenticide Act Amendments (1964) (Schierow 2004; EPA 2004a). The Amendment required the USDA to suspend the registration of pesticides that presented an 'imminent hazard' (EDF v Ruckelshaus 1971). Whilst the serving Secretary of the USDA suspended minor uses of DDT, he refused to suspend DDT use on cotton pests (EDF v Ruckelshaus 1971), suggesting that further studies were necessary. In the same year, the Secretary of the Interior issued a directive stating that the use of chlorinated hydrocarbons on Interior lands should be avoided unless no other substitutes were available (EPA 2003a). In November 1965, the Environmental Protection Panel of PSAC released another influential report entitled 'Restoring the Quality of Our Environment' (EPA 2003a), consolidating the need for change.

In many respects, the period 1969–70 was a watershed in both US society's and polity's attitudes to pesticides. Two additional publications were released by highly respected institutions, which cautioned the use of the most hazardous pesticides. These were the Committee on Persistent Pesticides, Division of Biology and Agriculture, National Research Council (NRC), the Agriculture Department in May 1969, and the Department of Health, Education, and Welfare's landmark report on the environmental health consequences of pesticides (Mrak Commission) (EPA 2003a). One of the key recommendations of both the PSAC reports and the NRC and Mrak reports was the phasing out of DDT (EPA 2004b), something that many states had already initiated as early as 1958. DDT use was already outlawed, except under emergency conditions, in Illinois, Massachusetts, New Mexico, New York, Rhode Island, Vermont and Wisconsin (EPA 2003a). Alaska, Arizona, California, Colorado, Connecticut, Ohio, Utah, Virginia and Washington had placed some limitation on the use of DDT (EPA 2003a). By this time, the traditional belief that, if properly used, pesticides were not a serious threat to human health and well-being had been replaced by an acknowledgement that at least some pesticides were hazardous (Meiners and Morriss 2001).

In addition, growing public concern over the impact of pesticides prompted President Nixon to establish the Environmental Quality Council as well as a complementary Citizen's Advisory Committee on Environmental Quality in 1969 (EPA 2004c). In the same year, the National Environmental Policy Act (NEPA-1969) recast the government's role from conservator of wilderness to the protector of earth, air, land and water. The law declared congressional intent to 'create and maintain conditions under which man and nature can exist in productive harmony', and to ensure for all Americans safe, healthful, productive, aesthetically and culturally pleasing surroundings (EPA 2004c).

Trying to stem uninformed public concerns about health hazards and increasing political support for banning DDT, the secretary of Health, Education, and Welfare established a commission to produce a report on DDT. The report, prepared primarily by the National Cancer Institute, noted suspicions about DDT but found there was no firm evidence that it was carcinogenic to humans. More testing was recommended (Dunlap 1981). Consequently, environmentalists attacked the report, and a wide range of groups pressed to have DDT banned because of its carcinogenicity. Smelling victory, the EDF sued the USDA and was allowed by the courts to challenge the USDA's registration of DDT. The federal appeals court took the opportunity to excoriate the Department of Agriculture for its clumsy handling of the matter. Moving quickly by Washington standards, federal agencies back-pedalled and began announcing that DDT use would be phased out (Dunlap 1981). The USDA quickly initiated action to cancel all DDT registrations for use against pests of shade trees, aquatic areas, the house and garden and tobacco. USDA further announced its intention to discontinue all uses non-essential to human health and for which there were safe and effective substitutes (EPA 2003a). The Secretary of the Interior also banned DDT from use on federal lands, and the FDA asserted its jurisdiction should DDT appear in any food product. If it did, the agency would feel free to limit or ban DDT use, an action opposed by the USDA.

In 1970, the USDA cancelled federal registration of DDT products for use on 50 food crops, beef cattle, goats, sheep, swine, seasoned lumber, finished wood products and so forth (EPA 2003a). This wave of environmentally conscious activity coincided with the first Earth Day, when an estimated 20 million Americans gathered together on 22 April 1970, to participate in a spectacularly well-publicised environmental demonstration (EPA 1990). In the same year, the use of 2,4,5-T was banned on all food crops except rice (Boardman 1986).

However, as concerns grew regarding USDA's administration of FIFRA (Rogers 1970), Congress eventually shifted responsibility for pesticide regulation from the USDA, and the FDA's programme for setting pesticide tolerances (FDA 2003a), to the newly created Environmental Protection Agency (EPA) in 1970. Politics between and within government agencies, and between Congress and the Nixon administration, and between the chemical industries and environmental activists dominated the debate (Anderson et al. 1999). The EPA's creation coincided with the culmination of the DDT debate, which had been kicked off by Carson who intimated that humans also faced DDT-induced risks (EPA 2004a). However, at its creation, the EPA not only inherited the function of pesticide registration from the USDA, but also the staff that served that function. The USDA economic entomologists who designed the pesticide registration process in the first place preached the advantages of effective pesticides and minimised discussion of debatable health risks. The same staff that had backed USDA Secretary Clifford Hardin's claim that DDT was not 'an imminent hazard to human health or to fish and wildlife' (Dunlap 1981) provided the EPA's administrator (Ruckelshaus) with the same counsel (EPA 2004a). However, whilst the new EPA may have been institutionally handicapped, judicial interpretations of FIFRA realigned rights and responsibilities. For example, whereas the initial judicial interpretations of FIFRA intimated that suspensions should be reserved for pesticides that caused substantial and irreparable harm to the public, after 1970, interpretations set a precedent that legal action to suspend a given pesticide could be made if possible harm to the public could be scientifically substantiated. This interpretation effectively placed the burden of proving the safety of a pesticide on the registrant, and was made explicit in suspension proceedings for the pesticides heptachlor and chlordane (discussed later). EPA regulations placed the burden of going forward on the agency and the burden of persuasion on the registrant (EDF v Ruckelshaus 1971). This precedent was first upheld by the District of Columbia Circuit Court of Appeals, which insisted that the EPA's administrator had a duty to suspend a registration when a substantial question of an imminent hazard existed (EDF v Ruckelshaus 1971).

In January 1971, following another lawsuit by the EDF, a tribunal of the US Court of Appeals in the District of Columbia ordered the EPA's administrator to begin the process of suspending DDT's registration, and to consider suspending DDT's registration immediately (EPA 2004a). DDT's high-profile cancellation hearing began on 17 August 1971 (EPA 2004a). By this time, the focus of public concern had shifted from general environmental damage to carcinogenicity. Indeed, in April 1972, after seven months of testimony, Judge Edmund Sweeney stated that 'DDT is not a carcinogenic hazard to man. The uses of DDT under the regulations involved here do not have a deleterious effect on freshwater fish, estuarine organisms, wild birds or other wildlife. The evidence in this proceeding supports the conclusion that there is a present need for the essential uses of DDT' (Sweeney 1972). The EPA's administrator reported that he had found no good reason to suspend DDT registration immediately. This and several other pesticides including 2, 4, 5-T (Agent Orange), dieldrin, aldrin and mirex did not appear to constitute imminent health threats (EPA 2004a). The conclusion of the review process prompted accusations that the outcome was overly biased in favour of agrichemical interests (EPA 2004a). Indeed, Mississippi Congressman Jamie Whitten was accused of using his position as chairman of the agricultural appropriations subcommittee of the House of Appropriations Committee to make Ruckelshaus conform to the interests of the agrichemical lobby (EPA 2004a).

In 1972, Ruckelshaus, who had allegedly never attended a single day's session in the seven months of EPA hearings, and, who admittedly, had not even read the transcript of the hearings, eventually gave way to the immense pressure (EDF 1999) to suspend DDT and overturned Judge Sweeney's decision. Ruckelshaus declared that DDT was a 'potential human carcinogen' and banned it for virtually all uses (Ackerly 1981). This decision was based on DDT's environmental persistence and carcinogenic risk to humans (EPA 2004a), and was likely bolstered by the EPA's interpretation of the Delaney Clause to mean that pesticides could be considered a potential risk to humans if they caused an estimated 1–1 000 000 additional probability of developing cancer (WSU 2003a).

Interestingly, in the years preceding the ban on DDT, the National Academy of Sciences (Handler 1971), the American Medical Association, the US Surgeon General (Steinfeld 1971), the WHO (WHO 1971) and the FAO (New York Times 1969) had been among those who spoke out in support of the continued use of DDT as a disease fighter and crop protectant. The principal crops affected by this action were cotton, citrus and a number of vegetables (EPA 2003a). In effect, the use of DDT became authorised only for emergencies (EPA 2003a), despite the fact that alternative products were more expensive and less efficacious (Anderson et al. 1999). Concerns were soon raised over how farmers could successfully protect their crops, particularly cotton, against insect attack (EPA 2003a). Because DDT and related chemicals were so cheap and effective, pesticide use increased fivefold between 1950 and 1970 (Rogers 1994). When individual States imposed bans or restrictions, DDT's use in the USA declined rapidly, from a peak of approximately 80 million pounds in 1959 to between 12 (EPA 2003a) and 13 (EPA 1972a) million pounds in the early 1970s. Between 1970 and 1972, over 80% of DDT was applied to cotton crops, with the remainder being used predominantly on peanut and soybean crops (EPA 2003a).

The decision to ban DDT met with major legal challenges from pesticide manufacturers (EPA 1972a). The decision was also not enough for the EDF. Indeed, it sued to have all production halted so that DDT could not be used for export or even used in case of domestic public health emergency, such as an outbreak of malaria (Dunlap 1981). DDT for export increased rapidly from 12% of total production in 1950 to 67% in 1969. Between 1970 and 1972, export of

DDT, primarily for use by U.S. Agency for International Development (USAID) and the UN, fell from 70 million pounds to 35 million pounds (EPA 2003a). Environmental organisations continued to press for bans on DDT production across the world, and bans were imposed in many nations. By the mid-1970s, the transformation of DDT from a public good to a public bad was complete. There it remains today (Anderson et al. 1999).

In response to wider congressional concerns over the effects of pesticides on farm workers and wildlife, particularly non-target insects and birds, and on food consumers, FIFRA was amended in the form of the Federal Environmental Pesticides Control Act (FEPCA) 1972 (EPA 2003a). In response to growing concerns over older pesticide chemistry, particularly in light of improved safety standards, FEPCA directed the EPA to re-register approximately 35 000, older products. This process was accelerated by reviewing generic pesticide groups sharing the same ingredients, instead of individual pesticides. Many older products were not re-registered (Schierow 2004). Approximately 50 000 pesticides were registered under the permissive pre-1972 requirements (Glicksman et al. 2003). Examination of all these pesticides proved impossible by the 1976 deadline for re-registration. Indeed, the EPA estimated that re-registration would take over a decade. In fact, the EPA only managed 25 pesticides a year (Glicksman et al. 2003).

In addition, FEPCA required pesticide registration before it could be legally distributed, sold, or offered for sale or shipped or received in any state (AMA 2003). FEPCA federal control extended to the application of pesticide to control misuse, including financial penalties (EPA 1972c). Historically, FIFRA intended that all products be periodically reviewed, that is re-registered. FEPCA reiterated this goal (within 4 years) (WSU 2003a). Under FEPCA 1972, all pesticides were to be classified for general use or restricted use. Restricted use refers to high-risk chemicals applied or supervised by certified applicators (EPA 1972c). The Act also contained provision for greater civil and criminal penalties (EPA 1972c), and endowed the EPA with the responsibility to establish pesticide packaging standards, regulate pesticide and container disposal, conduct research on pesticides and alternatives and monitor pesticide use and presence in the environment (EPA 1972c). Use in violation of product label became a crime (WSU 2003a).

FEPCA also required that the risks and benefits of pesticides be considered at registration, restricted registration, suspension and cancellation (Glicksman et al. 2003). However, FEPCA required the EPA to indemnify producers, distributors and others for economic loss when the EPA cancelled or suspended a registration. This obligation had far-reaching financial ramifications. For example, claims for two suspended and cancelled pesticides, 2, 4, 5-t/Silvex and EDB, totalled US$30 million (Glicksman et al. 2003), and estimated claims for dinoseb were valued at approximately US$40 million (Senate Committee on

Agriculture, Nutrition, and Forestry 1998; Glicksman et al. 2003). The EPA also bore the cost, estimated at US$195 million (Glicksman et al. 2003) of disposing of the three pesticides listed above. In addition, Section 136d (a) (1) of FEPCA authorised the EPA to permit continued sale and use of existing stocks of pesticides whose registration has been suspended or cancelled (Glicksman et al. 2003). FEPCA (1972) also insisted that registrants demonstrated that a pesticide could perform intended functions without unreasonable adverse effects on the environment (WSU 2003a). Indeed, the 1972 FEPCA amendments included: 'any unreasonable risk to man or the environment taking into account the economic, social and environmental costs/benefits of the use of any pesticide' (Boardman 1986). This had moved on from 'will not cause any unreasonable adverse effect on the environment' (Boardman 1986). Amendments in 1972 increased the role of the USDA and created a scientific advisory body so that the EPA could not use a lack of pesticide knowledge as a basis for banning it (Boardman 1986).

Passed in 1973 (USDA-ERS 2004), the Endangered Species Act of 1972 (FWS 2003) amended FIFRA to define imminent hazard to include situations involving unreasonable hazard to the survival of a species declared by the Secretary of the Interior to be endangered or threatened (FWS 2003). In addition, the Federal Water Pollution Control Act (FWCA), Clean Water Act and Consumer Product Safety Act passed in 1972 (USDA-ERS 2004), added a further tier of pesticide restrictions. For example, the FWCA provided for citizen participation in the enforcement of federal standards. Aggrieved private citizens could seek judicial relief against any polluter for violations of an effluent standard or limitation, or administrator if he fails to perform an act required of him under law (EPA 1972b).

2.4.8 Pesticide exposure and human health

Rachel Carson's critique of pesticide use is seen by many as the pivotal text that provided individuals and organisations with the impetus to question not only the use of pesticides but the industrial model of productivist agriculture itself. Instead of a constant stream of pro-technology rhetoric from national and regional governments, the media and the private sector, the public were exposed to the growing, and increasingly negative, rhetoric associated with the human health, nutrition and environmental impacts of pesticide use (Dittus and Hillers 1996). Concern over pesticide residues was not limited to one's own health or to one's children (Winter 1996) but also to those that formulate or apply pesticides (farmers, farm labourers and pesticide factory workers) (Macfarlane 2002). In a survey conducted by Macfarlane (2002), 80% of US shoppers indicated that they were concerned about pesticide residues.

Presence of children in the household also increases concern over food safety. The public often quoted cancer as a major concern (Macfarlane 2002).

Although chemical pesticides differ markedly in their toxicity to humans and their mobility, most people are not exposed to dangerous levels of pesticides in the air that they breathe or on things that they touch, unless they live near areas being sprayed or sites of pesticide production. The people who are at the greatest risk of exposure are those who work directly with pesticides themselves, particularly the undiluted concentrates. These include factory and farmer workers manufacturing or applying pesticides. They are exposed to pesticides on things they touch. This is especially dangerous when pesticides are lipid soluble and are able to pass through their skin. Conversely, many pesticides can also be dangerous when inhaled or ingested, either deliberately or by accident.

Overexposure to several pesticides can cause severe poisoning or death. In 1969, a study found a higher incidence of leukaemia and liver tumours in mice fed DDT compared to unexposed mice (Tarjan and Kemeny 1969). When WHO investigated the 1969 mice study, scientists discovered that both cases and controls had developed a surprising number of tumours. Further investigation revealed that the foods fed to both mice groups were mouldy and contained aflatoxin, a carcinogen (Edwards 1978). The tests were repeated using uncontaminated food and neither group developed tumours (Facts versus Fear 1998). Furthermore, in 1978, after 2 years of tests, a National Cancer Institute report concluded that DDT was not carcinogenic (Efron 1985). After comprehensive studies, there was also inconclusive evidence linking DDT and DDE to breast cancer (Kolata 1997). DDT was also accused of thinning eggshells of birds of prey; however, this is also contested (Enderson and Berger 1968). In the UK, a 3-year government study noted that the decline of peregrine falcon numbers had ended, even though DDT levels were as abundant as ever. The British study concluded that 'there is no close correlation between the decline in population of predatory birds, particularly the peregrine falcon and the sparrow hawk, and the use of DDT' (Advisory Committee on Toxic Chemicals 1969). It is still unclear whether or not DDT was responsible or whether it was caused by mercury, PCBs or the effects of human encroachment and general agricultural industrialisation (Milius 1988). In addition, Scott and associates of the Department of Poultry Science at Cornell University found no tumours, no mortality, no thinning of eggshells and no interference with reproduction caused by high levels of DDT (Scott et al. 1975). In the 1950s, 1960s and 1970s, organochlorines were widespread in industrialised countries and high levels of exposure were common with all those involved in their manufacture and application, and, through pesticide residues, the public. Nevertheless, there is little evidence of serious ill health (Pretty 1995). The impact of DDT on human health remains hotly disputed (Hough 1998).

Whilst banning DDT may have played a significant role in helping the bald eagle and other birds make a comeback in the past three decades, it has also allowed malaria to make a comeback. The worry of scientists that there was no good, less toxic, cost-effective substitute for DDT to control mosquitoes and other pests is as true today as it was four decades ago. A disease that was on the way to being vanquished has returned with a vengeance. The hundreds of millions of sufferers of malaria and the millions of families that lose infants to malaria are being denied an effective solution (Attaran et al. 2000; Goklany 2000). When DDT spraying ended, malaria's incidence rises markedly. The disease has returned to areas in which it had been eradicated: urban areas of the Amazon Basin, Korea, Armenia, Azerbaijan and Tajikistan. In Sri Lanka, after the spraying stopped in 1961, malaria jumped back to 500 000 cases by 1969 (Hicks 1992). In 1977, USAID estimated that the WHO campaign 1955–70 had saved 15 million lives and prevented 2 billion cases of malaria (Bull 1982). This figure was significantly lower than that released by the US-based National Academy of Sciences in 1970, which declared that 'in little more than two decades, DDT has prevented 500 million human deaths due to malaria that would otherwise have been inevitable' (NAS 1970). In Ceylon, DDT spraying was attributed with reducing malaria from 2.8 million cases in 1948 to 17 in 1968. After spraying stopped in 1964, malaria began to rise and reached 2.5 million in 1969 (Gerberg and Wilcox 1977). The use of DDT also had a dramatic impact in Zanzibar, where prevalence of malaria dropped from 70% in 1958 to 5% in 1964. By 1984, it was back between 50% and 60% (Facts versus Fear 1998).

Along with the problems associated with organochlorines, a second level of acute public health problems associated primarily with OP pesticides emerged. Unlike DDT, some of these chemicals could cause severe health effects and death after exposure to small amounts over only a few hours. At the same time, another problem, the long-term health effects from the chronic exposure to relatively small levels of pesticides, was also being reported. Studies implicated pesticides in the rising rate of breast cancer, in neurological damage to workers and in a range of problems associated with endocrine disruption (Murray and Taylor 2000).

Herbicides were also beginning to cause concern. First, paraquat received significant attention due to its extremely poisonous nature. Second, phenoxy herbicides, particularly dioxins associated with the production of phenoxy herbicides (Boardman 1986), were thought to be associated with cancer (Pretty 1995), teratogenic effects and male reproductive problems (The Ottawa Citizen 2003). However, several comprehensive literature reviews have disputed these claims suggesting that phenoxy herbicides posed only a negligible or minimal risk (Boardman 1986). In the USA, the 1969 report linking 2,4,5-T with teratogenic effects in rodents led to restrictions on the use of the compound in government spraying programmes (Boardman 1986). In 1970, 2,3,7,8-TCDD, which had been known about since 1950, and identified as a dioxin in 1965, was suspected as the causal agent (Boardman 1986). By the late 1970s, and the improvement of pesticide assessment techniques, the EPA reversed its earlier recommendation, suggesting that 0.1 ppm of the pesticide actually posed an unacceptable risk to health (Hay 1982). Discussions involving the EPA, US military, Dow and anti-pesticide activists deepened the interpretation and judgement based conflicts (Boardman 1986).

In 1972, the United Nations Conference on the Human Environment took place in Stockholm. The Conference recommended that programmes, to be guided by WHO, should be undertaken for the early warning and prevention of harmful effects of chemicals to which humans were being increasingly exposed, and for the assessment of the potential risks to human health (WHO 2003b). In 1973, the WHO made the first global estimate of pesticide poisoning. It suggested that there were 500 000 cases of acute pesticide poisonings annually, which resulted in approximately 5000 fatalities (WHO 1973). At that time, it was considered to be an unacceptably large problem. Pesticides, principally insecticides, caused most of the poisonings. These poisonings originated from the three main classes of pesticides: organochlorines, OPs and carbamates. Because of DDT's notoriety, organochlorines were one of the first families of pesticides to receive attention from researchers, the mass media and public. Indeed, in 1973, the WHO's Expert Committee on the Safe Use of Pesticides established the idea of a standard worker exposure protocol, with the intention of it becoming a definitive reference for those concerned with the safety of workers applying new organophosphorus pesticides, which by that time were beginning to replace DDT (Hough 1998). However, despite the widespread use of organochlorines throughout the 1950s, 1960s and early 1970s, there was little documented evidence of pesticide-related ill health in developed countries. This lack of evidence was primarily due to three related phenomena. First, because of the institutionalisation of industrial agro-food systems and the accompanying belief in the virtues and irreproachability of modern science, the systematic application of modern synthetic pesticides had been accepted as a normal constituent of environmentally benign and healthy modern farming systems. Second, as a result of the perceived human health neutrality of pesticides, there was little public sector institutional capacity to assess the potential dangers inherent in modern pesticides. Third, as there was either little or no formal incentive for pesticide companies to assess the human health and environmental risks of their products, many pesticides in use during the 1950s, 1960s and 1970s did not have toxicological profiles. Where toxicological profiles existed, they were generally the property of pesticide companies. This meant two things. First, access to these profiles was generally restricted. Second, the accuracy of in-house assessments released into the public sphere was understandably treated with suspicion.

At a regional level, since 1956, the Council of Europe's Social Committee has had a body that has the specific task of investigating dangerous chemical substances. Established by the Sub-Committee on Health Control of Foodstuffs, the Working Party on Poisonous Substances in Agriculture (known as the Sub-Committee on Pesticides after 1965) (Boardman 1986) developed guidelines on pesticide registration, first issued in 1962 (Boardman 1986). In 1967, a Council Directive was passed on the approximation of laws, regulations and administrative provisions relating to the classification, packaging and labelling of dangerous substances (Boardman 1986).

2.5 Summary

In less than a decade after the Second World War, productivist agriculture and the urban-centred industrial agro-food system had been institutionalised in the Western world, and, increasingly, in many developing countries, as the modern paradigm for food production, distribution and consumption. Over the last 40 years, world output has increased by 25%, even though the land base of agriculture has only increased by 10% and world population has increased by 90%. Food prices, in real terms, have decreased by 40%, so that rich consumers, such as Americans, only spend 14% of their household incomes on food. For many years after World War II, governments needed to demonstrate the availability of surplus food as a political tool. Even in the late 1950s and early 1960s, when problems with pesticides became increasingly apparent, agricultural production had to be maintained for political reasons. Rich farmers, the chemical industry and food consumers all wanted the benefits of the system in spite of the public concern over the human health and environmental impacts (Dent 2000).

In contrast to the preceding Period, the Productivist Period witnessed significant advances in pesticide science and application technologies. So much so that, by the 1970s, pesticides provided farmers with a formidable arsenal of products with which to prevent, control and even eradicate all agriculturally important pests and diseases. With few exceptions, chemical control became a panacea for farmers throughout the developed world and as well as becoming increasingly popular with farmers and other actors involved in agricultural production in developing countries. However, aside from several notable exceptions, even though the application of chemical products grew increasingly cost-effective from the farmers' point of view, agriculture, indeed the whole food system, it attracted increasing amounts of criticism, particularly on environmental, food safety and human health grounds.

2.5.1 The resilience of urban-centred industrial agro-food systems

Although urban-centred industrial agro-food systems received significant criticism from the media, and disaffected scientists (particularly ecologists and environmental economists) in the 1960s and 1970s, it managed to remain both intact and relatively stable. Even with support from a growing section of the scientific community and the relentless and often savage persistence of influential mass media, concern for the environment and public health throughout the 1960s and 1970s remained subservient to the needs of politicians and proponents of the growing industrialised agro-food sector. The techno-policy model of productivist agriculture remained entrenched, almost enshrined, as the only model for modern/advanced economies and, indeed, for the remaining underdeveloped countries of that time. As with other components of productivist agriculture, the production and use of pesticides remained institutionalised. Part of this resilience of the industrial agro-food system can be explained by the growing political and economic strength of stakeholders who had received significant benefits from the industrial agro-food system in the past and expected to continue to do so in the future. These stakeholders include farmers, farm input manufacturers and suppliers, processors, wholesalers and retailers of farm output. It was the intervening years between World War II and the first signs of tension within modern farming systems during the 1960s and 1970s that allowed these interests to become highly institutionalised and exceedingly influential in a plethora of policy develop processes, which directly or indirectly affected their economic well-being or political status.

Over time, a deepening convergence and correspondence developed between the aims, objectives of government and agribusiness stakeholders with regard to the development of agricultural policy. This occurred to such an extent that, in many cases, making a distinction between the two factions became increasingly difficult. Another key characteristic of these so-called agro-policy communities during this Period was their conservatism and robustness, specifically their ability to resist the infiltration of new stakeholders who did not share their own aims and objectives. This attribute meant that even though vocal members of the public, media and scientific community had much to say regarding industrial agro-food systems, their ability to influence the agricultural and food policy development process was extremely limited. This virtual 'closed shop' attitude was critically important in the maintenance of the productivist agriculture and the industrial agro-food system. Agro-food policy communities were so influential that were even able to delay the evolution of new governmental institutions with environmental remits, and often manipulate the scope of their remit so that their own interests were not compromised. It should also be noted that, publicly sponsored research, development and education remained aligned with furthering the interests of productivist agricultures and industrial agro-food systems.

That said, even as the industry sought to discredit its critics, some within the chemical manufacturing sector were recognising that changes in the way pesticides were promoted and used were necessary and inevitable (Murray and Taylor 2000). In 1967 (Hough 1998), the pesticide industry formed the

Groupement International des Associations Nationales de Fabricantes de Produits Agrochimiques (GIFAP), to lobby on behalf of the pesticide industry. GIFAP was a network of national agrochemical associations. The national associations of the five main pesticide-producing and -exporting nations (USA, Japan, Germany, UK and Switzerland) are full members, and in all, over 1000 companies and 90% of the world's pesticide producers are represented (Hough 1998). GIFAP undertook a range of lobbying and promotional efforts aimed largely at influencing international development and regulatory policies such as those being pursued by the FAO of the UN (Murray and Taylor 2000). GIFAP acquired liaison status with FAO in 1967 (Hough 1998). The relationship between the two institutions has always been strong, reaching a high point in the 1970s, when the Industry Cooperation Programme (ICP) was established as a joint bureau. ICP organised seminars in developing countries, outlining the importance of chemical technology in furthering development (Hough 1998), and the need to promote 'new and better ways' to distribute and use pesticides in the developing world. However, GIFAP's uncontested collaboration with the FAO was short lived, as, after political pressure from various NGOs, the FAO was obliged to curtail its ICP (Hough 1998). As criticism aimed at pesticides grew, the FAO's dominant role in international pesticide policy (Murray and Taylor 2000) formulation was increasingly contested from within the UN system as well as from environmental and anti-pesticide NGOs.

In the USA, it was, in part, the recognition of the pesticide industry's influence over pesticide regulation, and the ineffectiveness of the regulatory process that eventually led to. (1) Creation of the EPA in 1970 (Murray and Taylor 2000). (2) The subsequent transfer of jurisdiction for pesticide regulation from the United States Department of Agriculture to the EPA in 1972. While this transfer was hotly contested by the industry, it did, in some ways, bring greater credibility to the industry among the American public. However, while the EPA's remit was to balance public health, environmental risks and economic impacts, its political legitimacy was extremely fragile. Pesticide use was supported by a powerful constituency of industry, agricultural associations and much of the agricultural research establishment (Glicksman et al. 2003). While the USDA's and EPA's judgements on the regulation of pesticides often seemed irrational to neoclassical economists and unconstitutional to many citizens of developed countries, they were consistent with political economy approaches. Indeed, political economy theory suggests that costs are often left out, or roughly approximated, because decision/policymakers lack the resources and/or reliable methods to measure or compare them. Costs are also left out when influential persons or groups, acting in self-interest, do not have to bear them. From the point of view of the self-interested decision maker, the costs are truly external (Thompson 1971). This was certainly the case for interest groups such as farmers, agribusiness and environmentalists.

Chapter 3

The evolution of the pesticide industry in the Divergent Period (1974–2017)

3.1 Key characteristics of farming systems and broader economic, social and political contexts

Many of the key structural trends evident at the beginning of the Productivist Period were still evident at the start of the Divergent Period. Population levels continued to rise throughout the Productivist Period and had reached nearly 4 billion by 1974 (Sarre and Blunden 1995). However, by this time, population growth had become highly skewed, with very high levels of growth in developing countries, particularly in the tropics, and either relatively static or even negative population growth rates in many developed countries. In 1974, the population growth rate in developed countries was approximately 0.75%, compared to approximately 2.5% in developing countries (Islam 1995). By the start of the Divergent Period, 37.8% (UN Habitat 2004) of the world's population was urbanised and severely limited with respect to food production and/or collection opportunities. Industrial principles of agricultural production and management had deepened significantly and the use of industrial inputs, particularly pesticides, over much of the developed and developing world had become relatively prophylactic. The urban-centred industrial agro-food system had become institutionalised as 'the' model for food production, processing, distribution and consumption over most of the globe. However, across developed countries, concerns had grown regarding the impacts of productivist agriculture on: food surpluses; the environment; food quality and safety and human health.

The evolution of pesticides and the pesticide industry during the Divergent Period was principally underpinned by a so-called collapse of Fordism and the Second Food Regime and the search for stable social, economic and political arrangements in which capitalism could thrive once again.

3.1.1 The Collapse of Fordism and the Second Food Regime

It is suggested that the Fordist Regime of accumulation collapsed sometime during the early 1970s (Ward 1993; Levidow and Bijman 2002; Friedmann 2005; McMichael 2009b; González-Esteban 2017). In other words, the compromise between capital and labour, which had been brokered by nation state governments, and the supporting social, economic and political institutions that formed a stable mode of accumulation, had been fractured, seemingly, beyond repair. Although debate still rages over whether or not the Fordist Regime of accumulation actually collapsed during this Period, or, whether, in fact, it was able to adjust or realign itself, there is little doubt that, as a result of emerging contradictions and growing structural instability, Fordism slipped into crisis. It is suggested that crisis in Fordism arose due to a growing state of economic uncertainty, which first emerged in the world's most important economy, the USA, during the late 1960s, spreading to Western European economies during the early 1970s. Economic uncertainty took two principal forms; (1) a slowdown in productivity and (2) declining profitability throughout most industrial sectors.

3.1.1.1 A slowdown in productivity

It is suggested that a slowdown in productivity was rooted in the Fordist labour process, particularly the inability of mass production methods to realise further productivity gains within manufacturing, as well as their limited applicability to areas of the economy such as services. In this view, the limits of Fordism had been reached, both in a technical sense and in relation to a mass collective workforce whose resistance to the dull, repetitive rhythms of work had peaked. From a technical standpoint, increasing returns to scale from ever-larger mass production complexes were said to have been frustrated by the difficulties experienced in balancing what is produced with what is in demand at any one point in time. Fluctuation in the pace of demand, shifting patterns of taste, delays in re-starting the line and the time taken to re-tool machinery, all contributed towards a trend of declining productivity. In addition, from the point of view of labour, the excessive bureaucratisation of control, coupled with the tedium of work, had led to increased absenteeism and strife among the workforce. The very characteristics, therefore, which had once contributed to the success of the Fordist Regime, were now blocking the accumulation process.

3.1.1.2 Declining profitability throughout most industrial sectors

Declining profitability throughout most industrial sectors concerns a shift in the post-war pattern of global demand and a significant slowdown of post-war economic momentum. This, it is argued, occurred partly in response to the steady erosion of the US dollar as an international regulatory currency.

As strong economies such as Japan and Germany, and the EEC [now the European Union (EU)], challenged the hegemony of the USA in the world economic order. Indeed, the impact of increased competitiveness fuelled global instability, threatened trade wars, disrupted traditional patterns of trade with developing countries and enhanced the prospect of international recession (Atkins and Bowler 2001).

The overall result was further disruption in the post-war economic momentum. Compounded by the oil shock of 1973, which led to a dramatic increase in energy costs in the western nations, the long post-war boom, which had lent credibility to the notion of Fordism as an industrial era, appeared to be at an end. An important note is that, many developing countries, particularly those with divided societies and weak institutions for conflict management, suffered the most significant declines in economic growth (World Bank 2003). This phenomenon was compounded by the breakdown of oligopolistic pricing methods within national markets due to increased internationalisation.

Thus, within the spheres of both production and consumption, the national and international dynamics that had sustained Fordist economic growth were, in this view, beginning to crack, and with them the forms of social regulation that supported the Regime, affecting workers, capital and governments. Workers were increasingly dissatisfied with their lot. The compact that underpinned annual improvements in their standards of living and work conditions began to erode, leading to increased occurrences and intensities of worker militancy and less disposable income. On the other hand, capital became increasingly frustrated with slow economic growth and low rates of return on capital investments, and thus embarked on a quest to develop new, more profitable arrangements. Lastly, capital, and to a certain extent workers, across much of the world lost faith in the ability of nation state governments to facilitate economic development that would satisfy the requirements of both capital and labour. Since the crisis of Fordism in the early 1970s, capital, organised labour and governments have endeavoured to construct a new, and lasting, arrangement (i.e. a new mode of regulation) that had the capacity to balance their contemporary needs. However, after more than 40 years, debate as to whether or not Fordism has been superseded by a new and superior organisational superstructure is still ongoing. Currently, there are those who proclaim the possible emergence of a successor to Fordism in the form of Post-Fordism or a Corporate-Environmental Food Regime (Friedmann 2005) or the emergence of neo-Fordism or a Corporate Food Regime (McMichael 2005). At the same time, there are those that insist that Fordism did not collapse and that what we see today is a solution to the inherent problems of the Fordist Regime through minor adjustments (Pritchard 2009). Whichever argument you believe, it is undeniable that the face of capitalism in the new millennium is greatly different from that of the 1970s.

3.1.1.3 Crisis in the Second Food Regime

Just as the Fordist Regime of accumulation reached a crisis in the early 1970s, so did the urban-centred industrial agro-food systems and techno-policy model of productivist agriculture of the Second Food Regime. This crisis, often referred to as the international farm crisis, was attributed to two principal events. The first event took the form of a sudden and sharp increase in oil prices; brought about by a reduction in global oil supplies caused by a strategic restriction of oil supplied by the world's largest oil cartel the Organisation of Petroleum Exporting Countries. In the agro-food sector, this had the immediate effect of increasing agricultural input prices, particularly fuel, fertiliser, pesticides and machinery and increasing food transportation and processing costs further down the food chain. The second event took the form of an almost total elimination of surplus international wheat stocks (the world's principal food staple at the time) due to US-USSR grain deals. The combined effects of these two events was to plunge the global economy into recession, place a final nail in the coffin of the Bretton Woods System of exchange rate management, produce a significant increase in world grain prices and rekindle worries over international and domestic food security. It is ironic that, until this time, institutionalised food surpluses, a stagnant world market, growing evidence of the negative impacts of industrial agriculture on human health and the environment, and escalating costs of national farm support programmes had been the most likely candidates to undermine the techno-policy model of productivist agriculture (Redclift 1989).

In the sphere of food production, distribution and consumption, some analysts, such as McMichael (2005), have already proclaimed emergence of the Third Food Regime, or at least the beginning of a transition phase. However, after more than 40 years after the alleged breakdown of the Second Food Regime, many analysts remain to be convinced. Indeed, if the Third Food Regime has in fact emerged, the new regime could only be defined as highly differentiated and full of contradictions and tensions. Therefore, many scholars have indeed suggested that western capitalist societies (not to mention those in Eastern Europe) are still engaged in an enduring search for a new institutional compromise. For example, Amin (2000) stressed that 'reliance on sharp distinctions between phases . . . denies the key aspect of history as a complex and heterogeneous project of many determinations'. In addition, Peck and Tickell (2000) argue, 'it is premature to talk about a post-Fordist regime of accumulation because, while some significant experiments are under way in the production sphere, a coherent post-Keynesian mode of social regulation has yet to stabilize'.

What seems increasingly likely is that, throughout the last 40 years, agro-food capitals have explored a range of strategies that sought to maximise accumulation over heterogeneous social, economic and political landscapes.

Where strategies have been coordinated or where they have been compatible with, and/or complementary to, the strategies of other key actors in the food chain, agro-food capitals have successfully secured high and stable returns. Indeed, in the contemporary landscape, it may be possible to discern a range of new micro-food regimes that coexist and overlap each other in time and space. Indeed, just as the First and Second Food Regimes failed to encapsulate the whole world, why should we, for the sake of argument, endow the process of globalisation with the capacity to promote a single omnipotent regime of accumulation with its associated mode of social regulation? Because of the difficulties associated with the search for the Third Food Regime in the contemporary phase, many theorists have momentarily stepped down from the high-order macro-levels of analysis and begun to focus on meso-level explanations of food systems and, indeed, wider economic growth and development. Indeed, Goodman and Watt's provocative and sharply worded critique of Neo and Post-Fordist food systems literature of the 1970s and 1980s (Goodman and Watts 1994) significantly eroded the credibility of traditional Regulationist explanations of food system development. In its place, meso- and micro-level neo-regulationist frameworks of analysis, namely: (1) Conventions Theory; (2) Commodity Systems Analysis and (3) Actor-Network Theory (ANT) have emerged.

3.1.1.3.1 Conventions theory

Based on the earlier work of Boltanski and Thévenot (1991), Conventions Theory has become widely utilised by Neo-Regulationists since the mid-1990s in attempts to explain what Salais and Storper (1992) termed as the 'micro-foundations' of capitalisms reflexive turn. According to Storper (1997), Conventions Theory suggests that economic activity is choreographed by conventions, namely specific 'practices, routines, agreements, and their associated informal and institutional forms', which bind individuals and their actions together through 'mutual expectations' (Salais and Storper 1992).Where different combinations of conventions coalesce, they result in distinct 'worlds of production' or logics of production. Salais and Storper (1992) identified four worlds of production: Industrial World; World of Intellectual Resources; Market World and Interpersonal World.

- The 'Industrial World', is characterised by the logic of mass production of standardised and generic commodities, and by the logic of mass consumption. Standardised commodities are those that are produced using widely diffused methods, which adhere to accepted standards, norms, objectivised rules and testing procedures, so that competition comes to be inevitably centred on price (Storper 1997). Similarly, the generic commodity is attributed with recognisable inherent qualities that allow it to be easily traded on stable and predictable existing markets.

- Conversely, the 'World of Intellectual Resources' is characterised by the production of specialised products. Because the specialised product is made using restricted technology and know-how, competition is centred on quality, and is generally associated with recognised brands and trademarks.
- The 'Market World' is characterised by the production of standardised and dedicated products. Dedicated products are those targeted towards specific sets of clients. While efficiency and reliability remain important attributes, the recognition of trademarks and brands and attachments to place and tradition are also important, as are ecological, health and safety values. The 'Market World' is governed by domestic and civic conventions. Domestic conventions are based on trust of people, places or brand names, and the importance of the home, family and community. They stress the role of face-to-face relationships, loyalty and trust in maintaining a functioning society. Civic conventions rely on the adherence of a group of actors to a set of collective principles for the common good, and the expectation that individual interest will be subordinate to the common good: Fair Trade is good example of civic coordination.
- Lastly, the 'Interpersonal World' is characterised by the production of specialised and dedicated products, where the market is composed of interpersonal negotiations. The 'Interpersonal World' is also dominated by domestic and civic conventions, although to a much greater extent than the 'Market World'. Domestic and civic conventions in the 'Interpersonal World' are particularly based on trust, equity, collective social responsibility and ecological and economic harmony.

Conventions Theory, therefore, exemplifies the micro-social regulations that define economic actors' collective identities and coordinates their modes of economic participation (Storper and Salais 1997). Conventions themselves are seen to represent the codes, channels of interaction and ways of organising and coordinating behaviours (Storper 1992). According to Storper (1997), conventions coordinate the actions of actors and actants (non-human entities, i.e. food material, and machinery) within an 'action framework', taken to mean commodity system/chain/filière or actor network. In the field of agriculture, food systems and the environment, Conventions Theory, often in conjunction with Commodity Chain/Commodity Systems Analysis and ANT, has facilitated more rigorous explanations of agro-food system tendencies. On the one hand, towards standardisation and homogenisation, and, on the other hand, countertrends towards organic food, local food systems, local food labelling, an emphasis on quality, and so on that are leading to widespread restructuring of regulatory practices along the entire span of food chains. The work of Marsden and associates

(Marsden 1999, 2000) reflects particularly clearly this new, more localistic, consumption-focused regulation tradition in the sociology of agriculture.

3.1.1.3.2 Commodity Systems Analysis

Commodity Systems Analysis has been one of the most widely utilised conceptual and empirical frameworks for characterising the links between the production, distribution and consumption of goods and services, particularly in relation to Neo-Regulationist political economies of agro-food systems (Friedmann 1993; Watts and Goodman 1997; Watson 1998). Friedland (1984) initially developed Commodity Systems Analysis in the early 1980s. Since that time, various derivations have emerged. Essentially, Commodity Systems Analysis focuses on the progressive movement of a commodity through the sequential phases of production, distribution and consumption. In agro-food systems studies, researchers have focused on detailed empirical analyses of particular agricultural commodity systems. Their emphasis has been on: (1) The structures and strategies of multinational/transnational corporations. (2) The role of nation states. (3) The power asymmetries and contestation within chains. (4) The modes of organisation. (5) Determinations of the loci, and mechanisms, of institutional power used to allocate economic resources and wealth (producer-driven or buyer-driven Vorley 2002). (6) The importance of means of coordination versus means of production. (7) The dynamics of cross-border commodity chains. (8) New divisions of agri-food labour. (9) Agricultural industrialisation. (10) The marginalisation of labour and environmental concerns.

3.1.1.3.3 Actor-Network Theory

Actor-network analysis largely derives from the anthropology of science and technology ('technoscience') of Bruno Latour and Michael Callon (Latour 1987; Callon 1986). Their most important contribution was to bridge the previously intractable impasse between macro- and microanalyses and between society and nature. Indeed, the rise of actor-network approaches has arguably been the most significant trend in agrarian sociology over the last 40 years, and has been widely utilised in Neo-Regulationist agri-food studies and economic geographies (Whatmore and Thorne 1997) that have sought to develop flatter, less hierarchical and less deterministic theories of the economy (Thrift and Olds 1996; Watson 1998). Whilst acknowledge as an ontology, rather than a theory, ANT deploys the metaphor of the network to economic activity in order to exemplify how 'different kinds of nodes (people, firms, states, organisations, etc.) are connected to one another in complex and multi-stranded ways' (Thrift and Olds 1996; Watson 1998). In a similar fashion to Conventions Theory, ANT has been particularly useful in representing the organisation of social, cultural and biophysical elements in economic activity, and illuminating relationships between production, distribution and consumption.

ANT has brought increased conceptual and empirical robustness to recent agro-food analyses, particularly traditional commodity chain approaches that emphasised the unidirectional, and mechanistic, linearity of links within commodity chains. While still acknowledging the power asymmetries between sets of actors (or nodes), associations between varied actors and actants (non-human) are viewed as complex 'webs of interdependence' rather than fixed, vertical and unidirectional relationships. These webs of interdependence connect farmers, merchants, processors, distributors, wholesalers and retailers, and a range of other key actors and actants including scientists involved in agricultural R&D, NGOs and consumer groups, through vertical commodity exchange relationships and horizontal associations (based on 'multi-directional flows of information and materials') that either support or seek to undermine or reshape commodity exchange relationships.

From ANT's perspective, the globalisation of commodity networks is simply the extension of supportive associations over a greater spatial scale. Network extension can be achieved either by adding actors and actants to the network or by extending the spatial reach of existing actors and actants. In their study of the globalisation of rapeseed production networks, Busch and Juska (1997) suggest that it is contemporary technologies (new actants) that 'permit new forms of action at a distance'. Whatmore and Thorne (1997) also develop this notion of 'network lengthening' in tracing the connections of fair trade coffee networks, insisting that 'the power associated with global reach has to be understood as a social composite of the actions and competences of many actants'.

3.2 Summary

The following chapters start with the premise that crisis in the Fordist Regime of accumulation and in the Second Food Regime led to the evolution of a complex array of both complementary and contradictory drivers that created heterogeneous development contexts for the pesticide industry. In turn, this led to the emergence of a wide range of pesticide industry strategies that aimed to strike a balance between their own corporate needs and those of society within these contexts; and in so doing, achieve relatively stable conditions for continued profit generation. Each chapter organises these drivers into coherent themes, and discusses their impacts on the pesticide industry and on urban-centred industrial agro-food systems. In turn, each chapter outlines the pesticide industry's responses to these drivers over the past 47 years.

Chapter 4, entitled 'Pesticide industry evolution under Residual Productivism', addresses how the pesticide industry responded to continued, if

Table 3.1 Conventions associated with competing food system paradigms (1974–2017) after Boltanski and Thévenot (1991) and Thévenot (1998)

Conventions	Residual-productivist paradigm	Neo-productivist paradigm	Post-productivist paradigm	Sustainability paradigm
Commercial conventions (e.g. evaluation by price)	X	X		
Industrial conventions (e.g. focus of efficiency and reliability)	X	X	X	
Public conventions (e.g. recognition of trade-marks and brands)			X	X
Domestic conventions (e.g. attachments to place and tradition)			X	X
Civic conventions (e.g. importance of ecological, health and safety issues)			X	X

weakened, state protection of the productivist techno-policy model of Fordist agriculture in Western Europe, the USA and Japan. It emphasises the continued adherence to 'industrial' and 'commercial conventions (see Table 3.1), particularly with regard to the production and processing of staple food crops (see Table 3.3), and the conservative intransigence of powerful actors within key commodity systems/networks (see Tables 3.3–3.6 for a summary of food system characteristics). Chapter 5, entitled 'Pesticide industry evolution under Neo-Productivism', discusses the pesticide industry's role in the realignment of agricultural production in both the developed and developing world in light of Neo-Fordist developments. Its focus is on the drive for increased pesticide industry profits through company restructuring and product diversification and the production of technologies that endow farmers with the chance of reducing unit costs against static or falling output prices. Discussion is centred on the establishment, and widespread acceptance, of 'industrial' and 'commercial' conventions (see Table 3.1), and, ultimately, intellectual property conventions, and their protection, that exploit the emerging 'World of Intellectual Property' (see Table 3.2), particularly with regard to staple food and fibre commodities across the world (see Tables 3.3–3.6 for a summary of food system characteristics). Conversely, Chapter 6, entitled 'Pesticide industry evolution under Post-Productivism', discusses the pesticide industry's role in the realignment of agricultural production in both the developed and developing world in light of the Post-Fordist developments. This primarily takes the form of customer and retailer demand-driven adjustments in relation to key commodity

Table 3.2 'Action frameworks' and 'Worlds of production' associated with competing food system paradigms (1974–2017) after Salais and Storper (1992, 1997)

Conventions	Residual-productivist paradigm	Neo-productivist paradigm	Post-productivist paradigm	Sustainability paradigm
Industrial world (e.g. standardised/generic production)	X	X		
World of intellectual resources (e.g. specialised/generic production)	X[a]	X[a]	X[b]	
Market world (e.g. standardised/dedicated production			X	
Interpersonal world (e.g. specialised/dedicated production)			X	X

[a] Production of patent-protected pesticides and input trait biotechnologies enabled pesticide/life science corporations to differentially price products based on the cost/benefit ratios faced by their customers in relation to different crops in different countries. In food systems governed by commercial and industrial conventions, competition in relation to the final product was centred on price.

[b] Output trait biotechnologies are expected to be either: (1) differentially priced, based on the cost/benefit ratios faced by their customers, varying according to different output traits in different crops and different countries; or (2) farmers may be contracted to grow commercially viable output trait biotechnologies by life science/food processor alliances, where price with ultimately be centred on actual or perceived quality in food systems governed by domestic and civic conventions.

systems/networks, which are increasingly choreographed by public, domestic and civic conventions (see Table 3.1). This chapter explores the mechanisms through which the pesticide industry sought to exploit emergent food and fibre production/action frameworks, namely 'World of Intellectual Property', 'Market World' and 'Interpersonal World' (see Table 3.2). See Tables 3.3–3.6 for a summary of food system characteristics. Chapter 7, entitled 'Pesticide industry evolution and the Sustainability Paradigm', addresses how the pesticide industry responded to the constraints and opportunities associated with the integration of so-called sustainability goals into agro-food system regulation. This chapter explores the mechanisms through which the pesticide industry sought develop and define/redefine the role of pesticides and biotechnologies in sustainable food and fibre production systems, and to profit from so doing. The Sustainability Paradigm hinges on 'public', 'domestic' and 'civic' conventions (see Table 3.1) and the 'Interpersonal World' (see Table 3.2). See Tables 3.3–3.6 for a summary of food system characteristics.

Table 3.3 Structural characteristics associated with competing food system paradigms (1974–2017)

Residual productivist paradigm	Neo-productivist paradigm	Post-productivist paradigm	Sustainability paradigm
Urbanisation, concentration of wealth and separation of business ownership and control	Urbanisation, concentration of wealth and separation of business ownership and control	Urbanisation, counter-urbanisation, concentration of wealth and separation of business ownership and control	Counter-urbanisation and sustainable communities
Mass production of standardised commodities (e.g. wheat, maize, soybean and cotton) – production specialisation – usage of a narrow range of crop species	Mass production of standardised low value commodities (e.g. wheat, maize, soybean and cotton) – production specialisation – usage of a very narrow range of crop species with little genetic diversity	Batch production of standardised, high value-added, region speciality food products (e.g. Swiss cheese, English lamb and Scottish salmon) – production specialisation – usage of a relatively narrow range of crop species	Small-scale production of a wide range of locally derived, non-standard, food and fibre crops – production de-specialisation
Mass consumption of low-cost, standardised, often highly processed food products	Mass consumption of low-cost, standardised, often highly processed food products	Niche markets for differentiated high-value, authentic, natural and healthy food products	Local markets for locally produced authentic, natural and healthy food products
Mass advertising	Mass advertising	Targeted advertising	Local information networks

Table 3.4 Farm-level production characteristics of competing food system paradigms (1974–2017)

Residual productivist paradigm	Neo-productivist paradigm	Post-productivist paradigm	Sustainability paradigm
Increased commoditisation of agricultural production	Increased commoditisation of agricultural production	Increased commoditisation of agricultural production and the environment	De-commoditisation of agricultural production and the environment
Increased industrialisation of agricultural production	Increased industrialisation of agricultural production	Decreased industrialisation of agricultural production	De-industrialisation of agricultural production
High-input/high-output farming systems	High-input/high-output farming systems	Lower-input/high-output farming systems	Lower-input/lower-output farming systems
Significant reliance on off-farm inputs: pesticides, machinery, fertiliser, seeds and input trait biotechnologies, capital, and labour-little consideration of farm agro-ecosystem and local sociocultural context of production	Almost total reliance on off-farm inputs: pesticides, machinery, fertiliser, seeds and input trait biotechnologies, capital, and labour-little consideration of farm agro-ecosystem and local sociocultural context of production	Reduced reliance on off-farm inputs: pesticides, machinery, fertiliser, seeds, capital, and labour-much greater consideration of farm agro-ecosystem and local sociocultural context of production	Minimal reliance on off-farm inputs
Farm wages are regulated by the state	Farm wages are regulated by the market	Farm wages are proportional to the idiosyncratic skills of individual workers	
Farm workers possess medium skills levels in defined skill areas	Farm workers possess low skill levels in defined skill areas	Some farm workers possess high skill levels in a range of skill areas, while others possess low skill levels in defined skill areas	Farm workers possess high skill levels in a range of skill areas

Table 3.5 Regulatory and trade characteristics of competing food system paradigms (1974–2017)

Residual productivist paradigm	Neo-productivist paradigm	Post-productivist paradigm	Sustainability paradigm
National food self-sufficiency	National food self-reliance	National food self-reliance	Local food self-sufficiency
Regulation of national agricultural production the state in cooperation with business, farmers' representatives and the trade unions	Deregulation of agricultural production and usurpation of the power of organised labour	Re-regulation of agricultural production to include expansion of the sphere of economic activity – particularly the production of authentic, original and natural food products, biodiverse agroecosystems, beautiful landscapes, and the preservation of rural history and culture	Local control over food and fibre systems
Agricultural trade regulated by national governments and supranational governmental apparatus (European Commission)	Agricultural trade regulated by the market but strongly influenced by the World Trade Organization	Agricultural trade regulated by the retail sector and strongly influenced by national governments regulation of trade based on food quality, and the environmental and animal welfare aspects of food production	Agricultural trade and quality standards are principally unregulated as the exchange of food and fibre products are based on trust and producer pride and civic responsibility
Intellectual property rights associated with industrial agriculture are protected	Intellectual property rights associated with industrial agriculture are considered essential and fiercely protected through vertically integrated corporations	Intellectual property rights associated with industrial agriculture are protected through alliances between corporations	Relies on systems of common ownership
Increased subsumption of regional farming systems within deepening urban-centred agro-food complexes	Increased subsumption of global farming systems within deepening urban-centred agro-food complexes	Increased subsumption of global farming systems within deepening retailer dominated urban-centred agro-food complexes	Autonomous local agricultural production balanced with local consumption
Regional markets of Western Europe, the USA and Japan are targeted by pesticide corporations	Global markets are targeted by pesticide corporations	Pesticide corporations target high-value retailer product supply chains to negotiate the role of pesticides	Pesticide corporations struggle to define a role for pesticides and biotechnologies

Table 3.6 Environmental and human health characteristics of competing food system paradigms (1974–2017)

Residual productivist paradigm	Neo-productivist paradigm	Post-productivist paradigm	Sustainability paradigm
Environment regarded as a backdrop to agricultural production	Environment regarded as a backdrop to agricultural production	Environment regarded as an important component of agricultural production	Food production is harmonised with local environmental and sociocultural context
Consumer health externalities associated with industrial agriculture, particularly pesticide residues are redressed through government regulation	Consumer health externalities associated with industrial agriculture, particularly pesticide residues are not redressed	Buyer-driven agricultural product supply chains attempt to internalise consumer health externalities associated with market-based agriculture by making food system adjustments, particularly to agricultural production	Consumer health and nutrition is paramount
Agricultural worker health externalities associated with industrial agriculture are redressed through government regulation	Agricultural worker health externalities associated with industrial agriculture are not redressed	Agricultural worker health externalities associated with market-based agriculture are redressed through private sector (retailer) and government regulation	All efforts are made to internalise agricultural worker health externalities
Major agricultural productivity threatening environmental externalities associated with industrial agriculture are redressed by government regulation	Environmental externalities associated with industrial agriculture are not redressed	Environmental externalities associated with market-based agriculture are redressed by private sector (retail) and government regulation – in many cases, environmental costs and benefits are fully internalised	Environmental externalities associated with agricultural production are fully internalised

Chapter 4

The evolution of the pesticide industry under Residual Productivism (1974–2017)

4.1 Introduction

By the end of the 1970s, it was clear that the Fordist compromise between capital and labour, underpinned by its unique mode of social regulation, had ended. However, despite the demise of Fordism as a relatively coherent mode of regulation, Fordist principles of production and consumption remained embedded in many economies across the world, albeit in an emasculated form. Due, in part, to their unique characteristics and fundamental importance, urban-centred industrial agro-food systems, based on productivist agriculture, remained as residual elements of the Fordist era, particularly in Western Europe, the USA and Japan. This was especially the case where Keynesian-styled protection for food production remained firmly institutionalised, and a market still existed for low-cost, mass-produced, standardised and generic food products that had been grown, processed, distributed and retailed using industrial methods.

Chapter 4 argues that juxtaposed to the rapidly changing economic climate in Western Europe, the USA and Japan, where the Friedman styled supply-side monetarist policies of Reagan and Thatcher were implemented, sustained Keynesian-styled state support for industrial agro-food systems continued to provide a context that strongly influenced the strategic direction of the pesticide industry.

4.2 Key characteristics of urban-centred industrial agro-food systems and broader economic, social and political contexts

In many respects, this chapter may, at first sight, seem at odds with conventional analyses of corporate development over the past 40 years, as its focus is distinctly regional than global. While this chapter does not attempt to suggest that the globalisation of food systems is unimportant, it does stress the significance of national and regional contexts that framed the evolution of the pesticide industry throughout the 1970s, 1980s, 1990s, and the early twenty-first century in Western Europe, the USA and Japan. Indeed, globalisation is a central theme in the following chapters. So what are the unique characteristics that set Western European countries, the USA and Japan apart from other nations? In terms of economic development, these are the most prosperous democratic capitalist economies in the world, endowed with substantial financial resources. The financial capacity to underwrite agriculture, combined with the political will to do so, remained the cornerstone of pesticide industry growth. Population was another important factor. Even though population levels in Western Europe, the USA and Japan had begun to plateau or even decline, food security through food self-sufficiency remained an important political goal. Indeed, between 1975 and 2000, the average annual population growth rate in high-income OECD countries was only 0.6% (UNDP 2002), and easily matched per capita food supplies (Islam 1995). It is also important to note that, even though concerns over the human health impacts of industrial agriculture had grown, the majority of food production remained based on conventional high-input/high-output agricultural systems, characterised by the relatively prophylactic use of industrial inputs, particularly pesticides. The urban-centred industrial agro-food system remained institutionalised as 'the' model for food production, processing, distribution and consumption. In other words, agriculture in Western Europe, the USA and Japan remained protected inside a state-sponsored bubble. By default, this also implies the continued existence of well-established corporate agribusinesses, including strong domestic and multinational agrochemicals businesses.

4.3 Key external drivers of pesticide evolution

This section begins by outlining the specific combination of external drivers, which conditioned the evolution of the pesticide industry in Western Europe, Japan and the USA within the Residual Productivist context. Due to the influence and complex interdependency of agro-food system components, external drivers are defined here as driving forces outside the direct control of the pesticide industry. This definition therefore includes drivers over which the pesticide industry does have some control, namely continued political and financial support for urban-centred industrial agro-food systems. The Residual Productivist context of pesticide industry development was principally brought about by the combination of six key external drivers. (1) Falling real unit prices for staple food commodities and stagnant world markets. (2) Biological resistance to a growing range of pesticide AIs. (3) Expiry of pesticide patents. (4) Increased pressure to reduce negative human health impacts of pesticide use. (5) Continued political and financial support for the farming sector. (6) Scientific and technological advances, particularly in molecular biology. The first four drivers presented significant obstacles to the pesticide industry, whereas, the last two provided the means by which they could potentially overcome them. As this chapter focuses on the survival of productivist agriculture, and is therefore primarily focused on the economics of pesticide use and direct human impacts of pesticides, broader environmental externalities associated with industrial farming systems and pesticide use are addressed in Chapters 5 and 6.

4.3.1 Falling real unit prices for staple commodities and stagnant world markets

Aside from the brief hike in food prices, particularly wheat, during and immediately after the food and oil crises in the early 1970s, the falling real unit prices for staple commodities and stagnant world markets meant that most farmers were becoming increasingly subject to a growing cost-price squeeze (Evans 1998).

For a number of reasons embedded within the technological treadmill concept (see Chapter 2), most farmers were initially reluctant to explore alternative income-generating strategies (both agricultural and non-agricultural). Therefore, one of the key strategies employed by farmers during the Divergent Period was to remain on the technological treadmill, increasing agricultural inputs per area in an attempt to increase agricultural outputs per area and maintain or enhance their incomes, particularly as their respective governments continued to offer substantial output-based subsidies. Throughout this Period, most farmers still perceived pesticides to be one of the key productivity- and profitability-enhancing inputs, and an obvious route to survival, or even success. To explore alternative input-reducing, and potentially yield-reducing, strategies was to risk lower levels of profit or financial loss, even against the backdrop of financial support from their respective governments. In addition, one of the few potentially profitable avenues for growers of staple food crops was the production of high-quality commodities in the hope that they would receive higher premiums and higher financial returns. This strategy was often more, rather than less, dependent on pesticides. Indeed, lower cosmetic value (i.e. blemishes, in otherwise edible produce) is especially important in markets in North America, Japan and Western Europe.

For the pesticide industry, this meant that their products, at least under conditions of high pest and disease pressure, must be able to show a return on investment for the farmer. In addition, as the agrochemical sector continued to grow, corporate decisions increasingly needed to take into account shareholder demands for high rates of return on capital investments. This balancing act became increasingly difficult for pesticide manufacturers to achieve. Not only was the pesticide R&D process (including costs incurred through pesticide regulation) becoming increasingly expensive, pesticide companies were faced with both the expiry of pesticide patents and a growing biological resistance to a range of pesticide AIs. This meant that pesticide companies, as well as farmers, were increasingly subject to a growing cost-price squeeze.

4.3.2 Biological resistance to a growing range of pesticide active ingredients and expiry of pesticide patents

As if the challenge faced by pesticide companies to provide efficacious pesticides at a price that maintained farmer solvency, whilst generating substantial returns on investments for agrochemical company shareholders, was not enough, pesticide companies faced increasing biological resistance to a growing range of pesticide AIs, both old and new. The repeated use of a given pesticide often led to the development of pesticide resistance in insect pests, pathogens and weeds. Where resistance builds up, the pest control options available to farmers decline, leading to extra costs for the development of new options, and possibly to greater losses in yield due to pests and to the use of less efficient or more harmful AIs (DEFRA 2000).

Much of the early work on resistance builds up concentrated on insect populations (DEFRA 2000). Resistance to insecticides was first noted at the turn of the twentieth century. However, while still manageable, insecticide resistance became much more of a problem after the 1940s and the introduction of synthetic organic insecticides (Georghiou 1986). The first case of resistance to DDT was detected as early as 1947 (Georghiou 1986), only 8 years after its introduction. During the 1950s, approximately 50 insect species developed resistance to modern synthetic insecticides (Georghiou and Taylor 1976). Insecticide resistance continued to be a problem throughout the 1960s and 1970s. In the period 1965-77, the number of resistant insects and mites was estimated to have doubled (Allen 1980). By 1980, there were 432 species of arthropods with resistance to at least one pesticide (Bull 1982), and by 1986, 447 species of insects and mites had developed resistance to insecticides, including most major pests (Georghiou 1986). By the mid- to late 1980s, resistance had developed in all insecticides groups and at least 480 species of insect, mite or tick were recorded as resistant to one or more compounds (Georghiou 1986). Almost 50% of arthropod species could resist compounds in two, three, four or five classes of chemicals; 17 insect species were resistant to all five classes (Georghiou 1986). By 1990, 504 insect species had developed resistance to at least one insecticide (Pimentel et al. 1993), and at least 23 species, including Colorado beetle (Georghiou 1986), had developed resistance to pyrethroids, the most recently introduced class of insecticides. Whereas the presence of resistance was a rare phenomenon in the 1950s, it was the fully susceptible population that was rare in the 1980s (Georghiou 1986). The creation of new pest species also became a major problem. Bull (1982) suggested that, while there were currently only 5000–15 000 pest species, pesticide overuse could potentially create ten times that number, particularly in the tropics and subtropics (Bull 1982).

To a lesser extent, resistance began to appear in fungal and weed species. The first cases of fungal resistance occurred in 1940 (Georghiou 1986), but the number of resistant fungal strains dramatically increased throughout the 1970s and early 1980s (Georghiou 1986). In 1976, there were 67 different pathogens resistant to at least one chemical, almost double the number (34) in 1974 (Bull 1982). By 1979, 70 species of fungal pathogen were reported resistant, 59 species were resistant to the systemic fungicide Benomyl and 13 species were resistant to thiophanate (Georghiou 1986). By 1986, 100 species of fungal pathogen (Georghiou 1986) were resistant to fungicides, increasing to 150 (Pimentel et al. 1993) fungi and bacteria by 1990 (WRI 1994).

While not seen as a problem before the 1970s (DEFRA 2000), herbicide resistance in noxious weed species became a serious problem during the 1970s, 1980s and 1990s (DEFRA 2000). By 1976, 19 weed species had developed resistance to 17 herbicides (Bull 1982), and by 1984, between 41 (Georghiou 1986) (28 dicots and 13 monocots) and 48 (Georghiou 1986) weed species had developed resistance to at least one herbicide. Most instances of resistance involved the triazine herbicides, atrazine and simazine. In addition, at least seven weeds were resistant to other herbicides including the phenoxies, trifluralin, paraquat and the ureas (Georghiou 1986). By the late 1980s, 107 weed species had developed herbicide resistance. Of these, 81 were resistant to more than one chemical formulation (DEFRA 2000). By the early 1990s, between 113 (Pretty 1995) and 273 (Pimentel et al. 1993) weed species could withstand one or more herbicide product. Resistance in noxious weeds became more noticeable when certain species were exposed to intensive herbicide spray regimes (Georghiou 1986).

In an attempt to redress problems related to pesticide resistance, the Global Crop Protection Federation established the Insecticide Resistance Action Committee in 1984 to provide a coordinated crop protection industry response to the development of resistance to insect and mite pests, and to develop resistance management strategies in crop protection and vector control (CropLife International 2002).

Problems related to growing pesticide resistance were further compounded by the expiry of pesticide patents. In effect, this removed transient monopolies over patented materials and significantly reduced the profitability of individual pesticides on a per volume basis. According to ETC (2015), the lion's share of R&D expenditures were allocated to preserving sales of old chemical products facing patent expiration.

4.3.3 Growing pressure to reduce negative human health impacts of pesticide use in Western Europe, the USA and Japan

4.3.3.1 USA

Even as the political astuteness of pesticide reform groups improved, they could not hope to match the political experience and influence of the pesticide corporations. Because of this skewed distribution of power, the evolution of pesticide regulation in the USA has been extremely protracted and, according to most anti-pesticide activists, extremely disappointing (see Fig. 4.1). However, the environmental lobby made head way when Congress passed the Safe Drinking Water Act in 1974 (EPA 1992b). This Act initiated a process in 1975 to set national standards or Maximum Contaminant Levels for a range of hazardous contaminants, including several pesticides (EPA 1992b).

Growing public concern over the potentially dangerous effects of pesticides in the 1970s and early 1980s forced the EPA to issue Registration Standards for a number of pesticides. Early success in the campaign to eliminate the most hazardous pesticides included the removal of heptachlor. Heptachlor was first registered in the USA in 1952 for use as a broad-spectrum insecticide. In November 1974, the US EPA issued a Notice of Intent to cancel all registered uses of heptachlor except those for subterranean termite control and dipping of non-food plants. Heptachlor is of moderate toxicity in short-term animal feeding studies, and has been placed in Toxicity Category II for this effect. (Toxicity Category I indicates the highest degree of toxicity, and Category IV the lowest.) Heptachlor is mildly irritating to the skin, and is placed in Toxicity Category IV for dermal effects. Heptachlor is a Class B, probable human carcinogen, based on long-term animal feeding studies in which it produced benign and malignant liver tumours in mice. Heptachlor epoxide is also a Class B carcinogen, as it produces liver tumours in mice and female rats (BCERF 1998). Mutagenicity studies show that heptachlor and heptachlor epoxide affect DNA synthesis in certain human cells (EPA 1992a). Heptachlor and its metabolite, heptachlor epoxide, were known to persist in soil for many years and to bioaccumulate throughout the food chain (BCERF 1998). In 1975, both heptachlor and chlordane (another

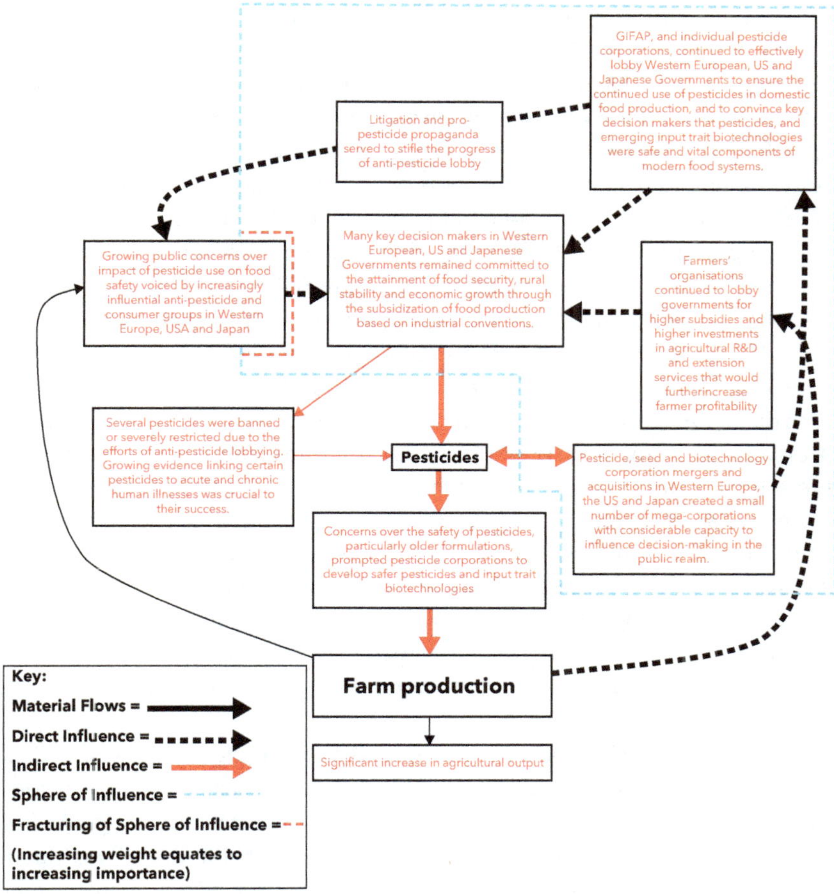

Figure 4.1 Pesticide manufacturers and the Residual Productivist model of agriculture in the Divergent Period.

organochlorine pesticide) were banned because of their links with cancer (EPA 1975b). Eventually, most commercial, agricultural and domestic uses of heptachlor were phased out in the USA between 1974 and 1988. All applications of heptachlor were banned by the EPA in 1988, except for the control of fire ants in power transformers (BCERF 1998). In 1989, all heptachlor tolerances (residue limits in foods) were revoked and replaced with action levels (limits set for unavoidable residues in foods resulting from environmental contamination, which are lowered as actual residue levels decline) (EPA 1992a).

In 1977, the EPA initiated a Rebuttable Presumption Against Registration (RPAR) review, now called a Special Review, against the organochlorine Lindane. The RPAR was triggered because of questions related to oncogenicity, fetotoxicity, teratogenicity, reproductive effects, its potential to cause blood dyscrasias and acute toxicity to wildlife (EPA 2003f). The EPA published Position Documents in 1977 through to 1983, resulting in the cancellation of certain uses of Lindane (EPA 2003f). Short-term exposure to Lindane can cause high body temperature and pulmonary oedema. Long-term exposure can cause liver and kidney damage. EPA's maximum concentration level (MCL) for Lindane is set at 0.2 ppb (EPA 2003d). Lindane is also persistent in the environment. In the USA, Lindane has not been produced since 1977, although it was imported in multiple forms for pharmacological and industrial use. However, the EPA restricts its use, and it can be applied only by certified pesticide applicators (Cecchine et al. 2000).

In 1981, the EPA issued a Registration Standard for endosulfan (another organochlorine pesticide). This was primarily due to endosulfan's associated toxins (chlorinated dibenzo-p-dioxins and dibenzofurans). This was followed by a Guidance Document in 1982 (EPA 2003e). Endosulfan has been shown to have a high acute oral and inhalation toxicity as well as slightly toxic dermal toxicity. It is an irritant to the eyes. Endosulfan is neither mutagenic nor carcinogenic. Endosulfan primarily affects the nervous system, and causes neurotoxic effects due to over-stimulation of the central nervous system (EPA 2003e). There is also evidence that endosulfan acts as an endocrine disruptor. Endosulfan poses risks for terrestrial and aquatic organisms, and can cause reproductive and developmental effects in non-target animals, particularly birds, fish and mammals (EPA 2003e).

While the pesticide industry was forced to concede the loss of several important pesticides, primarily old off-patent organochlorine insecticides, it was able to stifle attempts by anti-pesticide activists to regulate profitable, patent-protected, products. The power of the pro-pesticide lobby was clearly demonstrated by the amendment of FIFRA in 1975 (FWS 2003), which required that the EPA administrator consider the impacts of regulatory actions on production and prices of agricultural commodities and to notify the Secretary of Agriculture in advance of related rulemaking (FWS 2003). This allowed the pesticide industry scope to argue the case for a range of controversial pesticides and secure a key role for its traditional ally, the USDA. However, in 1975, during the Amendments first test case, the EPA refused to allow the use of DDT on cotton to control tobacco bud worm on the grounds that 'no substantial new evidence had arisen which may materially affect the 1972 order with respect to the human cancer risk posed by DDT, the environmental hazards of DDT and the need to use DDT on cotton' (EPA 1975a). In July 1975, the EPA prepared a report, 'DDT, A Review of Scientific and Economic Aspects of the Decision to Ban Its Use as a Pesticide', for the Committee on Appropriations of the US House of Representatives (EPA 2003a). In the same year, the EPA issued RPAR, stipulating that pesticide companies were required to prove that a pesticide was not dangerous (Boardman 1986).

In 1976, the EPA adopted a cost-benefit approach to pesticide registration (Glicksman et al. 2003). However, EPA's application of cost-benefit analysis was cause for concern. In one particular case, the EPA refused to cancel the insecticide chlorobenzilate for rust mite control on citrus crops in Florida, Texas and California because its withdrawal would have resulted in significant adverse effects on fruit size, appearance, crop yield and tree stock stamina (NAS 1980). Subsequently, the USDA, charged with supplying benefit data, failed to forward 'eight additional studies of the effect of chemical miticides on citrus yields. None of these additional studies found a statistically significant difference in yields between sprayed and unsprayed plots that could be attributed to rust mites' (NAS 1980). In addition, the National Academy of Science Committee, responsible for the allegations, concluded that estimated increased costs from cancelling the use of chlorobenzilate were between US$2.4 and US$9.2 million a year compared to the EPA's US$7 million a year (NAS 1980).

FIFRA was further amended in the form of the 1978 Federal Pesticide Act (FWS 2003), when Congress allowed the EPA to grant conditional registration for pesticides that are 'identical or substantially similar' to registered pesticides and that do not significantly increase the risk of any unreasonable adverse effects on the environment (Glicksman et al. 2003).

In 1979, the EPA banned the spraying of 2,4,5-T in Oregon forest rights of way after local miscarriages were found to be correlated with the use of 2,4,5-T (Boardman 1986). After the UK's response, the US EPA concluded that the available data did not support cancellation of 2,4,5-T, and that its use on rice and in rangelands was acceptable. Eventually, the question mark over 2,4,5-T was resolved in 1983 when Dow Chemical said that it was no longer economical to participate in costly negotiations with the EPA over restrictions on USA sales. Shortly afterwards, the EPA announced that the sale of 2,4,5-T would be illegal once existing supplies were depleted (Boardman 1986).

In 1982, the EPA cancelled most of the uses of toxaphene after determining that it had the potential to cause central nervous system effects including restlessness, hyper-excitability, tremors, spasms or convulsions when people are exposed to it at levels above the MCL for relatively short periods. Long-term effects include liver and kidney degeneration; central nervous system effects; possible immune system suppression and cancer (EPA 2003c). The EPA set toxaphene's Maximum Contamination Level at zero. Toxaphene is also very persistent, remaining in the soil for up to 14 years.

In 1983, the EPA ordered an immediate emergency suspension of EDB as a soil fumigant (EPA 1983). EPA cited significant new evidence that EDB was contaminating groundwater supplies in a number of states. Laboratory tests have shown that EDB is a carcinogen and mutagen and that it causes reproductive disorders in test animals (EPA 1983). In addition, short-term exposure to EDB can cause liver, stomach and adrenal cortex damage. It is also linked to reproductive system toxicity, particularly the testes (EPA 2003i). Long-term exposure to EDB can cause damage to the respiratory and nervous systems, and the liver, heart and kidneys (EPA 2003i). However, EDB could still be used in USDA's Japanese Beetle Control Programme (EPA 1983).

In the early 1980s, 45 000 different pesticides were on the market in the USA, formulated from 1400 AIs; only 37% had been adequately tested. Of the remaining pesticides, many had inadequate information of health effects, and 37% of AIs lacked any safety data (Clunies-Ross and Hildyard 1992).

4.3.4 Continued political and financial support for farming

For the growing pesticide and agricultural inputs sector, continued political and financial support for industrial high-input/high-output agriculture was essential. In many respects, it was the political will and financial resources that enabled agricultural support in the post-1970s food and oil crises era to continue (see Fig. 4.1). It was at this time that other key democratic capitalist economies, notably Australia and New Zealand, could no longer afford to maintain previously high levels of support for agriculture.

Agricultural subsidy equivalents (percentage of total value of agricultural production) in selected countries in 1989 were: Japan 71%; EU-12 30.1%; US 18.5%; Brazil 8.1% and New Zealand 3% (OECD 1993; Vorley and Keeney 1998). Total agricultural subsidy support in 1993 to OECD countries was US$180 billion, approximately US$16 000 per full-time farmer equivalent, or US$179/ha. Japanese farmers received US$13 billion/year to receive a price approximately ten times above the world price for rice (Vorley and Keeney 1998).

Food production in Western European countries, the USA and Japan, remained a key issue in these countries for two reasons. First, even though food production continued to increase in Western Europe and Japan, and populations were beginning to stabilise or decline, these countries were not self-sufficient in food and remained vulnerable to disruptions in food trade. It has been suggested that, particularly in Western Europe and Japan, food shortages both during and immediately after the Second World War had not been forgotten, and that increasing food production in search of food self-sufficiency was still a legitimate political goal. Many actors within national governmental administrations in Western Europe, the USA and Japan remained convinced of the value of the techno-policy model of agricultural development and maintained strong support for its continuance. Agricultural support acted as a welfare system for farmers, enhanced domestic food security and was believed to underpin the social, political and economic fabric of rural societies. Industrial agriculture continued to be the chosen model for agricultural development as demand remained for a system with the capacity to increase the production of relatively cheap, mass-produced, homogenised foods.

Second, food production in the USA vastly outstripped food consumption. Political support was maintained for farming primarily because agricultural production and agro-food exports were extremely important to the USA's economy. This continued emphasis on agricultural productivity was given further zeal in 1977 when the US National Academy of Science suggested that crop losses would need to be reduced by 20% in order to satisfy the food requirements of the additional 476 million more people predicted by global population models (Boardman 1986). Powerful actors in urban-centred industrial agro-food systems were able to secure significant levels of public finances, particularly for farm-level production, and private sector agro-food systems actors, including representatives of the pesticide industry, successfully resisted political pressures to scale down or even eliminate protection and financial support for industrial forms of agricultural production. This continued political and financial support for agriculture is one of the principal characteristics that set the pesticide industry development context of Western Europe, the USA and Japan apart from the rest of the world. However, it is important to note that, during the Divergent Period of pesticide evolution, stakeholder and governmental dynamics across the world began to change.

4.3.5 Scientific and technological advances

As the pesticide market throughout Western Europe, the USA and Japan grew increasingly competitive, as pests and diseases became increasingly resistant to pesticides and as public opinion towards pesticides became increasingly hostile, governments, in partnerships with pesticide companies, began to explore new and innovative alternatives to conventional pesticides. Scientific and technological advances generated throughout public research sectors continued to provide pesticide companies with useful spillovers, particularly in the field of molecular biology. For example, in the USA, public funding of US universities tended to favour biotechnology compared to alternative technologies, such as Integrated Pest Management (IPM) and organic production (see Fig. 4.1). By the early 1970s, publicly funded molecular genetics research, begun in the 1960s, was beginning to generate knowledge and products that offered profitable market opportunities for hi-tech companies (Goodman and Redclift 1991).

In the USA, the private sector attempted to profit from the potential of new biotechnologies. For example, in 1972, biochemist Dr Herbert W. Boyer and geneticist Stanley Cohen pioneered a new scientific field called recombinant DNA technology, by synthesising the first recombinant DNA molecule (USDA-ERS 2004). Excited by the breakthrough, venture capitalist Robert A. Swanson placed a call to Boyer and requested a meeting. Boyer agreed to give the young entrepreneur 10 min of his time. Swanson's enthusiasm for the technology and his faith in its commercial viability were contagious, and the meeting extended from 10 min to 3 h; by its conclusion, the company concept of Genentech was born. By 1976, Genentech, the world's first biotechnology company, was founded (Genentech 2003). During the late 1970s and early 1980s, several small biotechnology companies emerged in the USA. As with Genentech, university researchers, with the support of venture capital, established many of the early biotechnology companies. Entrepreneur Norman Goldfarb and Raymond Valentine established Calgene, one of the most important emerging biotechnology companies, in 1980. Raymond Valentine was an acting professor of plant biology at the University of California-Davis (University of Barcelona 2003).

The commercialisation of new biotechnologies, particularly those originating from the public sector, was strongly supported by the US Government through a number of Federal Acts, including: the Stevenson-Wydler Technology Innovation Act (1980) (USDA-ARS 2004b), the Bayh-Dole Patent and Trademark Amendments Act (1980), and the Federal Technology Transfer Act (1986) (USDA-ARS 2004b). The Federal Technology Transfer Act, the most significant amendment to the Stevenson-Wydler Innovation Act, was passed in order to accomplish the following. (1) Improve the flow of technologies with potential commercial application from federal laboratories to the private sector. (2) Empower directors of government-operated laboratories to enter into cooperative agreements and negotiated licensing agreements for inventions made at the laboratories. (3) Make technology transfer a responsibility for all federal laboratory scientists. (4) Create a charter for the Federal Laboratory Consortium for technology transfer. (5) Create royalty sharing with federal scientists and engineers (NTTC 2003). These Acts were designed to promote the commercialisation of federally sponsored R&D through cooperation between the research community, industry and both state and local governments (NTTC 2003).

By the early 1980s, pesticide corporations became convinced of the profitability of biotechnology and began ardently supporting biotechnology research both directly and by assuming control of the public research agenda in agro-biotechnology by persistent and effective lobbying of public officials (Thompson 1971). This had the effect of subsidising pesticide company research and helped to increase their markets. By this time, government and private sector stakeholders, that had forged the emergent biotechnology policy community, had institutionalised the role of biotechnologies – mirroring the case of pesticides and industrial farming methods in the 1940s and 1950s. Once fully institutionalised, there was little any opponents of biotechnology could have done, even if they had wanted to. Even an NRC report in 1989 that criticised the USDA/Land grant administrators for failing to investigate alternatives to the mechanically and chemically intensive technologies of conventional agriculture, had little impact on refocusing the direction of agricultural research (Thompson 1971).

In Western Europe, large pesticide corporations had also been working closely with their governments in an attempt to generate marketable biotechnologies and catch up with biotechnology sectors of the USA and Japan (Goodman and Redclift 1991). In 1982, a group of researchers from the University of Gent established the biotechnology company Plant Genetic Systems (PGS) (Bijman 2001a). In 1983, PGS was one of the first companies to successfully generate a genetically modified (GM) plant (Bijman 2001a). In the UK, the Government's Agricultural Food Research Council established the Agricultural Genetics Company (AGC) in 1983, with the aim to commercialise public sector research. AGC stakeholders, including Ciba-Geigy, could expect first rights to exploit whatever new technologies emerged (Goodman and Redclift 1991). In 1984, Rhône-Poulenc, the French-Government-owned chemicals corporation producing agricultural pesticides established its first molecular and cellular plant biology laboratory in Lyon (France). Its work was reinforced by a joint collaboration between Rhône-Poulenc and the Centre national de la Recherché Scientifique in 1986 (Assouline and Pierre-Benoit 2001).

4.4 Key characteristics of pesticide development processes and manufacturers

Throughout the period 1974–2017 in Western Europe, the USA and Japan, there can be no doubt that, the Residual Productivist-styled context of pesticide industry evolution presented numerous constraints and opportunities. In many respects, it is possible to view the evolution of the pesticide industry as a case of necessity being the mother of invention. Alternatively, many read more into the changes and suggest that, in a similar fashion to the processing and retail components of the food chain, the pesticide industry took advantage of the crisis of Fordism, and the role of the nation state, and thoroughly exploited their position to create private sector power never before witnessed. Whichever perspective is taken, there is no doubt that developments throughout this Period have been both breathtaking and groundbreaking. How did the pesticide industry in Western Europe, the USA and Japan take advantage of the opportunities of continued financial support of their domestic farming sectors and exploit new science and technologies? How did they achieve this whilst, at the same time, addressing the constraints of falling farm output prices, increasing pressures to reduce the negative human health impacts of pesticides, increasing resistance to pesticide AIs, the expiry of patents and the increased competition that patent expiry brought? This section details the strategies adopted by pesticide companies in Western Europe and the USA that aimed to address the key agricultural production and human-health-based drivers outlined above.

Pesticide corporations needed to respond to increased competitiveness in pesticide markets, which was brought about by falling farm output prices, increased R&D costs, additional costs imposed by tighter pesticide regulation and increasing demands from pesticide company shareholders for high rates of return on investments. In an attempt to increase efficiencies and appropriate increasingly higher levels of value added from the state-sponsored urban-based industrial agro-food systems of Western Europe, the USA and Japan, most pesticide corporations did two things: engaged in horizontal mergers and acquisitions (M&A) within the pesticide industry, and increased control over the agro-food sector through vertical M&A.

In response to growing pesticide resistance, expiration of pesticide patents and increasing pressures to reduce the negative human health impacts of pesticides, pesticide companies have utilised key advances in science and technology to develop new and safer pesticides, and input trait seed-based biotechnologies. In addition to the development of safer pesticides and crop protection solutions, the pesticide industry launched a series of well-funded, sophisticated rear-guard actions against anti-pesticide campaigns.

4.4.1 Horizontal mergers and acquisitions within the pesticide industry

During the Divergent Period, pesticide companies in Western Europe, the USA and Japan aimed to increase profitability through M&A of competitive firms (see Fig. 4.1). Known as horizontal integration, this process was widespread throughout the agro-food system as well as other sectors of the world economy during the mid- to late 1970s, and continued throughout the 1980s, 1990s and into the twenty-first century. Table 4.1 summarises some of the most strategically important acquisitions and mergers that occurred between Western European, USA and Japanese base pesticide corporations. Table 4.2 provides a snap shot of the top 15 pesticide corporations (by sales) in 1988 and 1992. This is as they were before the mega M&A of the early twenty-first century.

4.4.1.1 Rationalising horizontal concentration in the pesticide industry

Before attempting to rationalise horizontal concentration in the pesticide industry in response to Residual-Productivist-styled drivers, it is important to quickly recap important evolutionary changes during the 1950s and 1960s when the pesticide industry experienced a phase of unprecedented growth and

Table 4.1 Key acquisitions and mergers in the agrochemical industry in Western Europe, the USA and Japan, 1973-2017

Company name	Date of merger/acquisition and company/companies involved
ChemChina	In April 2017, ChemChina acquired Syngenta. Syngenta emerged in 2000 through a merger of the agrochemical operations of AstraZeneca (which had been formed in turn by the merger of Astra and Zeneca in 1999) and Novartis (formed by the merger of Ciba-Geigy and Sandoz in 1996). Novartis acquired Ishirara Sangyo Kaisha Ltd. and the crop protection business of Oriental Chemical Industries (OCI) in 1998. Zeneca (formerly ICI) acquired the US company Stauffer Chemical BV in 1987. Ciba-Geigy (formed by the merger of Ciba and Geigy corporations in 1970) launched its US agrochemicals division in 1984, and acquired the Austrian agro-chemicals company Maag AG in 1990. Sandoz acquired the agrichemical division of Zoecon Corp & Velsicol Chemical Corps' (USA) agricultural businesses in 1986. Novartis acquired Merk & Co.'s, (USA) Crop Protection business in 1997, and in 2001, Syngenta acquired Tomono Agrica (Japan).
Bayer	In September 2016, Bayer acquired Monsanto. Previously, Bayer had acquired Aventis in 2001/2. Aventis formed in 1999 through a merger of Hoeschst and Rhône-Poulenc. In 1994, Hoechst and Schering merged to form AgrEvo. Hoechst acquired Celanese Corporation (USA) in 1987 and called it the Hoechst Celanese Corporation. Nor-Am was formed by the merger of Nor-Am Chemical Company (subsidiary of Schering) and TUCO (division of Upjohn Company) in 1985. Previously, Nor-Am Chemical Company had been formed by a merger of Nor-Am Agricultural Products and Boots Farm Corporation (BFC) in 1984. Before that, BFC had been formed by merger of Boots/Hercules America (BHA) and Fisons (UK) in 1981. Rhône-Poulenc (including May & Baker) acquired Union Carbide Agricultural Products in 1987. Before that, Union Carbide Agricultural Products acquired Amchem Products in 1977. Monsanto was acquired by Pharmacia in 1999. Pharmacia divested Monsanto in 2002.
BASF	In 1986, BASF acquired Wyandotte Co. (USA) and established the BASF Corporation in the USA sometime before 1992. BASF acquired Sandoz US and Canada Corn in 1996, Microflo in 1998 and American Cyanamid in 2000. American Cyanamid acquired Shell International's agrochemical business in 1993.
DowDuPont	In August 2017, Dow AgroSciences and DuPont merged to create DowDuPont. Dow AgroSciences (formerly DowElanco, which was formed by the merger of Elanco - containing Eli Lilly - and Dow - containing Murphy Chemicals - in 1989), Union Carbide (USA) in 1999 and Rohm and Haas (USA) in 2001. DuPont had previously acquired Royal Dutch/Shell's USA agrichemicals' interests in 1986.

prosperity. This growth and prosperity was primarily based on the development of important new pesticides, and the significant expansion of pesticide sales, particularly in domestic markets. Annual sales growth rates averaged 10% during the 1960s (Agranova 1999). By the early 1970s, pesticide companies, either as independent corporations or divisions of larger diversified corporations,

Table 4.2 Top 15 Agrochemicals Companies (by sales in US$ 000 000) in 1988 and 1992, ranked by 1992 sales

Company name	Agrochemical sales 1988	Agrochemical sales 1992
Ciba-Geigy	2140	2940
ICI	1960	2362
Rhône-Poulenc	1630	1939
DuPont	1440	1930
Bayer	2070	1921
Monsanto	1338	1647
DowElanco	1420	1580
Hoechst	1020	1359
BASF	1000	1185
American Cyanamid	-	1000
Sandoz	-	874
Schering	-	818
Shell	940	700
FMC	-	455
Kumiai	-	437

had evolved into large and successful businesses. However, whilst the 1950s and 1960s had been highly profitable for pesticide manufacturers, the rapid growth in demand for pesticides, and the capacity and willingness of farmers to pay high prices for them, had permitted the pesticide industry to evolve in such a way as to institutionalise relatively inflexible and inefficient systems of production and distribution.

In the aftermath of the oil/food crisis of 1973, when the pesticide industry faced increasingly competitive pesticide markets and calls for tighter pesticide regulations (which translated into increased R&D costs), many companies were subject to falling profitability at the same time as shareholders demanded ever higher returns on investments. As the cost of developing pesticides continued to rise throughout the 1970s and 1980s (including R&D and meeting regulatory requirements), pesticide companies struggled to allocate sufficient resources for R&D. Indeed, R&D costs per pesticide in the 1970s, estimated at US$10–20 million (Boardman 1986), rose to US$140 million in 1992 (Vorley and Keeney 1998) and $256 million in 2015 (ETC 2015). In 1976, Dow Chemicals estimated the costs of regulatory compliance across the whole pesticide product range to be US$186 million (Boardman 1986). By 2015, regulatory compliance associated with releasing just one new pesticide was estimated at nearly US$90 (ETC

2015). Global pesticide research expenditure increased from US$740 million in 1987 to US$1900 million in 1992, and product defence costs increased from US$263 million to US$482 million over the same period (Agrow 1993).

The financial capacity to develop and register pesticides, and the significant time lag between starting the pesticide development process and receiving a return on investment with the first sales, became substantial obstacles for the pesticide industry, particularly the smaller companies. However, as most corporations were highly diversified, and have been since their inception or at least since the 1960s, the overall performance and evolution of agrochemical companies has generally been greatly affected by the successes and failures of their parent institutions. Basic chemicals and plastics have been subject to cyclical swings, while pharmaceuticals, part of the speciality chemicals, cosmetics and varnishes were not. Pharmaceuticals are the most profitable products. This meant that Bayer was less dependent on market swings than ICI or BASF.

M&A offered a potential mechanism to regain the levels of profitability experienced throughout the 1950s and 1960s, in what was now a mature industrial sector (Wield et al. 2010). Between 1973 and 2017, all of the major Western European, USA and Japanese pesticide corporations engaged in horizontal mergers and/or acquisitions with their competitors. There are several discernible trends associated with pesticide corporation M&A that have been affected by the Residual Productivist context. These trends include the search for: (1) greater economies of scale; and (2) increased share/control of the key agrochemicals markets in Western Europe, the USA and Japan, including the acquisition of new pesticide technologies (Wield et al. 2010).

4.4.1.1.1 Economies of scale

Since the early 1970s, one of the key motives behind M&A was the search for profit maximising efficiencies associated with economies of scale. Economies of scale became an important goal as R&D expenditures and regulatory costs (associated with food and environmental safety tests) continued to rise, and pesticide sales began to slow or even fall. As R&D expenditures and regulatory costs are both sunk costs, they are a potential source of economies of scale. Through utilising economies of scale, pesticide companies can lower average costs, allowing pesticides to be sold at more competitive rates whilst maintaining acceptable levels of return on investments. The new, more competitive, environment provided a clear incentive for pesticide companies to expand (Fulton 1997; Hayenga 1998). Indeed, those that did not get large were vulnerable to being driven out of the market by larger and more cost-efficient firms.

Economies of scale have been used to justify consolidation within the pesticide industries of Western Europe, the USA and Japan (Ollinger and

Fernandez-Cornejo 1998). It is also important to note that key pesticide companies not involved in mergers or acquisitions between 1974 and 1992 were equally convinced of the benefits of economies of scale. Indeed, Ciba-Geigy, one of the most important pesticide companies in 1992, had already merged in 1970. This merger was the culmination of a complex series of events that incorporated both international and regional drivers. First, shortly after World War I, Swiss agrochemicals companies (Ciba, Geigy and Sandoz) formed the Basel AG cartel. This was in response to increased regional competition from the recently formed German I.B. Farben cartel. However, after World War II, the I.B. Farben cartel was dissolved and only the Basel cartel remained. Dissolution of the Basel cartel quickly followed in 1951, as protection from foreign companies was no longer deemed necessary.

Ironically, it was to be the increasing levels of international competition, and competition from Swiss based Sandoz, that initiated the merger between Ciba (founded in 1859 by Alexander Clavel) and Geigy in 1970 (Novartis 2003b). ICI, one of the most important pesticide manufacturers at the time, already had significant economies of scale prior to 1974. While ICI finished the 1980s well, the early 1990s recession caused ICI to divest and close 40 non-strategic and/or unprofitable businesses. ICI pioneered the restructuring process in Europe, cutting staff numbers by 23%. Part of its restructuring involved the divestment of its agrochemicals division as Zeneca. Zeneca, as with most agro-biotechnology companies, invested in the development of high-throughput screening facilities to identify new products (Steinrucken and Hermann 2000). These processes use combinational chemistry to provide large numbers of chemicals for testing, and genomics provided insights into plant biology to enable rational targeting and design of pest control products (Evans 2000). In 2000, Zeneca was able to screen 100 000 new chemicals per year compared to the previous 15 000, which resulted in one successful product every 1–2 years (Tait 2001). This was a key strategy to get improved products to market in order to increase or at least maintain market share (Tait 2001).

Rhône-Poulenc had incorporated May & Baker sometime between the First and Second World Wars. It is important to note that Ciba-Geigy, ICI and Rhône-Poulenc were the three largest pesticide companies (by sales) in 1992 (see Table 4.2). It is most likely therefore that economies of scale have been beneficial to these companies and not detrimental. It is interesting to note that, between 1974 and 1992, all key M&A involving key pesticide manufacturers were between domestic rivals or pesticide companies based in Western Europe, the USA and Japan. Thus, rapid and formalised consolidation of the pesticide industry took place within the protective cocoon of the three largest pesticide markets. These economies offered the highest levels of agricultural protection and were home to the world's most intensive agricultural systems.

4.4.1.1.2 Increased market share, market power and the acquisition of new pesticide technologies

Closely allied to increased economies of scale, increased market share was another key motive behind M&A. By default, acquisitions of, and mergers with, competitive companies automatically expanded market share through the addition of production plants, patented pesticides and important distribution networks. Pesticide companies endeavoured to capture even larger market shares, and greater control over agricultural inputs, by focusing on sales volumes and market share rather than short-term profits on pesticide products. In the 1970s and 1980s, pesticide company M&A were about market presence and distribution (Horstmeier 1996b). Multinational pesticide corporations could not survive without good penetration of pesticide markets in the USA, EU and Japan (Vorley and Keeney 1998).

Ultimately, it was the lower average costs of pesticide production, gained through increased economies of scale that enabled pesticide companies to market competitively priced pesticides and capture ever-larger shares of pesticide markets. Increased market share enabled pesticide companies to increase profits by capturing high levels of return on competitively priced pesticides.

4.4.2 Increased control over the agro-food sector through vertical mergers and acquisitions

Since the 1970s, pesticide companies secured increased control of the food system through mergers with, and acquisitions of, seed and biotechnology companies. Known as vertical integration, this process was widespread throughout agro-food systems as well as other sectors of the world economy.

Vertical integration can be defined as a process whereby successive stages in the production and distribution of a product are progressively controlled by a single firm. Classic examples of vertical integration include Cargill's and ConAgra's control of key commodity chains in the USA. It has been estimated that between 120 and 1000 seed companies were acquired by pesticide companies throughout the 1970s and 1980s (Clunies-Ross and Hildyard 1992). All the major pesticide companies engaged in mergers and/or acquisitions of seed and/or biotech companies between 1974 and 2017 (see Table 4.3). Table 4.4 provides a snap shot of the top ten seed companies (by sales) in 1988. This is as they were before the mega M&A of the 1990s and early twenty-first century.

4.4.2.1 Rationalising acquisitions, mergers and developments in seed and biotechnology industries

In parallel to the explanation of horizontal concentration in the pesticide industry, vertical concentration through acquisitions, mergers and new developments

Table 4.3 Key acquisitions, mergers and developments in the seed and biotechnology industries in Western Europe, the USA and Japan, 1974–2017

Company name	Date of acquisition/merger/development and company/companies involved
DuPont	DuPont acquired Nickerson Seeds (USA), part of Royal Dutch/Shell, in the 1970s. In 1986, DuPont entered into a joint venture with DNA Plant Technology Corporation (USA). Interestingly, DuPont acquired Conoco Oil in 1981 to take control over one of the pesticide industries' major inputs. DuPont acquired Pfister Hybrid Corn Co. (USA), and Protein Technologies International Inc., 20% of Pioneer Hi-Bred (USA) in 1997. In 1998, DuPont acquired Cereals Innovation Centre (Dalgety-USA), Griffin Corporation (USA) and Hybrinova S.A. (France). In 1999, DuPont acquired the remaining shares in Pioneer Hi-Bred.
Monsanto	Monsanto acquired Hybritech Seed International (USA) in 1982, and Jacob Hartz Seed Co. (USA) in 1983. Monsanto acquired part of Calgene Inc. (USA) in 1995 and Asgrow Seed Co. (USA) in 1996 – formerly acquired by Upjohn in the 1970s. Monsanto acquired part of Dekalb Genetics (USA), Holden's Foundation Seeds (USA), Corn State Hybrid Service Inc. (USA), Corn States International (USA) and remaining Calgene shares in 1997. Monsanto acquired Cargill's International Seed business (USA), remaining Dekalb Genetics shares, and PBI Cambridge (UK) in 1998. In 1999, Monsanto acquired Delta Pine Land Co., Plant Genetics Inc. and part of Mendel Biotechnology. Monsanto acquired Seminis in 2005.
Syngenta	In 1998, Novartis acquired Sturdy Grow Hybrids, Agritrading (Italy), Agra (Italy), Agrosem (France) and Koipesol Semillas (Spain). Zeneca had previously acquired Mogen (NL) in 1997. Ciba-Geigy acquired Funk Seeds (USA) in 1974, and established a Japanese research centre in 1987 and Japanese research laboratory in 1990. Sandoz acquired Rogers Seeds (USA) in 1975, Northrup King (USA) in 1976, Sluis & Grut (S&G) in 1980 and Hilleshög NK in 1989. By 1987, Sandoz had acquired 60/40 interests in Repligen, as well as three other agrochemicals R&D laboratories in the USA. VanderHave B.V. acquired Interstate Seed (USA) in 1983 and Payco Seed Company in 1988. In 1991, Interstate Seed and Payco Seed Company merged to form Interstate Payco. ICI acquired Garst Seed Company (USA) in 1985, Sinclair McGill (UK) and Societé Europèene de Semences (France) in 1987 and a technical centre in Japan in 1989, followed by another in Canada in 1990.
Dow AgroSciences LLC	Dow AgroSciences acquired Mycogen, including Cargill's Hybrid Seeds- (USA), and established Advanced AgriTraits in 1998. In 2009, Dow acquired a number of strategic seed companies such as Hyland Seeds (Canada), MTI (Austria), Pfister Seeds (USA) and Triumph Seed (USA).
Bayer	Bayer acquired Molecular Diagnostics sometime before 1986, and Miles Inc. and Cutter Biologicals sometime before 1992. Bayer acquired Agro Stefes in 1995 (as AgrEvo) and Plant Genetic Systems (as AgrEvo) in 1996. Bayer acquired Sunseeds Corporation (USA) in 1997, and Biogenetic Technologies (NL), Rio Colorado Seeds (California-USA-as Nunza) and the diagnostics division of Chiron Corp (USA) in 1998. Hoechst acquired a minority share of KWS, Germany's largest agricultural seed company in 1985, and acquired Nunhems Seeds, a Dutch vegetables seed company, in 1986. In 1987, Hoechst acquired a Japanese biotechnology centre and production plant. Rhône-Poulenc acquired agricultural R&D and production capacity in the USA, and Japan before 1988.
BASF	BASF established an agricultural research centre sometime before 1992.

Table 4.4 Top ten Seed Companies (by sales in US$ 000 000) in 1988

Company name	Seed sales 1988
P'oneer Hi-Bred Int.	735
Sandoz	507
Limagrain	370
Upjohn	280
A trois	257
ICI	250
Cargill	230
Shell	200
Dekalb-Pfizer	174
Ciba-Geigy	150

Source: Tansey and Worsely (1995).

in seed and biotechnology industries was primarily driven by the growing internal demands of pesticide company shareholders for high rates of return on investments. Pesticide companies invested in seed and biotechnology companies based on three key assumptions: (1) Higher profits were expected from synergies between seed and pesticide businesses. (2) New biotechnologies, principally delivered through the seed, offered greater public legitimacy and promised much higher rates of return than pesticides. (3) Greater control over food systems offered a way to attain efficiencies and appropriate increasingly higher levels of value added from the state-sponsored urban-centred industrial agro-food systems of Western Europe, the USA and Japan.

4.4.2.1.1 Expected synergies from the amalgamation of pesticide and seed businesses

During the 1960s and 1970s, the seed industry of North America, Western Europe and Japan boomed. This was primarily due to the introduction of high-yielding hybrids, based on the development of hybrid technologies in the 1930s (World Bank 2003), particularly hybrid maize in North America, hybrid sugar beet in Europe and hybrid vegetables in Southeast Asia (UNESCO 2003). For example, new maize hybrids were capable of achieving four times the yield, compared to non-hybrid maize in the 1930s. High yields, and the associated increase in profitability, enabled hybrid maize to capture 95% of the North American maize seed market by 1975 (Kloppenburg 1984). Corporate interest in hybrid technology was high because, due to declining yield performance of farmer-saved hybrid seeds, farmers were obliged to purchase premium hybrid seed each year from the seed company. This enabled seed corporations to capture significant annual remittances from farmers and generate high returns on hybrid crop R&D (World Bank 2003). Subsequently, high seed company profits from hybrid maize seed sales during the 1970s attracted the attention of agrochemical, petrochemical and pharmaceutical companies (Goodman and Redclift 1991), which aimed to exploit possible synergies of the seed business with their own commercial interests. It was assumed that vertical integration would allow agrochemical companies to reduce transaction costs by eliminating the uncertainty that arises when research, seed production and seed distribution activities must be coordinated through market exchange mechanisms. The acquisition: of Northrup King (USA) by Sandoz AG (Switzerland) in 1976; of Funk Seeds (USA) by Ciba-Geigy AG (Switzerland) in 1974; of Nickerson Seeds (USA) by Shell (UK/Netherlands); and of Asgrow Seeds (USA) by the Upjohn Co. (USA) are examples of this type of vertical integration that took place during the 1970s (Sehgal 1996). See Table 4.3.

4.4.2.1.2 Greater public legitimacy and higher rates of return from new biotechnologies

CEOs in pesticide and seed corporations were quick to recognise the profit potential of biotechnologies. Much of the excitement in biotechnologies came from: (1) Technology foresight studies, which predicted good returns on investment. (2) Strong support from government (in the USA, Western Europe and Japan). (3) The interest of venture capital to invest (Tait and Chataway 2003). During the 1980s, 1990s and early twenty-first century, CEOs wanted to rapidly secure a leadership role in biotechnologies and capture a profitable size of the emerging market, either in input or output genetically modified organism (GMO) traits (Tait and Chataway 2003). Indeed, according to Tait and Chataway (2003), 'Monsanto managers had an almost unshakable faith' in biotechnologies.

Quick, low-cost, registration of new GM products in North America made biotechnologies a much more financially attractive option for large agrochemical-seed companies than developing new pesticides (Clapp 2002). Compared to $256 million required to develop a pesticide product, it typically cost US$136 million to bring a new GM crop to market (ETC 2015). Again, the relatively low cost of developing a new GM crop was assured through constant US Government lobbying to hold off tight legislation on biotechnologies, stating that regulation would act as a considerable disincentive for investment (Werner 1999). Whereas high returns on hybrid crop varieties drove seed company acquisitions in the 1960s and 1970s, it was to be the seed's vital importance for biotechnology that drove seed company acquisitions in the late 1970s and early 1980s (Bijman 2001a).

However, as pesticide companies had little internal competence in emerging biotechnologies they were faced with five options. First, pesticide companies could develop in-house biotechnology R&D capacity. Second, pesticide companies could contract biotechnology companies and universities to generate genetically engineered crop plants to meet their own specifications. This option became even more popular with European and Japanese pesticide corporations after the stock market crash of 1987, when biotechnology companies in the USA were starved of cash investments and subject to lower borrowing capacities due to decreased share values. Several pesticide companies entered into these so-called 'clientelist alliances'. Examples include Ciba-Geigy and Agri-diagnostics, and Rhône-Poulenc and Calgene. Third, pesticide companies could enter into joint ventures and limited R&D partnerships with biotechnology companies. Fourth, pesticide companies, or their parent diversified corporations, could underwrite new biotechnology companies with venture capital or equity participation. Indeed, initial venture capital investments in Calgene were subsequently supplemented by further capital inputs during the 1980s from a range of venture capital sources, including Mitsui-Japan (Mitsui 2003). Fifth, pesticide companies could merge with, or acquire, emerging biotechnology companies. As most directors of these early start-up biotechnology companies possessed little commercial experience, and were ill-equipped to finance the high absolute costs of R&D associated with the production of revenue-yielding products that were subject to long and unpredictable lags between laboratory trials and regulatory approval, fledgling biotechnology companies were extremely vulnerable to corporate takeovers. Throughout the 1980s, pesticide companies such as Monsanto, ICI and Hoechst implemented all five strategies concurrently (Goodman and Redclift 1991).

In the 1960s, Monsanto executives recognised the growing importance of commercial biology, particularly in association with their expanding agricultural business (Monsanto 2003b). However, it was not until 1978 that Monsanto initiated its own molecular biology programme (Monsanto 2003a). Monsanto authorised the construction of its Chesterfield Life Sciences Research Center in 1980, and by 1981, biotechnology became firmly established as Monsanto's strategic research focus (Monsanto 2003a). In 1982, Monsanto scientists celebrated their first genetically transformed plant cell. In the same year, in conjunction with the acquisition of Dekalb's wheat research programme, Monsanto Hybritech Seed International Inc., was established (Monsanto 2003a). In 1983, biotech scientists succeeded in growing plants with genetically engineered traits. In the same year, Monsanto acquired Jacob Hartz Seed Co. (USA). In 1985, Monsanto initiated a strategic restructuring in an attempt to consolidate its businesses around high value-added proprietary products (Monsanto 2003a). Emphasis was placed on the life science business. Non-strategic businesses, such as polyurethane film, Astra-Turf, Fome-Cor foam boards and so forth, were divested (Monsanto 2003a). In 1991, NatureMark was established to sell genetically engineered potatoes.

However, it was not until the mid-1990s that Monsanto accelerated its spending spree with its acquisition of part of Calgene Inc. (USA). In 1996, Monsanto acquired another leading biotech company Plant Genetics Inc., plus a number of important seed companies including the US seed company Asgrow Seed Co. (one of the largest soybean seed companies) and Plant Breeding International (UK) (Groundup 2003j). In 1997, Monsanto strengthened its biotechnology capacity through the acquisition of part of Dekalb Genetics (USA), and remaining Calgene shares. Additional seed company acquisitions included Holden's Foundation Seeds (USA) for US$1.2 billion, Corn State Hybrid Service Inc. (USA) and Corn States International (USA). Monsanto's acquisition of Holdens gave the company access to an estimated 25–35% of the maize acreage planted in the USA (ETC 1998). In 1998, Monsanto purchased Cargill's seed operations for $1.4 billion and the remaining Dekalb Genetics shares for $2.3 billion (Groundup 2003j). In 1999, it bought Delta and Pine land and subsidiaries for $1.82 billion (Groundup 2003j), plus Plant Genetics Inc., and part of Mendel Biotechnology. Delta and Pine land controlled 71% of the US cottonseed market (RAFI 2000a). It also bought Unilever's European wheat breeding business for $525 million.

While Monsanto was clearly at the forefront of plant biotechnologies, other corporations quickly followed suite. Bayer has also been incredibly active in the search for the access to new markets, particularly through the vertical integration of pesticide, biotechnology and seed companies. Bayer acquired Molecular Diagnostics (USA), and Miles Inc. (USA) and Cutter Biologicals (USA) sometime before 1992. Bayer (as AgrEvo) acquired Agro Stefes in 1995; PGS (Belgium) in 1996; Sunseeds Corporation (USA) in 1997; and Biogenetic Technologies (NL), Rio Colorado Seeds (California-USA-as Nunza) and the diagnostics division of Chiron Corp (USA) in 1998. By 1985, Hoechst had acquired a minority share of KWS, Germany's largest agricultural seed company, Nunhems Seeds, a Dutch vegetables seed company, in 1986, and a Japanese biotechnology centre and production plant in 1987. In 1989, Hoechst built its first production facility for genetically engineered human insulin (Aventis 2003).

Interestingly, Rhône-Poulenc continued to view crop protection as its core activity. As a result, it sold off its seeds subsidiary in 1994 and developed a global partnership with Limagrain instead. Plant biotechnology was seen as a knowledge tool for improving the process of discovery for new molecules for plant protection, and as a means of finding desirable agronomic traits for major crops (Assouline and Pierre-Benoit 2001). Major industrial crops were targeted by Rhône-Poulenc and the need to fill gaps in complementary pesticide products (herbicides and insecticides). Niche markets such as strawberries and raspberries were seen as unprofitable (Assouline and Pierre-Benoit 2001).

Rhône-Poulenc stated that 'the future of crop protection passes through the combination of innovative chemistry with different biotechnology' . . . . Rhône-Poulenc's strategy consists of focusing on innovation and investing in the development of genes that provide a clear competitive advantage (Assouline and Pierre-Benoit 2001).

While Bayer's short-term strategy focused on crop protection, its long-term strategy focuses on crop production and food production (Domain-B 2003). Much of Bayer's interests in biotechnology came through the acquisition of Aventis CropScience in 2001, particularly the AgrEvo component (Wield et al. 2010). AgrEvo invested heavily in biotechnology in the early 1980s primarily for pharmaceuticals. With this background, it had in-house capacity to embark on plant biotechnology R&D, but still contracted out R&D to small specialised biotechnology companies. Indeed, glufosinate tolerance was contracted to PGS (Bijman 2001a). In 1996, the acquisition of PGS (for Euros 436 million) gave AgrEvo access to wide range of intellectual property rights (IPRs) through cross-licensing patents, allowing it to release weed control, insect resistance and hybrid vigour through products such as LibertyLink, StarLink and SeedLink (Bijman 2001a). According to Bijman (2001a), AgrEvo's strategy was to transition from a crop protection company to a crop production company. AgrEvo no longer wanted to sell pesticides but to become a provider of productivity enhancing solutions to the farmer. 'We have taken the strategic decision to develop from a plant protection enterprise to a company offering products for enhancing the productivity of plant production through innovations in chemistry, biotechnology and plant breeding' (Bijman 2001a). Biotechnology was seen as central to this new strategy and necessitated the purchase of seed businesses (Bijman 2001a). AgrEvo established its Seeds/Crop Improvement product group in 1997, after the acquisition of PGS and the shift of Nunhems Seeds from parent company Hoechst to AgrEvo (Bijman 2001a). Main crops included canola [oil seed rape (OSR)], maize, soybean and vegetables (Bijman 2001a).

In addition to Monsanto, other US pesticide corporations have been actively acquiring biotechnology and seed companies. DuPont entered the biotech race in 1986 when it entered into a joint venture with DNA Plant Technology Corporation (USA). In 1997, DuPont acquired Pfister Hybrid Corn Co. (USA) and Protein Technologies International Inc. (USA) plus a 20% share of Pioneer Hi-Bred International (USA), the world's largest seed company, for US$1.7 billion (Reuters 1999a). The two companies jointly invested US$400 million in an agricultural research effort and formed a new company, Optimum Quality Grains (Lipin et al. 1999). This was followed in 1998 by the acquisition of the Griffin Corporation (USA), Hybrinova S.A. (France) and the remaining shares in Pioneer Hi-Bred for US$7.7 Billion in 1999 (Lipin et al. 1999). In 1998, Pioneer-Hi-Bred controlled 39% of the US maize market and 17% of the soybean market (Kalaitzandonakes and Hayenga 2000).

Alongside Monsanto, DuPont is also investing in wheat at all levels of the agricultural supply chain. In January 1998, DuPont purchased the Cereals Innovation Centre (specialising in wheat-based food ingredients) from the UK-based food processor Dalgety. In May 1998, DuPont acquired Hybrinova, the hybrid wheat business of Lafarge, S.A. (France), and announced plans to extend its hybrid wheat business in the future (ETC 1999).

Dow AgroSciences has also been active. In 1998, Dow AgroSciences acquired Mycogen - including Cargill's Hybrid Seeds - (USA) (Fulton and Giannakas 2001). Dow stated that the acquisition of Cargill's Hybrid Seeds would result in a 'larger and more efficient platform from which to launch biotechnology products'. The purchase included all seed research, production and distribution assets of Cargill Hybrid Seeds in the USA and Canada, except Cargill's Inter Mountain Canola, Goertzen Seed Research and the western Canadian seed distribution business. Dow AgroSciences established Advanced AgriTraits.

BASF entered the mainstream biotechnology race in mid-1998, and intensified its involvement in biotechnology in 1999 through its investment in BASF Plant Science (BASF 1998). In 2000, BASF announced that it intended to make a major long-term commitment to plant biotechnology (Bullock 2000). This was to include: (1) The allocation of more than 700 million euros (>US$680 million), over 10 years, for research in plant biotechnology. (2) Establishment of flexible organisation (BASF 2000). (3) The allocation of additional funds for acquisitions, including seed companies and other appropriate companies. (4) The formation of a holding company, BASF Plant Science GmbH, to oversee the company's seed operations and plant biotechnology efforts.

BASF saw the application of biotechnology to agriculture as providing the company with 'new and attractive growth opportunities in a "new agromarket" that is arising from the increasing coalescence of crop protection, seed production and plant biotechnology' (BASF 2000). In its efforts to play catch-up in the agricultural biotechnology arena, BASF had already founded two autonomous biotech-related companies: Metanomics and SunGene (BASF 2000). BASF had also acquired a 40% ownership stake in Swedish seed company Svalof Weibull, which formed the basis of the BASF Plant Science GmbH biotechnology research effort (Bullock 2000). In addition to the two research joint ventures in Germany, BASF owns three research companies in Sweden, Canada and the USA. 'The research objectives of BASF Plant Science GmbH involve improving low temperature tolerance and drought tolerance'. If BASF is successful, these optimised plants could potentially constitute a major advance for the agriculture of many countries that are regularly confronted with crop failures due to extreme climate conditions. All told, about 300 new research jobs were expected to be created by BASF to focus on its agricultural biotechnology effort (Bullock 2000).

Finally, while assuming the guise of a silent partner, Syngenta's commitment to the development of biotechnologies has rivalled Monsanto's (Tait 2001). However, while Syngenta has devoted significant financial resources and has developed a substantial range of innovative technologies, and acquired seed production and distribution networks, it has taken a much more conservative approach to the introduction of biotechnology. As ICI, the company acquired Garst Seed Company (USA) in 1985, Sinclair McGill (UK) and Societé Européene de Semences (France) in 1987, and a technical centre in Japan in 1989. By early 1990, ICI was selling 15 000 products in 150 countries, and was the world's third largest chemical group, second largest agrochemical producer and fifth largest seed producer (Corporate Watch UK 2002b). After an unsuccessful takeover bid in 1991, ICI reduced its vulnerability to take over bids by splitting the company into a chemicals company and a bioscience group, called Zeneca in 1993 (Corporate Watch UK 2002b). To date, Syngenta (formerly Zeneca) acquired biotechnology and seed interests through the internalisation of Ciba-Geigy. Ciba-Geigy had previously diversified into seeds businesses in 1974 with the acquisition of US-based Funk Seeds International, and established a special biotechnology unit in 1980 (Novartis 2003b). Its internalisation of Sandoz also endowed Syngenta with biotechnology and seed capacity. Sandoz moved into the seeds market with the acquisition of US-based Rogers Seed Co., in 1975, followed by Northrup King (US-based maize seed breeder and supplier) in 1976, the Dutch Zaadunie Group in 1980 and the Swedish Hilleshög company (sugar and fodder beet breeder and supplier) in 1989 (Novartis 2003b). By 1987, Sandoz had acquired 60/40 interests in Repligen, as well as three other agrochemicals R&D laboratories in the USA. Latterly, Syngenta (formerly Zeneca) focused more on 'output traits than input traits. Its research included nutritional characteristics of cereal crops, and incorporation of effects that it could sell as beneficial to consumers and their health' (Wield et al. 2010).

As Novartis, Syngenta acquired Mogen (NL) in 1997, and, by default, Sturdy Grow Hybrids, Agritrading (Italy), Agra (Italy), Agrosem (France) and Koipesol Semillas (Spain). Syngenta also inherited Novartis' Biostein Agrobiological Research Centre in 1996 (Groundup 2003m). In 1998, Novartis announced the creation of the Novartis Institute for Functional Genomics (Novartis 2003b), and its intention to invest US$600 million over 10 years on the world's biggest gene mapping project, involving 180 scientists (Groundup 2003l). Novartis aimed to respond to lower financial returns and pressure on distribution channels, due to low farmer profitability, and the need for cost savings and efficiencies, plus new challenges with precision farming and biotechnologies, and the growing influence of food and feed processors and retailers and consumers.

A New Agribusiness Strategy aimed to strengthen leadership and achieve cost savings by implementing new product lines, improving product mix, optimising asset utilisation, reducing purchasing costs and eliminating approximately 1100 jobs worldwide. Focusing on pillar staple crops, Novartis' new strategy focused on: (1) The food and feed chain and crop innovations and (2) Integrating breakthrough technologies, and exploring joint synergies between crop protection and seeds (Wield et al. 2010). Global teams were formed to define overall strategies for pillar crops (maize, vegetables, cereals and rice) and important crops (oilseeds, sugar beet, cotton, fruits and grapes). Products showing a poor return were withdrawn from unprofitable markets or rights were sold to competitors (Tait 2001).

4.4.2.1.3 Greater control over food systems offered a way to attain greater efficiencies and appropriate increasingly higher levels of value added from the state-sponsored urban-based industrial agro-food systems of Western Europe, the USA and Japan

Over time, it became apparent that acquisition was the favoured strategy of pesticide companies in their search for high and stable returns on investment. The belief in the long-term profitability of integrated pesticide, seed and biotechnology companies stems from a number of assumptions, namely: (1) Synergies between pesticide, seed and biotechnology companies would yield greater efficiencies and profitability. (2) Relatively cheap stock prices in seed and biotechnology companies would quickly increase in value when new technologies were marketed. (3) Higher levels of appropriation would be achieved due to greater control over distribution networks of new technologies. (4) In order to achieve high returns on investment, IPRs associated with the development of new technologies must be protected.

Synergies between pesticide, seed and biotechnology companies were expected to lead to the production of technologies that yielded greater cost efficiencies and profitability for both pesticide corporations and farmers. The use of new technologies was expected to reduce farmers' unit costs of production, and, therefore, by default, they would become increasingly important components of farming systems, relative to the declining importance of pesticides. New technologies were also expected to increase pesticide company profits through the capture of new and profitable markets, initially through the production of high-yielding hybrid seeds, and latterly through the production of herbicide and seed germplasm packages, marketed through extensive international seed marketing networks (Goodman and Redclift 1991). New technologies could also potentially reduce reliance on pesticides and contributed to more environmentally friendly, and consumer friendly, crop production methods.

The relative cheapness of emerging biotechnology companies compared to their assumed future worth was another key justification for pesticide corporations to merge with, purchase equity in, or acquire outright, emerging biotechnology companies. The relative newness of biotechnologies, their long and unpredictable lead times to market, and the stock market crash of 1987, all contributed to relatively low stock values of biotechnology companies.

Another key justification of vertical concentration was the expected higher levels of appropriation, due to greater control over distribution networks of biotechnologies. Their commercial strategy was to capture profits along the agribusiness chain from the laboratory to the farm (Sehgal 1996). As the cost of biotechnology R&D increased, pesticide/biotechnology corporations faced considerable losses if they are unable to recoup their research investment through sales of germplasm products.

Vertical integration allows the risks associated with individual activities to be greatly reduced. Thus, the research division is ensured of an outlet for its products, the seed production division is ensured a steady supply of high-quality biotechnologies, as well as a reliable outlet for the commercial seed it produces, and the marketing and sales division is ensured an adequate supply of quality products. Second, because research, seed production and seed distribution activities are carried out within the same corporate entity, profits and losses can be distributed in a way that maximises the objectives of the firm, even if this means cross-subsidising certain inherently unprofitable activities.

Many pesticide corporations were concerned over transaction costs and the capacity of, and incentives for, seed companies to efficiently and effectively undertake plant breeding, multiplication and marketing of GM germplasm. They were convinced that by increasing horizontal and vertical concentration in the farm inputs sector, pesticide companies would be able to exert more power in the system and extract greater amounts of surplus value through the provision of crop varieties with inbuilt pest and disease resistance, through herbicide and germplasm packages.

By 2004, Monsanto supplied seeds for 97% of the total global GM maize area (46.4 million acres). Monsanto also supplied seeds for 91% of the worldwide GM soybean area (109 million acres), 63.5% of the global GM cotton area (14.1 million acres) and 59% of the global GM canola area (6.3 million acres) (ETC 2005). A quote from Robert Farley, president of Monsanto's Ceregen division, eloquently captures the expansionary sentiments of the industry. 'What you are seeing is not just a consolidation of seed companies; it's really a consolidation of the entire food chain. Companies like ours, who want to continue to be in the food and feed production business, are all trying to secure our spot along that chain' (Horstmeier 1996a).

Lastly, vertical concentration was perceived as a necessity for the protection of IPRs associated with the development of new biotechnologies. In order to realise significant returns on their investments, pesticide/biotechnology companies needed to not only create mechanisms/pathways to get their products to farmers, but also protect the extremely valuable intellectual property contained within them, as well as the scientific know-how that went into producing them. Again, they had two options. They could either license genes to seed companies or vertically integrate the supply of new biotechnologies through M&A with/of seed companies. However, because of the vulnerability of IPR associated with licensing of either genes to seed companies or through alliances or subcontracting arrangements with independent biotechnology companies, many pesticide corporations opted to acquire biotechnology and seed companies (Dow 2003b).

This strategy did not work for all new entrants. First, the time required to convert biotechnologies into products took much longer than originally envisioned. Second, there was a conflict between the entrepreneurial management style of the comparatively smaller seed companies and that of most big chemical corporations. Third, the learning phase was longer and more complex than expected, which led to poor financial performance. Finally, unlike chemicals, seeds could not be marketed globally, but only in agroclimatic regions similar to where they are developed. The culmination of these problems persuaded some companies to abandon biotechnologies altogether. For example, Shell began divesting their seed and biotechnology businesses in the early 1990s (Sehgal 1996).

4.4.3 Development of new and safer pesticides and input trait biotechnologies

Since the late 1970s, scientific advancement continued to facilitate development of a steady stream of new and generally less toxic and persistent pesticide products (see Fig. 4.1). However, the average rate of discovery for new AIs between 1980 and 2000 was only 12.7 compounds per year (Pesticide News 2000), significantly lower than the rate of discovery during the 1950s (18), 1960s (19) and late 1970s (16) (Boardman 1986). Between 2000 and 2012, the number of new AIs in pesticide corporation R&D pipelines decreased by 60% (ETC 2015). Part of the explanation for this decreased rate of AI discovery was that most of the chemicals suspected of possessing pesticidal activity had already been discovered or developed. In 1956, the number of products screened per market product was estimated to be 1800. By 1972, this had increased to 10 000 and to 12 000 by the late 1970s (Boardman 1986).

Pesticide producers were continually under pressure to develop new chemicals to replace older AIs, particularly insecticides, to which resistance had developed. As well as being safer, new pesticides were also designed to be more selective (Vorley and Keeney 1998). That is to say, they were increasingly targeted at specific pests, with little or no activity against beneficial/non-target pests (NRC 2000). Application rates for new pesticides were also reduced. For example, the typical application rate for herbicides dropped from approximately 3 kg/ha in the 1960s to 100 g/ha in the late 1980s (Aigner et al. 1998); insecticide and fungicide application rates also dropped from approximately 2.5 to 25 g/ha and 1.2 to 100 g/ha, respectively (Aigner et al. 1998). The decline in application rates was made possible primarily because new pesticide products contained more potent and concentrated AIs (Aigner et al. 1998).

As governments imposed more stringent pesticide regulations, and pesticide R&D costs escalated, pesticide corporations increasingly enforced strict selection criteria for potential pesticides early in the product development process (Bijman 2001a). The cost of developing agrochemicals increased fourfold between 1975 and the early 1990s (Hartnell 1996) and increased by a further 118% between 1995 and 2005 (ETC 2015). This contributed significantly to declining pesticide industry profitability in the 1980s.

Between 1983 and 1993, world pesticide R&D expenditures increased 3.2% per year, while the pesticide market only grew by 1.5% per year (Roberts and Begley 1994). Most additional expenditure was in getting regulatory approval, increasing from 17.5% in 1972 to 47% of total costs in 1989 (Ollinger and Fernandez-Cornejo 1998). Other factors contributing to declining pesticide industry profitability included the length of pesticide R&D and regulatory processes, which often took up to 8–10 years, eating into the pesticide patents life (Clapp 2002). In addition, brand names were increasingly under threat from cheaper generic products (Clapp 2002).

The application of new science enabled the development of a string of new biotechnologies. The first wave of agricultural biotechnologies aimed to increase crop yields, simplify crop management, reduce pesticide use and increase profits for all those involved in the food chain from the laboratory to the field (UNESCO 2003). Company entry into the crop biotechnology industry is suggested to have peaked in the early 1980s, with production innovation throughout the 1980s (Kalaitzandonakes and Hayenga 2000).

Ultimately, Monsanto led the commercialisation of GM seeds for major crops. Roundup Ready [herbicide tolerant (HT)] soybeans, maize and cotton were released in 1996 (Deutsche Bank 1999). Soon after, Monsanto released Roundup Ready (glyphosate-, glufosinate-ammonium- and bromoxynil-tolerant maize), canola, sugar beets and more (Genewatch 2003a). By 1997, the area of land sown with HT soybean was already around 17% of the US soybean area, growing to 56% in 1999, 68% in 2001, 81% in 2003 and 94% in 2016 (Feedstuffs 2002; USDA-ERS 2004; USDA-ERS 2017). By 2003, HT cotton had expanded to 59% of the cotton hectarage, and, by 2016, HT maize was grown on 89% of the maize area in the USA (USDA-ERS 2017).

Monsanto also launched a range of insect-resistant GMOs in 1996, namely, Bt maize and Bt cotton. These were resistant to insect attack, especially the important European corn borer and rootworm (BIO 2004b). Bt maize quickly increased to 8% of the US maize area in 1997, 26% in 1999 and 29% in 2003 (ERS 2004). Likewise, Bt cotton expanded rapidly from 15% of US area in 1997 to 32% in 1999, 37% in 2001 and 41% in 2003 (Fernandez-Cornejo and McBride 2002; USDA-ERS 2004).

Whilst not as high profile as Monsanto, Syngenta also released a number of GM crops in the USA as well as a number in Europe. In the USA, Syngenta marketed a number of soya and maize crop lines incorporating Liberty Link and Roundup Ready traits. It was also developing a range of Acuron HT GM crops (Corporate Watch UK 2002a). Syngenta introduced glyphosate-tolerant sugar beet (USA 1999); Bt insect- (including European corn borer) resistant maize (USA 1995); glufosinate-ammonium- and Bt-insect-resistant maize (USA 1996); Imazethapyr-tolerant maize (Canada 1996); and delayed fruit softening tomato (Genewatch 2003c). Syngenta produces two varieties of GM Bt maize YieldGuard and NK Knockout. NK Knockout is the only GM crop approved for commercial growing in the EU (Corporate Watch UK 2002a). By the turn of the millennium, 3% of Syngenta's sales came from GM crops.

DuPont also introduced a number of GM crops in the Canada and the USA. DuPont introduced imidazolinone-tolerant maize into Canda in 1994, followed by imidazolinone-tolerant OSR in 1995 (Genewatch 2003b). In 1996, DuPont introduced sulfonylurea-herbicide-tolerant cotton into the USA and introduced male sterile and glufosinate-ammonium-tolerant maize in 1998. DuPont also released GM crops in Eastern Europe and South America. For example, DuPont released Bt maize in Croatia and LibertyLink maize and Bt Maize in Hungary (Groundup 2003f). In 1998, DuPont released Bt-insect-resistant and glufosinate-ammonium-tolerant maize, and glufosinate-ammonium-tolerant maize and soybean in Argentina (1998). In 1997, DuPont also released Bt maize and LibertyLink maize in New Zealand (Groundup 2003f).

In turn, Bayer (as Rhône-Poulenc) introduced Bromoxynil-resistant tobacco in France and bromoxynil-tolerant cotton in the USA in 1995 and LibertyLink and Invigor Canola in Canada (Assouline and Pierre-Benoit 2001). Rhône-Poulenc introduced LibertyLink maize in the USA in 1997 and Canada in 1998. Soon after, Rhône-Poulenc released LibertyLink Cotton, Rice and Soybean (BIO 2004b). Lastly, Dow AgroSciences introduced Herculex 1 Insect Protection in maize and was developing Rootworm-resistant Corn and Insect-Protected-Cotton (BIO 2004b).

Figure 4.2 illustrates the adoption of GM crops, namely HT and insect resistant (Bt), in the USA between their release in 1996 and 2017.

According to Wield et al. (2010), Monsanto led the release of GM crops for two key reasons: First, 'it had less to lose from investment in biotechnology than the bigger agri-chemicals firms since it had a very narrow technological base with a large proportion of its profit coming from one chemical – its glysophate herbicide, Round-Up'. Second, 'the R&D strategy of developing GM

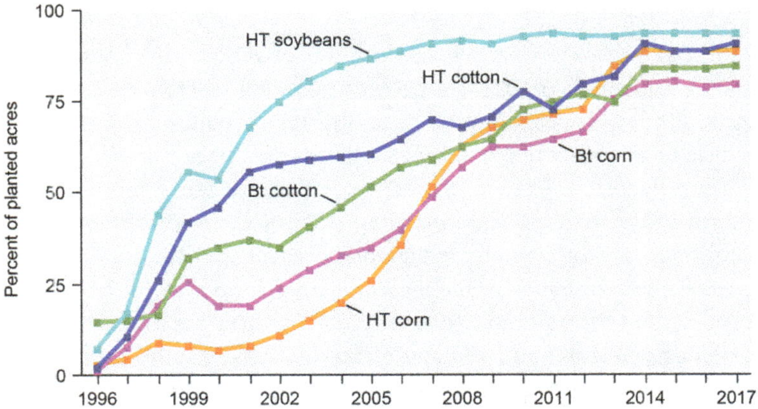

Figure 4.2 Adoption of genetically engineered crops in the USA, 1996-2017. Data for each crop category include varieties with both HT and Bt (stacked) traits. *Sources:* USDA, Economic Research Service using data from Fernandez-Cornejo and McBride (2002) for the years 1996-99 and USDA, National Agricultural Statistics Service, *June Agricultural Survey* for the years 2000-17.

crops resistant to Glysophate, and also developing GM induced insect resistance based on Bacillus Thurigeinisis (Bt) genes, fitted well with a company that had significant herbicide market share but little presence in insecticides'.

4.4.4 Rear-guard action against anti-pesticide campaigns

This section focuses on how the pesticide industry responded to increasing demands in Western Europe, the USA and Japan to remove, or restrict, the use of a growing number of pesticides that were believed to adversely impact human health (see Fig. 4.1).

4.4.4.1 Pesticide litigation

Litigation involving pesticides increased dramatically during the 1970s and 1980s in Western Europe and the USA. The amount, and success, of pesticide litigation tended to be based on two key criteria: (1) the changing attitudes of key environmental, consumer, pesticide industry and government actors towards pesticides and (2) the, closely associated, speed of institutional change.

Throughout the 1970s and 1980s, the perceived costs and benefits of pesticides underwent significant revision in several important sections of society. First, key conclusions of rapidly emerging pesticide cost-benefit analyses and health impact studies tended to strengthen the resolve of environmental and consumer organisations that pesticides were hazardous and that their use should be avoided if public health and environmental integrity were likely to be adversely affected. Second, individuals within governments and their legal apparatus have evolved from official positions of almost total unanimity in the acceptance and wholehearted support for pesticide use to a range of diverse positions. These positions ranged from the maintenance of strong support for pesticides and new technologies to the endorsement of programmes to drastically reduce the reliance of pesticides through the promotion of more integrated, and often organic, forms of agricultural production. Part of this heterogeneity can be explained by the cultural context of pesticide use in individual countries, the resilience of the industrial agro-food systems and the policy communities that support them (discussed below), and the difficulty of proving links between pesticides and negative effects on human health. Lastly, pesticide industry attitudes towards pesticides changed little throughout the 1970s and 1980s in Western Europe, the USA and Japan. Whilst scientists and sales staff may have harboured growing reservations over the safety of their pesticides, the official position of the pesticide corporations remained one that continued to argue the public goods nature of pesticides and that staunchly supported the use of synthetic pesticides in industrial agro-food systems, which underpinned economic and food security.

Whilst there is no doubt that certain pesticides can be extremely hazardous, the link between the negative effects of pesticide use on human health when used in accordance with manufacturer's instructions and GAP has been difficult to prove. When acute toxicity occurred, pesticide companies continued to claim that illness, or even death, arose from the blatant misuse of their pesticide products. For example, on occasion, individuals have drunk paraquat in order to commit suicide.

In the case of chronic poisoning from pesticides, it has been, and continues to be, notoriously difficult to establish, beyond reasonable doubt, the direct link between cause and effect in a court of law. Chronic poisoning from exposure to pesticides is difficult to unequivocally establish for several reasons. First, the symptoms of pesticide poisoning are easily confused with other unrelated illnesses and are often misdiagnosed. Second, if pesticide poisoning is suspected, it is exceedingly difficult to determine a scientifically valid, and consequentially a legally valid, link between cause and effect. Third, pesticide poisoning often affects different people in different ways. For example, children are generally more susceptible than adults. Effects can be gender specific (The Ottawa Citizen 2003), or specific to an individual's genetic composition (Cecchine et al. 2000). Protective clothing and differences in skin properties and integrity influence dermal exposure, and inhalation exposure may vary with ventilation or as a result of other factors, including properties of an individual's airway membranes (Cecchine et al. 2000). The rates at which pesticides are cleared in the body depend on amounts, genotype and activity of enzymes involved in their metabolism. Individual differences in cofactors modify the effect of pesticides. They may accelerate or slow a body's pesticide metabolism, or permit or inhibit toxic effects of pesticides. Cofactors, such as vitamins C and E, photochemicals and cholesterol (Cecchine et al. 2000) may contribute to clinical effects. Pesticides can often exacerbate existing medical complaints, making the pesticides role in the illness harder to prove. Fourth, individual pesticides can interact with one another in an additive, synergistic or antagonistic manner (Cecchine et al. 2000). Fifth, most biological tests on pesticides are not carried out on humans, therefore adding another tier of uncertainty (Pesticide News Briefs 2003). Finally, the number of documented chronic pesticide poisonings in Western Europe, the USA and Japan has been extremely low.

As the attitudes and/or political standpoints of key actors involved in the pesticide debate evolved, institutional change followed suit, albeit at different speeds and with a diverse range of outcomes. In the UK and USA, for example, the rate of institutional change was relatively slow. By default, this meant that pesticide companies in these countries found both their courts and political leaders generally sympathetic to the continued use of pesticides, making any change incredibly slow.

4.4.4.2 Pro-pesticide lobbying in the environmental and agricultural policymaking arenas

In contrast to the following three chapters, this chapter is unique as it sets the scene for the evolution of the pesticide industry by focusing on the Second Food Regime's urban-centred industrial agro-food systems' inherent capacity to resist change, despite overwhelming external pressures for it to do so (see Fig. 4.1). The continuation of Fordist-styled agriculture in Western Europe, the USA and Japan can be explained by the continuation of the intractable and protracted nature of the political mechanisms that regulate agriculture and the robust resilience of key interest groups, which remained the dominant actors in agro-food policy communities. These key actors include representatives of agrochemicals companies as well as fertiliser and machinery manufacturers, farmers and food processors and retailers. In concert with sympathetic pro-farm sector support actors, these interest groups have secured continued effective political and financial support to underwrite the key remnant components of Second Food Regime, which maintains enough coherence and durability to secure continued profitability. However, aggregate inability or reluctance to change is due to the activities of key interest groups rather than the appropriateness of the Fordist model of agriculture throughout this Period. Whilst growing public anxiety had been elevated into the political arena, proponents of tighter regulation, particularly of pesticides, were no match for political astute and well-connected agro-food interests, which aggressively attacked virtually all challenges to the continuation of industrial agriculture. Indeed, an excerpt from Bayer's 1981 annual report typifies industry's response to external criticism: 'in view of the challenges posed by world hunger, emotional attacks against conscientious agricultural chemical research are attacks against humanity'. This argument was still being used in 1993 when Ciba-Geigy stressed that 'without chemicals, you would have to feed pests to people' (Ciba-Geigy 1993).

Industry lobbies, including those subscribed to by pesticide corporations and their parent companies, and individual pesticide corporations and divisions, were able to promote institutional contexts that overtly favoured pesticides and biotechnologies, or at least stifle attempts to impose costs or restrict production and sales of their technologies. For example, according to industry analysts, biotechnologies were perceived by governments to be high-tech innovations with the potential to generate significant commercial benefits. The biotechnology industry was strongly supported by the US Government, most national governments in Europe, and by supranational institutions. These institutions have invested significant resources in order to pump-prime the foreseen biotechnology revolution (O'Mahoney 1999).

Conversely, commercial food systems actors have secured privileged and influential roles in food system regulation at both national and international levels. For example, at the 1982 CCPR meetings, the USA sent eleven representatives. Among this delegation, three were from the citrus industry and two from the agrochemicals sector. Similarly, the UK sent nine representatives, one from the Food and Industry Council and one from the British Agrochemicals Association (BAA). According to Boardman (1986), GIFAP has sent up to 30 members to Codex meetings. Indeed, GIFAP members can request to make statements and can be called upon by the chairman to comment or take part in various working groups, plus they have the opportunity for substantial networking (Boardman 1986). By 1983, GIFAP had chemical company members from 21 nations, many from the USA, Japan, Germany, Britain and Switzerland, only Brazil and India from the south, Argentina Philippines and Mexico were associate members (Boardman 1986).

Individual pesticide/chemical corporations were also active at a national level. For example, DuPont has been accused of attempting to weaken US laws designed to make polluters pay. One of the key pieces of legislation referred to was Comprehensive Environmental Response, Compensation and Liability Act also known as Superfund (passed in 1980); set up after the Love Canal disaster, when it was discovered that the Hooker Chemical Company had dumped more than 20 000 tons of toxic chemicals (Groundup 2003b). According to the Public Research Institute Group, DuPont and others 'have lobbied Congress to roll back the polluter pays principle and weaken clean-up standards at the nation's worst sites'. DuPont and others 'have fought efforts to expand the public's right to know about toxic chemicals used in the workplace, consumer products and communities' (Groundup 2003b). DuPont was also accused of influencing the eventual structure of the US Superfund law; by removing liability provisions, which, among other things, would have allowed victims of toxic wastes to sue chemical manufacturers in federal courts (Groundup 2003b).

4.4.4.3 Pro-pesticide education and public relations

Throughout the late 1970s and 1980s, distrust of the farming industry resulted in a growing scepticism towards safety messages emanating from the pesticide industry. Throughout the 1980s, the pesticide industry responded to criticism with the rhetoric of denial or neutralisation, continuing to imply that they were being unfairly blamed by an irrational and misguided public (Tombs 1993). In concert with many OECD governments, industry perceived and/or framed the problem as one of ignorance. Accordingly, this was a problem that could be redressed through the provision of accurate information about the benefits of pesticide use (see Fig. 4.1).

In 1990, the Agricultural Council of America (ACA) launched a US$50 million Food Watch Campaign to restore public confidence in the safety of their food supply (Vorley and Keeney 1998). The Crop Protection Association (CPA) also supplied free educational material for schools, featuring Benny Broccoli and his buddies, describing the significant benefits of fungicide use and so forth. Individual companies have developed their own programmes of education. Bayer received an award for its programme 'Making Science Make Sense', which supported the work of volunteers and promoted science literacy through a national public education campaign. This campaign included Allegheny Schools Science Education and Technology Inc. (ASSET), a non-profit organisation that promoted inquiry-based science education reform in Pittsburgh, 1992 (Bayer 2003h).

4.5 Key outcomes of pesticides and biotechnologies development and application in Western Europe, Japan and the USA

4.5.1 National food surpluses, national food security and farmer profitability

In the USA, Western Europe and Japan, national food surpluses continued to increase between 1974 and 2017 (FAO 2017), and were believed to have buffered national security (see Fig. 4.1). During this time, farmers' expenditures, chemical application and crop yields continued to increase. Between 1972 and 1980, the pesticide market grew by 5% per year (Bull 1982).

In the USA, pesticide use increased by 500% (Goodman and Redclift 2001) between 1950 and 1986 and by 250% (Glicksman et al. 2003) between 1964 and 1993. By 1993, there were 40 000 registered pesticides and 600 AIs in the USA (Glicksman et al. 2003). The USA accounted for 27% of the world market in pesticides - exporting 450 million pounds and importing 150 million pounds (Pimentel et al. 1993).

Benefit/cost ratios for the use of pesticides ranged from 3:1 (Bull 1982), 4:1 (Pimentel and Pimentel 1979) and up to 6:1, with most of the benefit passed on to consumers in the form of lower food prices (Zilberman et al. (1991). Estimates including indirect costs on humans and environment are approximately 3:1 (Pimentel and Pimentel 1979).

4.5.2 The battle over pesticide regulation

By the 1970s, in Western Europe, the USA and Japan, pesticide registration had become a prerequisite for pesticide release. Before release, all pesticides must have undergone a series of tests that assess a range of criteria, including their effects on human health through direct and indirect exposure to dilute and undiluted forms of the pesticide. In recent years, many pesticides have been subject to re-registration. This is a process whereby pesticides that met initial registration criteria but were suspected of falling short of new stricter guidelines were required to undergo further tests to prove that they were safe. The re-registration process also includes a selection of previously unregistered pesticides that were subject to grandfather rights during the initial registration process.

Despite the resilience of urban-centred industrial agro-food systems in Western Europe, the USA and Japan, the continued and relentless pressures to reduce the human health impacts of pesticides, from a growing number of environmental organisations and sympathetic individuals in key public positions, meant that a new trench of pesticide regulation was initiated throughout the 1970s and 1980s. This ultimately resulted in the establishment of increasingly stringent pesticide regulations, which, in turn, created increasingly higher costs for the pesticide industry. Additional costs usually came in the form of bans or use restrictions on existing pesticides, more extensive and stringent safety tests and a more protracted and comprehensive registration processes. Even though many actors, concerned with the negative impacts of pesticide use on human health, were less than overwhelmed with the progress made during the 1970s and 1980s, anti-pesticide activists have made considerable inroads. These inroads have primarily been made in several key areas: (1) Removal of some of the most hazardous pesticides from the market in Western Europe, the USA and Japan. (2) Pesticide use restrictions and limits to human exposure. (3) Pesticide residue limits on food products. (4) Labelling, formulation, packaging and repackaging, storage and safety standards. (5) Pesticide registration. (6) Pesticide taxes. (7) Pesticide reduction subsidies. (8) Voluntary regulation of pesticide use by manufacturers (including voluntary withdrawals, and actions to reduce some of the negative impacts of pesticide use).

As consumers became increasingly unconvinced of the utility or need for pesticide use, they also became increasingly unwilling to accept the health risks associated with pesticide use, through either direct exposure or ingestion in the form of pesticide residues on food, particularly when most of the benefits were perceived as accruing to industry (Macfarlane 2002). Changing attitudes towards pesticide use in Western Europe, the USA and Japan led to the initiation of policy processes aimed at addressing the public's concerns. Ultimately, this led to the imposition of increasingly stringent pesticide regulation. However, in the political arena, simultaneous struggles over the role of pesticides in agricultural production remained intense, and generally highly polarised. Opposing interests often fought to maintain the status quo and to re-legitimise pesticide use or to remove pesticides for sale and use completely. The public debate on risk was a battle to win over public opinion; all sides used scientific information to defend their positions. Many in industry and government considered that public reaction to pesticides continued to be based on emotion, rather than science. In turn, they believed that scientific rationality would prevail when the facts were known and the truth understood. Conversely, many environmentalists point to 'bad science' used by the proponents of pesticide technologies and insisted that their, less than supportive, views are more valid (Erlich and Erlich 1996; Irwin et al. 1997). The outcome of these battles was therefore highly heterogeneous, dependent on the political and media connections of pro- or anti-pesticide lobbies, and increasingly on their skilful contributions to the drafting process for new national legislation and use of erstwhile objective scientific evidence.

4.5.2.1 USA

By the end of the 1980s, and beginning of the 1990s, in roads into the regulation of some of the most hazardous pesticides had been made. In 1986, the Safe Drinking Water Act was successfully amended to include additional pesticides: dalapon, dichloromethane, dinoseb, diquat, endrin, glyphosate, picloram and simazine to the list of regulated water contaminants (EPA 1992b).

In June 1986, the EPA issued a Registration Standard for glyphosate, which is of relatively low oral and dermal acute toxicity. It was placed in Toxicity Category III for these effects (Toxicity Category I indicates the highest degree of acute toxicity, and Category IV the lowest). The acute inhalation toxicity study was waived because glyphosate is non-volatile and because of adequate inhalation studies with end-use products showing low toxicity (EPA 1993).

In December 1986, the EPA issued a Registration Standard for ethyl parathion, stating that the Agency planned to initiate a Special Review due to acute human and avian concerns (EPA 2000). Ethyl parathion was first registered as a pesticide in the USA in 1948, and is among the most highly toxic chemicals registered with EPA. It was placed in Acute Toxicity Category 1 (most toxic) for acute eye, skin and inhalation effects. It is a potent plasma, red blood cell and brain acetyl cholinesterase inhibitor. It is toxic by all routes of exposure and cholinesterase inhibition occurs following acute, subchronic and chronic exposures to low doses of ethyl parathion (EPA 2000). Ethyl parathion is very highly toxic to birds, fish, aquatic invertebrates and small mammals, poses a high acute risk to birds, mammals and aquatic invertebrates, and also poses a high chronic and reproductive risk to birds. This high acute, chronic and reproductive risk would also be expected to impact on many endangered species in the ethyl parathion use area. Ethyl parathion is very highly toxic to pollinating insects such as bees and has a well-documented history of bee kill incidents (EPA 2000).

By 1991, the Agency and the registrants of ethyl parathion reached an agreement under which the registrants agreed to limit use sites and restrict application and post-application practices (EPA 2000). Since 1991, the use of ethyl parathion in the USA has been limited to nine crops: alfalfa, barley, canola (rapeseed), corn, cotton, sorghum, soybeans, sunflowers and wheat. All of the technical ethyl parathion sold in the USA is produced by Cheminova Agro A/S, formulated at one location and sold under the Cheminova label (EPA 2000). It is formulated as a liquid and may be applied only using aerial equipment (EPA 2000).

One of the most celebrated achievements was the 1988 amendment of FIFRA. The FIFRA Amendments Act (1988) required the EPA to review and re-register all pesticides registered before 1984 (EPA 1988; Walth and Pulaski 2003). In addition, it authorised the collection of fees by the EPA to support re-registration activities (EPA 1988). Re-registration was comprised of five phases: (1) EPA published a list of AIs subject to re-registration, as well as asking whether or not pesticide companies intended to re-register their AIs. (2) Registrants were asked to formally agree to supply missing data. (3) Registrants were required to supply adequate data and flag studies that indicate adverse effects within 1.5–2 years after 1988 Amendments. (4) The EPA was required to respond by independently identifying gaps, and issuing requirements for registrants to fill gaps 2–4 years after the 1988 Amendments. (5) The EPA was required to have comprehensively examined all of the data submitted in support of pesticide re-registration and have either re-registered or taken other appropriate action within approximately 3–9 years after enactment of the FIFRA Amendment (EPA 1988). By 1988, 600 AIs required re-registration (WSU 2003a). To accelerate the process of re-registration, in 1988, Congress imposed a 10-year re-registration schedule. To pay for this, Congress directed the EPA to require registrants to pay re-registration and annual registration maintenance fees on pesticide ingredients and products (Schierow 2004).

Within this amendment, the EPA gained authority to cancel pesticide registration if new test data, aided by modern analytical techniques and equipment, and an improved understanding of the effects of pesticides, show that it causes unreasonable adverse effects on human health or the environment. Where unreasonable adverse effects could be shown, the EPA was authorised to take action to suspend the registration of a pesticide in order to prevent an imminent hazard (EPA 1988). Until the FIFRA Amendments of 1988, the EPA had been required to accept certain suspended and cancelled pesticides and dispose of them at government expense, and due to a pesticide industry derived indemnification provision, was required to reimburse holders of suspended and cancelled pesticides for financial losses suffered, up to the cost of the pesticide (EPA 1988).

In response to the FIFRA (1988) deadline, more than 40 000 studies were submitted to the EPA (Glicksman et al. 2003). However, the EPA made incredibly slow progress. By mid-1993, the EPA had only managed to register 250 pesticides (Glicksman et al. 2003). This allowed many potentially hazardous pesticides to remain in use (Walth and Pulaski 2003).

The media also continued to provide a forum for debate over pesticide use and, through the use of scare tactics, provide a useful tool to sell newspapers and increase TV ratings. After DDT, Alar has been one of the most publicised pesticides in the USA. In 1989, a '60 Minutes' report alleged that eating apples sprayed with the pesticide Alar (daminozide) could increase the risk of developing cancer. The report was based on a publication by the Natural Resources Defense Council, a consumer advocacy group, which suggested that Alar in apples created a 45 in a million cancer risk, and predicted that 6000 American school children could get cancer from Alar residues. The EPA's in-house assessment indicated that the hazard was 20–50 times higher than the negligible risk standard of one in one million lifetime cancer risk (Macfarlane 2002). Finkel (1995) calculated that this level of risk could have resulted in 25 additional deaths. Public reaction to the presence of carcinogenic residues on apples was one of outrage.

According to Erlich and Erlich (1996), several factors may have had influenced that reaction. First, the EPA had known about the potential dangers of Alar and chosen not to act. Second, Alar posed the highest risk to children. Third, the apple was seen as an icon of healthy eating. Fourth, the principal beneficiaries of Alar use were perceived to be growers and the food industry, rather than consumers (Erlich and Erlich 1996). Public uproar and extensive media coverage resulted in apples being removed from supermarket shelves and school canteens across the USA (Lee 1989), and caused apple sales across the USA to plummet.

Growing political pressure for the EPA to act culminated in a ban on the use of Alar in 1991. Apple growers and the pesticide industry responded by claiming that the Alar ban was an 'illustration of environmentalists gone awry and creating unnecessary panic' (Macfarlane 2002), and that the financial viability of farmers would be at risk if pesticides were continually removed from the market. The farming and pesticide industry insisted that the increased cancer risk from Alar had been calculated at 0.025% and therefore should be considered as trivial (Lee 1989).

By this time, science was being used to justify both the pro- and anti-pesticide positions. Consequently, this led to claims, from both sides of the argument, of bad science. Some simply did not believe the scientific evidence presented before them, either because of the credibility of the scientists that generated the data, the methods employed, and the disciplinary background of the both the scientist that undertook the research or the person evaluating the evidence. Consequently, when the science failed to clearly quantify and qualify pesticide risk, and became increasingly viewed as subjective, factors other than science became increasingly likely to inform public's opinion on pesticides (Freudenburg and Pastor 1992). At this stage, science itself began to lose its intrinsic and unquestioned objectivity, and, accordingly, its inherent legitimacy. Ultimately, apples were restored to school menus (Lee 1989).

As long as older, and cheaper, pesticides remain on the market, they acted as a disincentive to the development and adoption of newer lower-risk pesticides (Walth and Pulaski 2003). Indeed, it was not until the mid-1990s, after a series of pesticide and food residue scares in the USA, that the Clinton administration eventually drafted the Food Quality Protection Act (FQPA). The new legislation sought to reduce pesticides residues and examine the dietary impacts of pesticides, particularly in light of recent evidence indicating the enhanced impacts of pesticide residues on children compared to adults. The new legislation imposed the strictest pesticide restrictions ever witnessed in the USA. First, under the new law, farmers could no longer argue that the economic costs of banning a pesticide outweighed the health benefits. Indeed, if a pesticide was deemed as too dangerous, especially to children, it would be banned, regardless of the cost to farmers. Second, the EPA would be required to integrate an additional safety factor for children. In practice, this meant that the safe level of most existing pesticides would need to be reduced to 10% of their previously determined safe dose; just 0.001% of the amount a rat can stand. The mandate of FQPA was to ensure that tolerance levels were safe 'a reasonable certainty that no harm will result from aggregate exposure', to study the effects of endocrine-disrupting pesticides, and to consider exposures from all possible routes–food, drinking water, residential and others. In addition, pesticides with a common mechanism of action were to be evaluated as a group, to account for possible additive effects of exposure to many different chemicals, and all tolerances were to be reassessed by 2006 (WSU 2003a); this meant the re-evaluation of 9600 tolerances within 10 years (Walth and Pulaski 2003).

Initially, prospects for the Clinton Administration bill deteriorated when Republicans captured control of the House and the Senate in 1994. Indeed, the new Congress aimed to disempower, rather than empower, the EPA (Walth and Pulaski 2003). However, as voters grew increasingly cynical towards the Congress' environmental agenda, Republican's adopted the FPQA and pushed it through the House unanimously (Walth and Pulaski 2003). At the signing ceremony, President Clinton promised that 'if a pesticide poses a danger to our children, it won't be in our food'.

The FQPA (1996), which replaced the Federal Insecticide, Fungicide, and Rodenticide Act (FIFRA) and the Federal Food, Drug, and Cosmetic Act, was hailed as a landmark piece of pesticide legislation (Metcalfe et al. 2002). It provided the impetus to develop or adopt new ways to determine the health effects of pesticides, based on a growing understanding of the cumulative effects of pesticides on society's most vulnerable segments, and to promote policies to significantly reduce the negative human health impacts of pesticide use (Metcalfe et al. 2002). Now, the EPA must account for the aggregate exposure resulting from all pathways, including cumulative exposure to multiple pesticides through a common mechanism of toxicity (Metcalfe et al. 2002). The FQPA could have large impacts on both pesticide users and the rest of society. New safety standards targeted the OP and carbamate insecticide classes of pesticides and aimed to restrict or cancel several pesticides uses (Metcalfe et al. 2002).

The first milestone for this 10-year reassessment effort was 3 August 1999, when 33% (approximately 3207) of the total must have been completed (Walth and Pulaski 2003). Between 1996 and 1999, the US EPA completed re-evaluations of just 48 pesticides (or groups of similar pesticides).

4.5.2.1.1 Azinphos methyl

Azinphos methyl, better known by Bayer's brand name of Guthion, became the first pesticide examined under the FQPA 1996. By April 1997, internal EPA records indicated that Azinphos methyl posed a 'serious risk concern' (Walth and Pulaski 2003), and announced 'major reductions' in Guthion's use, something that received considerable acclaim from environmentalists, the media and sympathetic members of Congress (Walth and Pulaski 2003). Within days, a leaked copy of the memo on Azinphos methyl arrived by fax at the Washington, D.C. headquarters of the American Crop Protection Association, representing 57 pesticide manufacturers. Soon, word of the memo reached the 13 companies that produce most of the world's OPs. In response, the pesticide industry launched a massive lobbying campaign that, even by its critics' estimation, was devastatingly effective in its simplicity. It had one goal: tell Congress that if EPA was not stopped, a political price would be paid. Pesticide companies mobilised the big grower organisations and co-ops that represented their biggest customers. Farmers barraged Capitol Hill with letters. In congressional offices, the American Crop Protection Association had growers handed out red fly swatters. Leave things to the EPA, farmers told their representatives, and these fly swatters will be all we have to fight off pests.

Indeed, shortly after agribusiness lobbying began, Charles Stenholm, the highest ranking Democrat on the House Agriculture Committee, informed the Vice President Al Gore that Democrats from key farming states would suffer at the November polls over a restrictive pesticide policy. Stenholm also hinted that Gore's own 2000 presidential aspirations could be damaged (Metcalfe et al. 2002). Indeed, according to Walth and Pulaski (2003), Stenholm informed Gore that he did not think that anybody with aspirations for national office wants to be responsible for stifling agricultural technology. Two days before the EPA's deadline, Gore sent a memo to EPA ordering it to conduct an open, public process, something the agency had not previously done, and requiring that the EPA to cooperate with the USDA. In addition, Gore created an advisory panel to review the way EPA was carrying out the pesticide law. Gore, who had been a key supporter of the new pesticide law, was seen as selling out to the agribusiness lobby, which had grown increasingly concerned that the EPA might ban key pesticides before they had a chance to defend their interests.

Bayer differed in its strategy but finally struck a deal with the EPA (Walth and Pulaski 2003). Bayer favoured tests with human subjects, instead of the EPA's practice of relying on experiments with laboratory animals to find how pesticides act in the human body and measure health risks (Walth and Pulaski 2003). After conducting human tests on Scottish students, Bayer concluded that the 'use of Azinphos methyl did not pose an unreasonable risk to either workers or consumers'. Conversely, the EPA's computer model estimated that as many as 50 000 kids a day would get a dose the EPA considered too risky. In early July, EPA officials summoned executives of Bayer Corp. and the US Apple Association to hear an unwelcome message: Bayer had to propose dramatic cuts in the use of Guthion. If the company could no't find a way, the EPA might move to ban Guthion (Walth and Pulaski 2003). A voluntary agreement was reached to save the EPA from initiating an expensive and time-consuming cancellation hearing against Guthion, a move that probably would have raised questions about the chemical in other countries.

Most importantly for Bayer, the revised restrictions required only minimal cuts in Guthion's use. The new rules capped the total use of Guthion during a growing season at 4.5 pounds/acre, down from 6 pounds allowed previously. However, the allowable amount was almost twice the average 2.4 pounds/acre of Guthion that growers used nationwide (Walth and Pulaski 2003). In addition to capping the maximum annual applications of Guthion, Bayer supported lengthening the field re-entry phase for Guthion from 14 to 21 days, the same amount of time California now required after rolling back its emergency restrictions, but only if growers were allowed to use heavier Guthion doses (Walth and Pulaski 2003). The EPA agreed, and the new rules went into effect in January 1999.

Unlike the California authorities, however, EPA officials did not explicitly agree that Bayer's human-subject testing could be considered (Walth and Pulaski 2003). In Washington State, where 55% of Guthion is used, growers use 2.8 pounds an acre. EPA officials stated that lowering the per-acre cap will knock out some of the highest Guthion users, but they acknowledge that most growers will not be affected. The EPA also limited overall use to 1.92 million pounds – about a 10% cut from current levels, which vary from year to year. The cap prevents growers from increasing Guthion use if other bug-killers are banned in the future (Walth and Pulaski 2003).

Ultimately, EPA rules on Guthion's use were largely written to satisfy apple growers and the pesticide's manufacturer, Bayer Corp., both of whom argue that the EPA is exaggerating Guthion's risk to children. As a result, the EPA's ability to enforce the landmark pesticide act was severely hobbled. Canny lobbyists and bureaucratic dawdling, a leaked memo and presidential politics, pesticide-eating Scottish students and an odds-making computer programme named after a European gambling resort all combined to save a doomed pesticide (Walth and Pulaski 2003).

When FQPA was passed, 49 OP pesticides were registered for use in pest control throughout the USA. When the EPA released the Revised Organophosphate Cumulative Risk Assessment in 2002, 14 pesticides had already been cancelled or proposed for cancellation, and 28 others have had considerable risk mitigation measures taken through partial use bans (EPA). A few of the better-known cases include restrictions placed on the popular OP pesticides methyl parathion, chlorpyrifos, diazinon and the aforementioned azinphos methyl. In August 1999, the use of methyl parathion, considered to be one of the most toxic and widely used OP insecticides, was cancelled on many fruits and vegetable crops that constitute a large portion of children's diets (EPA 1999). In June 2000, restrictions were imposed on chlorpyrifos. These restrictions were widely contested by apple, grape and tomato producers, who projected the restrictions would have a significant impact on production and were upset that these effects were not considered by the EPA in the regulatory process (EPA 2000b). In January 2001, diazinon's use was banned for one-third of existing agricultural crops (EPA 2001).

In the case of Alar, Guthion and EDB, if the public overreacted, 'it was probably as much due to the reluctance of EPA to act on scientific evidence as to public misunderstanding of the risks' (Krimsky and Plough 1988). Mistrust of government and industry's management of both the Alar and EDB situations served to weaken the EPA's credibility and forge a divide between balanced and objective science for the public good and the selective abuse of science by government decision makers and the pesticide industry to meet their own policy goals (Macfarlane 2002). As Irwin et al. (1997) suggest, science was now controlled by government and industry. Claims relating to the abuse of science were given further credibility with a string of embarrassing disclosures of the unethical and unscientific practices of private sector laboratories in relation to pesticide tests. In 1983, Industrial Bio Test in the USA fabricated evidence for pesticide trials (Erlichman 1991). It was also believed that the US Government had known about this practice as early as 1977 (Wheelwright 1984). This scandal was suspected of stifling Codex negotiations for months if not years (Boardman 1986). Nearly a decade later, in September 1992, the president and three other employees of Craven Laboratories in Texas were found guilty of falsifying tests on pesticides in order to get them approved for pesticide companies. Craven's evaluations had been accepted by the US EPA, MAFF (UK) and others for a number of years (Hough 1998).

4.5.2.2 UK

The 1975 Health and Safety Regulations on worker safety (Bull 1982) and the Poison rules and poison list (Bull 1982) of 1978 served as the UK's principal command and control mechanism to regulate pesticides. In stark contrast to the USA, the UK's favoured approach was voluntary, non-statutory, control in the form of the Pesticide Safety Precaution Scheme (PSPS), which came into existence in 1957. The stated goal of PSPS was 'to safeguard human beings (whether they are users, consumers of treated produce, or other members of the public), livestock, domestic animals and wildlife, against risk from pesticides' (Boardman 1986). In essence, the PSPS was effectively a non-binding gentlemen's agreement between the government and pesticide manufacturers. PSPS policy was largely determined by the PSPS Advisory Committee, on which there was ample representation from the agrochemical industry, but little, or no, representation from consumer groups, environmentalists and farm workers.

Pesticide manufacturers proposing to introduce a new pesticide, or new use for an existing one, notified MAFF, provided specified information, and then refrained from introducing the product until agreement had been reached on the appropriate precautionary measures (Boardman 1986). Manufacturers provided MAFF with pesticide data and MAFF decided how to proceed. There were four stages: trials clearance, limited clearance, provisional commercial clearance and commercial clearance. Draft recommendations normally went through the Advisory Committee on Pesticides, resulting in recommendation sheets (Bull 1982).

The PSPS was designed to deal with clearance of safety aspects of pesticides (Boardman 1986). It covered toxicological testing, investigation of residues, environmental impact and day-by-day management of relations with industry (Boardman 1986). In addition, the Agricultural Chemicals Approval Scheme (ACAS) was another voluntary scheme created to test the efficacy of popular brands (Bull 1982). Under the terms of ACAS, a product containing a new AI could not be approved until its safety had been considered by the PSPS, and the necessary precautions recommended (Boardman 1986). However, PSPS officials were obliged to cooperate and collaborate closely with industry, and agreed to the release of potentially more hazardous pesticides, than they would have done if they had been a totally independent body with statutory powers (Boardman 1986). In parallel to PSPS and ACAS, pesticide safety in the UK was promoted by the circulation of advice on GAP (Boardman 1986).

Trade unions and political parties also played an influential role in pesticide regulation in the UK. Indeed, the trade unions and the Labour Party insisted that technical decisions made by experts lacked realism (Boardman 1986). Appeals to anti-chemicals prejudice and fear were prominent features of trade union initiated anti-pesticide campaigns (Boardman 1986). One such campaign was launched against the phenoxy pesticide 2,4,5-T. However, even though doubts about 2,4,5-T were raised (Bull 1982), in 1980, the Advisory Committee on Pesticides concluded that exposure to 2,4,5-T under GAP posed negligible risks (Boardman 1986). Indeed, despite further studies linking exposure to phenoxies to cancer (Boardman 1986) in 1982, the UK Government continued to permit their use (Boardman 1986). By this time, 2,4,5-T had already been banned in the USA, Sweden, Norway, Holland and Italy (Bull 1982). Germany had withdrawn registration of 2,4,5-T in 1981, amidst strong industry opposition. However, after industry used weaknesses in the scientific argument against 2,4,5-T, Germany allowed 2,4,5-T sale between 1982 and 1985 (Boardman 1986). Indeed, the UK ACP concluded that the Oregon 2,4,5-T study, used to support the ban on 2,4,5-T, was invalid on both scientific and statistical grounds (Boardman 1986) (i.e. studies were either not sufficiently documented or, where they had been fully documented, could not be sustained).

With the exception of arsenic and lead, the British Government had no formal scheme for the statutory regulation of maximum pesticide residue limits in foods (Boardman 1986). Residue limits were controlled by GAP (Boardman 1986), and residues detected by government chemists were generally below those recommended by Codex (Boardman 1986). By the early 1980s, 97–98% of all main crops, cereals and vegetables in Britain were sprayed at least once (Clunies-Ross and Hildyard 1992). Because of the UK's informal approach to pesticide regulation, by the mid-1980s, 41 pesticides were on sale in the UK that were banned or severely restricted in other countries. In 1985, the London

Food Commission found that 49% of pesticides in common use in the UK were possible carcinogens, 61% were suspected mutagens and 90% possible allergens (Clunies-Ross and Hildyard 1992).

The voluntary principle in UK pesticide regulation was also applied to pesticides sales, storage, transport and competence of staff. In 1978, at the suggestion of the British Government, the UK pesticide industry established BASIS, which defines itself as a 'self regulated scheme, in line with Government de-regulation policy, giving balanced and independent advice to registered distributors' and does not aim to 'emulate the role of any government enforcement agency'. The BASIS Registration Board consists of representatives of all trade associations with pesticide interests such as the CPA, National Association of Agricultural Contractors, National Farmers' Union (NFU), United Kingdom Agricultural Supply Trades Association (UKASTA) and Association of Independent Crop Consultants. The Board also has members elected by distributors as well as a representative from MAFF as an observer, and is headed by an independent Chairman. BASIS also administers the Professional Register for sales and advisory personnel employed within the agrochemical industry.

While many of the traditional working assumptions of regulatory approaches were questioned by a growing range of new pesticide actors (Boardman 1986), it was not until the UK's entry of the European Commission (EU) that the PSPS was held to task because of the conflict of interests involved in pesticide regulation. The EC was much keener on formal regulation, particularly as UK products destined for use in other EC countries were not regulated in line with EC protocols. The voluntary and industry-centred approach was too lax and open to abuse for Brussels (Boardman 1986). While the closed agricultural policy community had been able to resist pressures for change arising from environmental and consumer groups, the UK Government was obliged to draft formal pesticide legislation in order to plicate Brussels. Regulation came in the form of the Food and Environment Act 1985. In addition, as a direct result of the Control of Pesticides Regulation 1986, the BASIS Storekeeper, Field Sales and Technical Staff certificates were required, by law, by all those involved in the storage, sale and supply of pesticides. The statutory Code of Practice for Sale and Supply of Pesticides recommends that all companies have an independent annual assessment and names BASIS as an independent inspectorate. Almost every company distributing pesticides is registered with BASIS.

However, despite regulatory improvements, by 1987, some fresh water lakes in the UK contained over 50 ng/L of pesticide residue, tenfold the EC wide limit of 5 ng/L (Bowler 1992), and estimated acute poisonings due to pesticides were higher than many other OECD countries. Whilst MAFF began its phase-out of DDT in 1969, a MAFF study on pesticide residues on food in 1989 concluded that food was still contaminated with DDT, despite a ban on its use. Babies were likely to exceed WHO ADI levels for organochlorines if they drank baby's milk prepared from cows' milk and that residue levels of many pesticides exceed official MRLs. The report also found that 58% of infant rusks were contaminated with OP pesticides and many known carcinogens including Lindane, dieldrin, pentachlorophenol, Malathion, captan, parathion and others.

In 1992, a National Rivers Authority survey found that atrazine levels exceeded EU drinking water standards in 11% of samples. In 1993, 75% of breaches in drinking water standards were from pesticides, mainly atrazine and simazine (Bowler 1992). As a result, agricultural uses of atrazine were banned later that same year. A report by the UK Working Party on Pesticide Residues determined that up to 48% of potatoes, 32% of cereals and 29% of fruit and vegetables sampled contained pesticide residues (ARWPPR 1992). In 1992, 29 people suffered acute poisoning after eating cucumbers illegally treated with aldicarb (Hicks 1992).

While the existence of pesticide residues on food and water supplies was both an embarrassment and concern for the British Government, a legally defensible causal link between pesticides and ill health or even death has been difficult to prove, particularly when pesticides have been used as per manufacturers' instructions. Indeed, in 1985, the BAA was keen to stress that there had been no fatalities caused by pesticides during normal use on a UK farm since 1974 (Hough 1998). Furthermore, by the early 1990s, MRLs were only exceeded in 2% of samples in the UK, compared to 3–7% in Germany and 14% in Italy (estimated 48% in oranges) (Vorley and Keeney 1998).

Accordingly, greater protection for consumers, agricultural workers and the natural environment was called for (Boardman 1986). However, much anti-pesticide rhetoric calling for greater regulation of procedures for pesticide registration and use was discounted by government because it was emotionally based and did not conform to the scientific norm. Some called for involvement of the Health and Safety Executive, which had been established in 1974 to protect workers in their workplaces. Indeed, with the release of 2,4,5-T in the late 1970s, the Trade Union Congress (1980) called for the transfer of responsibility for the control of pesticides to the Health and Safety Executive (Boardman 1986). In practice, room for manoeuvre was limited because of the continued importance of traditional regulatory bodies, and by the government–industry relationship built up over many years through PSPS consultations (Boardman 1986).

4.5.2.3 EU

Initial work by the Council of Europe's Social Committee on pesticide regulation began in the late 1960s. In 1976, the Council of Ministers passed a Directive defining a range of maximum pesticide residues for fruit and vegetables (Boardman 1986). This was based on the fact that different member states had different maximum pesticide residue limits on their produce that could potentially create barriers to free intra-European trade (Boardman 1986) and hinder the free movement of goods. The work of the Social Committee prompted other committees such as the Public Health Committee (1978) to establish guidelines on topics such as safe handling, accidents and storage and disposal of pesticide containers (Boardman 1986).

The Harmonisation Directive of 1978 could be seen as a prelude to the registration Directive (first voiced in 1976), which caused significant concern in industry and in sympathetic governments. First, sovereignty was threatened. The UK and Denmark fought hard to keep national sovereignty over pesticide registration (Boardman 1986). Second, data handling and IPR was an issue. Third, concerns were voiced over the flood of US and Japanese imports unless reciprocal arrangements were made in the USA and Japan (Boardman 1986). The sixth Amendment of the dangerous substances Council Directive in 1979 constituted a far-reaching statement on toxic chemicals policy. The aim was to ensure that substances could not be placed on the market, on their own or in preparations, unless they were notified to the competent authority of one of the member states, and packaged and labelled in accordance with certain specified definitions (Boardman 1986). While industry was also keen to harmonise standards across the EEC, they aimed to harmonise downwards. After a tough fight, particularly on the basis of IPR, the pesticide industry managed to exclude pesticides from the Directive but in doing so created a divide between public and private sectors goals in relation to pesticide regulation in the EEC (Boardman 1986). Indeed, the pesticide industry claimed that the proposed standards ignored those laid down by Codex (Boardman 1986), and that they acted as a barrier to free trade.

4.6 Summary

It can be argued that evolution of the pesticide industry in Western Europe, Japan and the USA has been strongly influenced by the continuity of Fordist-styled principles of industrial organisation institutionalised in urban-centred industrial agro-food systems. The pesticide industry, in concert with the remaining non-farm components of the food system, changed significantly. In order to survive, remain competitive, or even prosper, non-farm components of the urban-centred industrial agro-food systems have undergone a protracted phase of restructuring, which has involved structural and operational changes. The most important structural change has been the rapid concentration with food systems, both horizontal and vertical. There were two key operational changes in pesticide industry. First, new pesticide products were developed that are inherently less hazardous to human health. Second, new pesticides and input trait biotechnologies were developed with the potential to generate significant profits for the pesticide industry, and allow farmers to remain financially solvent. By the end of the 1980s and beginning of the 1990s, the pesticide industry in the USA, Western Europe and Japan had grown significantly, in both sales and profitability and political power. However, despite the immense power of the pesticide corporations and their parent companies, the profitability and legitimacy of pesticides and new biotechnologies could not be taken for granted. By the early 1990s, the pesticide industry was in a difficult position. Pesticide corporations were constantly fighting, on the one hand, to resist the development of tighter pesticide regulation, unless it was related to cost-saving downward harmonisation of standards (Boardman 1986), while, on the other, fought to water-down and standardise regulation when it did arise. Ultimately, growing evidence linking pesticides with a range of human health and environmental problems necessitated new strategies from the pesticide industry.

Chapter 5

The evolution of the pesticide industry under Neo-Productivism (1974–2017)

5.1 Key characteristics of urban-centred industrial agro-food systems and broader economic, social and political contexts

Unlike the previous chapter, where the evolutionary context of the pesticide industry was primarily an extension of the previous Fordist Second Food Regime discussed in Chapter 2, this chapter outlines strategic responses of the pesticide industry in response to a combination of key drivers that created a Neo-Productivist/Neo-Fordist context of development. This concept echoes what McMichael (2009a,b) calls the 'Corporate Food Regime' or the 'food from nowhere regime', which he asserts emerged during the late 1980s. The term 'Neo-Fordism' has been used to describe a range of key social, political, economic, organisational and technological characteristics that define, what many have proclaimed to be, the beginnings of a new and stable regime of accumulation and mode of social regulation. Neo-Fordism embodies a range of characteristics in common with the Fordist Regime of accumulation, namely continued urbanisation; the concentration of capital; the separation of ownership and control; the mass production of standardised products; continuation of a mass consumption culture; mass advertising; and the growth of the professional, managerial and administrative 'service class'. However, as the prefix 'Neo' implies, new Fordism embodies a range of new characteristics that allegedly endow Neo-Fordism with the capacity to succeed where Fordism could not. These subtle differences include: (1) Restructuring of the labour process and (2) usurpation of the power of both national governments and organised labour in economic management.

5.1.1 Restructuring of the labour process

Although Neo-Fordism remains committed to the 'scientific management' of labour and the assembly line production of mass-produced standardised products, it does so in such a manner as to overcome the slowdown in productivity and declining profitability that allegedly signalled the end of the Fordist Regime. Neo-Fordism is suggested to overcome these problems by employing rather draconian measures.

First, Neo-Fordist principles are suggested to enhance labour productivity and therefore reduce unit costs in order to remain both competitive and profitable. To this end, the scientific management of labour and assembly line production is refined in such a way as to minimise costs and maximise productivity. It is suggested that this so-called 'lean production', 'Toyotism', 'Just-in-Time' (JIT) system is achieved by the elimination of unnecessary actions of people or equipment that do not contribute to added value. Enhanced productivity and profitability can be achieved by eliminating: over-production, time delays, multiple handling, unnecessary steps or work/procedures, unnecessary transportation, excess inventory (inputs and outputs), over-processing, production of a defective part or something that needs reworking, and by implementation of the concepts of continuous flow and consumer pull.

Second, Neo-Fordist principles directly reduce unit costs of production. It is suggested that, the centralisation of managerial control, combined with the Fordist break-up of the labour process into discrete elements, has enabled larger firms to move parts of their production processes to peripheral locations. During the 1970s, numerous manufacturing corporations relocated production in less developed countries such as Hong Kong, Taiwan, Singapore, South Korea and Brazil. The search for new locations was based on attempts to maintain productivity levels by tapping available pools of unskilled, cheap and non-unionised labour. The relaxation of employment regulations, due to instigation of neo-liberal free market economic policies, meant that labour savings could also be made in advanced economies. This was achieved through deskilling, multi-tasking, the use of temporary, part-time or contract workers, and greater flexibility for management in setting hours and tasks. Indeed, McMichael (2005) exclaims that neo-liberal policies have led to a world-scale casualisation of labour.

Third, in the search for increased profits, greater control over production and consumption was required. To this end, Neo-Fordist principles promoted concentration and integration within networks of production-consumption. This was achieved by the centralisation of bureaucratic and hierarchical managerial control with: vertical chains of command, mechanistic organisation, vertical and horizontal integration, global outsourcing of low-cost resources from the developing world by developed country's MNC, and 'structured flexibility', or quasi-vertical integration, which aims to effectively utilise both the merits of in-house production and contracting out to subcontractors.

According to Lipietz (1987), the spread of Neo-Fordist principles across the globe, albeit in an uneven and partial manner, represents an attempt to overcome the crisis of Fordism at the 'centre' – that is, in the western economies. In fact, the 'centre/periphery' distinction offers a useful way of summarising the directions of change that Neo-Fordism represents.

In the central or core western economies, new technologies are said to be creating a polarisation of the workforce. Therefore, while the highly skilled, technical elite amass advantages gained from increased automation, the majority are either deskilled or unemployed. Meanwhile, at the global periphery, routinised, labour-intensive methods of production predominate. The latter, often called 'Peripheral Fordism' (Lipietz 1987) embodying 'primitive or bloody Taylorization', should not be read as an entirely negative scenario, as this type of production has created the possibility for economic advancement among the newly industrialising countries; witness the Asian economies, for example (Lipietz 1987). However, the promotion of Peripheral Fordist industrialisation encourages the convergence of the techno-economic paradigm, comprising Taylorist and Fordist principles of production, without the corresponding social conditions that would make it possible to regulate labour relations, and without a Keynesian economic arrangement that would link workers' earnings to actual demand. This strategy is therefore called peripheral since it is Fordist in terms of production but flexible in terms of labour (Lipietz 1987).

According to McMichael, the 'Corporate Food Regime' maintained the divisions of labour and markets that solidified during the Second Food Regime. This primarily maintains the production of staple grains produced in the rich north, which are traded for high-value products, such as meats, specialty horticultural crops and off-season fruits, fruit juices and vegetables produced in the poor South (McMichael 2009a). In addition, McMichael (2013) also attributes the Corporate Food Regime with two parallel phenomena. First, the rapid increase in commodity value chains linking markets for high-value food and drink products in the North with low-cost production and processing platforms in the South, coordinated by growing corporate agro-food capitals. Second, the rapid expansion of land grabbing (by both corporations and governments alike) in order to secure bulk staples for animal feed or biofuel.

5.1.2 Usurpation of the power of both national governments and organised labour in economic management

As Chapter 3 suggested, the crisis of Fordism was not only an economic phenomenon but also a political, social and environmental one; this had two important impacts. First, confidence in the ability of developed nation states to manage the economy in a manner that satisfied the needs of the labour force and employers had been shattered. Once again, the business communities were to dissolve the post-war bargain with labour and pursue new, more profitable, economic strategies. In many developed economies, business communities were more able to pursue their aims by slowly dismantling Keynesian-styled economic and social regulation that depressed profits. Therefore, in countries such as the USA, the UK, Australia and New Zealand, governments were persuaded to pursue neo-liberal policies that allowed greater flexibility in the labour process and endowed greater operational freedom to businesses. Many nations replaced their Keynesian welfare policies with 'neo-liberal' policies to attract industry and investment and to placate the demands of international capital markets and agencies like the IMF. Under neo-liberalism, governments slashed social services as well as labour, environmental and consumer safety standards.

Second, while greater freedom and flexibility better suited the needs of business communities, footloose international corporations still required a strong international regulatory framework to ensure market stability and predictability. Thus, new compacts were made with international regulatory institutions, namely the IMF, World Bank and, most importantly, the World Trade Organisation (WTO; Atkins and Bowler 2001). Once corporate business was firmly entrenched in these regulatory institutions, its needs could be met through the pursuit of favourable economic and trade policies. Extraordinarily powerful new corporate configurations attempted to replace governments by engineering new mechanisms of monopoly control over resources and technology (ETC 2001a).

In 1999, for example, Monbiot highlighted concern over the influence of the European Round Table of Industrialists (ERT) with regard to the formulation of new EU policies (Monbiot 1999). The ERT is an alliance of the chief executives of Europe's largest companies, whose purpose is to formulate policies for adoption by the EC. According to Monbiot, to date, ERT has been astonishingly successful (Monbiot 1999). Indeed, Monbiot (1999) suggested that the Single European Act was framed not by the EC but by Wisse Dekker, the president of Philips and subsequently the chairman of ERT. Monbiot concludes by suggesting that Dekker's proposal became the basis of the EC's 1985 White Paper. According to Monbiot (1999), not only did the ERT insisted that EU expansion should follow the precise sequence agreed at Helsinki, but that all new entrants to the EU should be required to deregulate and privatise their economies and make significant investments in infrastructure designed for long-distance freight.

According to Monbiot, the ERT, and other trade bodies, has quietly been preparing for a single market with the USA, and that this Trans Atlantic Partnership aimed to eliminate regulatory barriers restricting the free exchange of goods and services between Europe and America. In practice, this would mean that once a product has been approved in one part of the trade block, it must be accepted everywhere. In this vision, an expanded Europe would become part of a single trading block with the USA, Canada and Mexico (NAFTA), which itself was expected to expand throughout the rest of the Americas and Caribbean. In addition, the US Senate had already signed a bill (the Africa Growth and Opportunity Act) forcing African countries to accept NAFTA terms of trade. At the same time, Russia and most of Asia were being obliged to deregulate by the IMF (Monbiot 1999). Monbiot concluded by suggesting that in time, the world would consist of a single deregulated market, controlled by transnational corporations (TNCs), in which no robust law intended to protect the environment or human rights would be allowed to survive (Monbiot 1999).

5.2 Key external drivers of pesticide evolution

The Neo-Productivist context of pesticide evolution arose during the early to mid-1970s, in parallel to the Residual Productivist context. Indeed, it shared many of the same drivers of Residual Productivism. These were: falling real unit prices for staple commodities and stagnant world markets; resistance to a growing range of pesticide AIs; expiry of pesticide patents; increasing pressure to reduce negative human health impacts of pesticide in Western Europe, the USA and Japan; and scientific and technological advances (see Chapter 3). However, the Neo-Productivist context of pesticide industry evolution differs from that of Residual Productivist context in several important respects. First, concern about the durability of political and financial support for high-input/-output industrial farming systems in Western Europe, the USA and Japan. Second, a reduction in production-based farm subsidies. Third, introduction of neo-liberal economic and trade policies. Fourth, increased globalisation. Fifth, industrialisation of farming systems in developing countries. Sixth, non-existent, or limited, regulation of pesticide production and use in developing countries.

5.2.1 Concern about the durability of political and financial support for high-input/-output industrial farming systems in Western Europe, the USA and Japan

The continuity of political and financial support for high-input/-output industrial farming systems in Western Europe, the USA and Japan was the dominant, and indeed, defining, driver in the Residual Productivist context of the pesticide industry's evolution. This driver was also to become an important Neo-Productivist driver. Whilst the discussion of the pivotal role of a single driver in two quite separate evolutionary contexts is a complex one, it is nonetheless central to the understanding of the strategies employed by, and the changes witnessed in, the pesticide industry. In the context of Residual Productivism, pesticide corporations visualised a future in which the continued political and financial support, achieved by relentless lobbying of key agricultural policymaking actors by agribusiness actors, could potentially extend the Fordist-styled Second Food Regime, allowing restructured pesticide companies to increase both profits and market share. This would be achieved by continuation of the relatively automatic acceptance by key actors of profitable new pesticide products and emerging biotechnologies.

In the context of Neo-Fordism, pesticide corporations visualised another potential future in which, despite the relentless lobbying of key agricultural policymaking actors by agribusiness actors, political and, more importantly, financial support for high-input/-output industrial farming systems could not be assured, and that the Fordist-styled Second Food Regime would undergo a phase of quite radical restructuring. In the Neo-Productivist context, the pesticide industry could no longer rely on the appropriation of high profits from production-based state-subsidised agriculture. Indeed, it was highly probable that, at some time in the near future, farmers in Western Europe, the USA and Japan would be increasingly exposed to free markets in agricultural commodities, and that both the consumption of pesticides, and per-unit profits on pesticides, might fall considerably. In this context, or conversely, under this scenario, the pesticide industry would need to dramatically restructure to maintain profits. Indeed, in order to substantially increase industry profits, pesticide corporations would need to both increase their share of existing markets, and, if possible, develop substantial new sales outside their traditional markets of Western Europe, the USA and Japan.

Ironically, the WTO, which is one of the central instruments of Neo-Productivism, has been described as a 'hangover from the 2nd Food Regime' rather than a 'putative successor'. If, as many suggest, the Neo-Productivist or Corporate Food System/Third Food Regime has actually materialised, it still carries with it much of the baggage from the Second Food Regime (Pritchard 2009).

5.2.2 Reduction in production-based farm subsidies

At a time when European farmers had turned a significant net EC import demand into net annual subsidised grain exports of 27 million tonnes, the need to subsidise the farm production in the EC was constantly being called into question (McCalla 1993). The issue of subsidised EC grain was brought into sharper resolve when US farmers began to lose global market share and witnessed low grain prices in the early 1980s. US farmers pressurised the US Administration to reduce subsidised EC agricultural production at their earliest convenience. Hence, liberalisation of agricultural trade was thrown into the spotlight during the Uruguay GATT talks of 1986.

However, it was not until the GATT/WTO accord of 1993 that the EC agreed to decouple (or unlink) subsidisation of agriculture and rural communities from agricultural production. This was based on the EC's own restructuring of the CAP in 1992. Indeed, in 1992, the EC had transformed traditional subsidisation from a per tonne basis to an area or headage basis. This was enough, to some extent, to plicate the US and other trade partners. However, as Pritchard (2009) puts it, 'the period from the inception of the Uruguay Round in 1986 until the Seattle meeting in 1999 represents a global politics of food in which elite northern interest strategised to create and use the WTO as a tool to preserve their own subsidy regimes, while at the same time enforcing liberalisation on the rest of the world'. Pritchard (2009) goes further by arguing, 'Northern countries have showed no desire to alter the paradigms in which their domestic food policies were framed'. Whilst it was not an abandonment of support to farmers, it brought enough uncertainty into the agro-food system. Namely, it was unsure how farmers would adjust their use of pesticide and other inputs associated with high-input/high-output agricultural production in the new subsidy system. Indeed, at the time, there was a great deal of discussion regarding the move towards lower-input and lower-output production systems, in which the use of pesticides was envisaged to be greatly reduced. See Chapter 6 for a discussion of agri-environmental reform.

5.2.3 Introduction of neo-liberal economic and trade policies

It was economic crisis in the 1970s that prompted a so-called 'New Right' liberal counter-revolution, triggered by the political hegemony of UK's Prime Minister Margaret Thatcher and the USA's President Ronald Reagan. These two important figures and their theorists became the vanguard through which laissez-faire thinking re-entered the realm of development thinking. At the Cancun Conference of 1981, Reagan and Thatcher directly attacked the Keynesian notion of a New International Economic Order (NIEO), expressing distaste for enhanced foreign assistance and commodity stabilisation (two planks of NIEO).

The call for free markets and liberal reforms was given a further push by the debt crisis of 1982 and the growing dominance of the IMF and World Bank and their Structural Adjustment Programmes. Indeed, the World Bank made deregulation and market liberalisation a condition of loan receipt. World Bank stipulations followed, what came to be known as, the Washington consensus. This stipulates that loan recipient countries adhere to the following criteria: (1) Fiscal discipline, redirection of public expenditure towards education, health and infrastructure investment. (2) Tax reform-broadening the tax base and cutting marginal tax rates. (3) Interest rates that are market determined and positive (but moderate) in real terms. (4) Competitive exchange rates. (5) Trade liberalisation-replacement of quantitative restrictions with low and uniform tariffs. (6) Openness to Foreign Direct Investment (FDI). (7) Privatisation of state enterprises. (8) Deregulation-abolishment of regulations that impede entry or restrict competition, except for those justified on safety, environmental and consumer protection grounds. (9) Prudential oversight of financial institutions. (10) Legal security for property rights.

The ideas favoured by the New Right were monetarist, in which Milton Friedman (1962) figured centrally. Abhorring forms of government regulation, they advocated a minimalist role for the state. The New Right claimed that free markets maximise human welfare: economically, markets efficiently distribute knowledge and resources; socially, liberal individualism would maximise moral worth; and politically, liberalism would maximise political freedoms since it rests on the most efficient distribution of resources and wealth (Johnston et al. 2000). By the late 1980s, Neo-liberal economic policies had spread across much of the globe (Johnston et al. 2000). Indeed, the pro-market position of neo-liberalism informed development theory and practice through the multilateral regulatory institutions of the WTO, the IMF and the World Bank (Johnston et al. 2000). The WTO, with over 122 member states, is the most powerful institution (World Bank 2000). Since 1993, the WTO (formerly the GATT) has been the principal regulatory mechanism for agriculture, food and trade. The liberalisation of agricultural production–consumption networks has been most noticeable in Australia and New Zealand.

Latterly, in response to paralysis in recent rounds of WTO talks, the Transatlantic Trade and Investment Partnership (TTIP) is seeking the following: (1) To dismantle remaining market protections. (2) To expand protections for investors over both consumers and farmers. (3) To severely restrict the ability of governments to use public policy to reshape food systems (Hansen-Kuhn and Suppan 2013). According to Hansen-Kuhn and Suppan (2013), TTIP, led by the USA, is a confidential corporate-focused initiative aimed at 'relaxing the EU's Precautionary Principle regarding food-associated chemicals (pesticides, packaging and additives), nano-technologies and genetically modified organisms (GMOs), and to regulate procurement at the possible expense of local participatory food democracy initiatives defined as "localization barriers to trade"'.

In agreement with Otero (2004), 'neoliberal globalism also depends centrally on the state and its attempt to impose the market as a self-regulating mechanism. The state, while changed in significant ways, continues to play a central role in neoliberal times'. Indeed, according to Beyond Pesticides (2017), agribusinesses have not found a need to penetrate the WTO directly. Instead, by flexing its muscles and leveraging influence at the highest political levels in the USA and Western Europe, it has been able to influence the WTO at a distance through the US Administration and EU representation. Much of this was achieved through the interchange of senior government and corporate personnel, known as *'revolving doors'*. In several cases, former government officials take up senior corporate positions. Their strong connections with government personnel ensure a clear channel through which to advocate their agenda. Conversely, many key corporate personnel move into government positions and are able to promote a corporate agenda (AHRP 2016; Cronin 2013; Beyond Pesticides 2017).

5.2.4 Increased globalisation

Globalisation is a complex concept, with varied interpretations. The interpretation used here is comprised of three dimensions (Higgot and Reich 1998). First, globalisation has a temporal dimension as an historical epoch. Thus, globalisation can be said to have emerged in the early 1970s, after the break-up of the Bretton Woods system of national economic organisation. The epoch of globalisation is one in which the world witnessed the ascendancy of a neo-liberal market-oriented approach to economic management at the expense of the Keynesian welfare model that had dominated for much of the post-World War II era. Globalisation can therefore be seen as the latest stage in the development of capitalism (UN 1998).

Second, globalisation is characterised functionally as an intrinsically related series of economic phenomena. These include the liberalisation and deregulation of markets, privatisation of assets, retreat of state functions (particularly welfare ones), diffusion of technology, cross-national distribution of manufacturing production (FDI) and the integration of capital markets. In its narrowest formulation, the term refers to the worldwide spread of sales, production and manufacturing processes, all of which reconstitute the international division of labour (Higgot and Reich 1998). While the process of globalisation focuses on the distribution of finance, production, technology, regulation and authority as indicators of change, it argues that many of these activities are not historically 'new' (Higgot and Reich 1998). Rather, their volume scope and clustering make them significant. They have never occurred with such depth, breadth and speed in the past; 1970 and beyond is not similar to the phase prior to the outbreak of World War I in 1914 (Perraton et al. 1997). Market reform and the retreat of the state may have occurred previously, but never in combination with the explosive growth of FDI, multilateral institutions and the spread of a single ideology. A corollary of this definition is that globalisation leads to convergence (Williamson 1996). While many point to the association between an open market and economic growth and suggest that with a 'growing cake' all the distribution problems will be easier to solve, others point out that this is by no means automatic (Novartis 2004). As power is shifted away from governments and towards the private sector, in particular towards MNCs and

TNCs, fears of corporate exploitation spread (International Council on Human Rights Policy 2002). Indeed, it was suggested that, market liberalisation and the standardisation of products and norms will increasingly lead to economic marginalisation of governments and the erosion of social cohesion and political consensus within national governments, reducing their bargaining strength in the international context. Most opponents claim economic globalisation lacks an ethical framework and political underpinnings (if not control) to ensure that economic performance is subordinated to essential human and social goals (International Council on Human Rights Policy 2002). Many fear that economic globalisation will precipitate a return to a type of global Manchester capitalism: when workers and governments in developed countries must compete ever more fiercely with workers and governments in developing countries. This, they fear, will lead to the worldwide erosion of social and environmental standards, more marked social asymmetries (UN 1998), even the reproletarianisation of labour, and to permanent damage to the environment (Novartis 2004). Many forecast the relative, if not absolute, impoverishment of whole groups of countries and the majority of their inhabitants (UN 1998).

Third, globalisation is characterised as the intellectual hegemony of western (Gosovic 2002), and predominantly, American values. It stresses the diffusion and assimilation of American technological capability, finance and institutions (political and economic ones in the public sector, and 'best business practices' in the private sector). In this definition, globalisation is a normatively good thing that represents the triumph of modernisation and democracy defined as industrialised economic development involving the characteristic features of a limited state apparatus, representative government and a liberal concept of freedom and choice. In this regard, current proponents of globalisation hardly sound distinct from their intellectual forebears in the modernisation tradition in American political thought according to which homogenisation of values occurs around the principles of capitalism and democracy. However, others denounce it because of the fear that it will lead to 'predatory capitalism' on a global level, with all its unsavoury features such as greed, ruthlessness and egoism (International Council on Human Rights Policy 2002).

As with the previous discussion of Residual Productivism, although US hegemony may have weakened, many institutions and regimes have either survived the decline of a hegemon or been created in the absence of one. Whilst numerous institutions have endeavoured to reconstitute or reframe the context for development after decline of the Bretton Woods System of international economic management, the three pillars of that system (World Bank, IMF and WTO) continued to play an exceedingly dominant role. As Hough (1998) suggests, both the GATT/WTO and World Bank have continued to function according to the same conditioning norms as before. The WTO and World Bank are classic examples of where the US maintains considerable influence, particularly if it can curry support from Western European nations and Japan (Hough 1998), which now rival the USA as major economic and political powers (Hough 1998). Its principal influence has been to perpetuate pre-1973 norms, or at least slow change, in a way that continues to favour US interests. This phenomenon may exist due to the USA possessing enough influence to favour development, which is in its interest (Hough 1998). Another explanation is that, while its dominant position may have weakened, the US Administration has been able to muster significant support from increasingly powerful MNCs and TNCs. Conversely, the opposite proposition, that the US Administration is dominated by corporate interests that display scant regard for anything other than the bottom line, is also plausible. The end result, however, has been the establishment of capitalist globalisation by western countries, characterised by the implementation of strategies for the global market and global governance. The global market is based on an economic system created, and increasingly controlled, by the TNCs. Conversely, global governance is legitimised by international policies imposed by the IMF, World Bank and the WTO (Pechlaner and Otero 2010).

5.2.5 Industrialisation of farming systems in developing countries

Whilst Chapter 2 outlined some of the key characteristics of processes involved in the industrialisation of farming systems in developing countries through the formative years of the mid-1960s and early 1970s, this section of Chapter 4 outlines the importance of agricultural modernisation/industrialisation in developing countries since the late 1970s for Western European, US and Japanese pesticide corporations. While the costs and benefits of pesticide use were hotly contested in the USA, Western Europe, Japan and a growing number of OECD countries, the imperative of economic growth and food security remained the dominant goals of most developing country governments.

Indeed, intergovernmental institutions and heads of key nation states repeatedly pressed the importance of these goals. For example, Boardman (1986) stated that 'pesticides were unequivocally an urgently needed tool of economic development' and that the goal was to assist developing countries in three ways. First, by obtaining the pesticides required for meeting urgent needs and for implementation of their plant protection programmes. Second, by proceeding to develop their own pesticide industries where feasible. Third, by strengthening their plant protection capabilities through training, consultant missions and supporting research activities (Boardman 1986).

Pesticides continued to be energetically promoted by the World Bank, the Consultative Group on International Agricultural Research and FAO (Plunkett 1991). For example, on the one hand, the World Bank issued guidelines in 1985 aiming to avert any damage to human health or the environment from pesticides when used in development projects it funded (Hough 1998). On the other hand, in the fiscal years 1993–5, the World Bank approved US$56.9 million worth of contracts for pesticides and agrochemicals to be supplied by France (38%), Germany (27%), the UK (15%), the USA (11%) and Japan (10%). Six companies' names are associated with US$3 million or more in World-Bank-approved agrochemical sales over the 3-year phase 1993–5. These were Rhône-Poulenc, BASF, Zeneca, Sumitomo, FMC Corp. and Helm (Germany). Another five were to receive US$1–3 million: Bayer, Roussel Uclaf, Cyanamid, Air Lloyd (Germany) and Hoechst. The company at the top of this list, Rhône-Poulenc, was the big winner in terms of sales. Between 1993 and 1995, it stood to make US$18.6 million or 33% of the value of all World-Bank-approved contracts benefiting the G-7 agrochemical industry, over US$12 million more than BASF, the first runner-up.

In addition, under policy guidance from the World Bank and IMF, and with support through agricultural projects and structural adjustment lending, developing countries were increasingly emphasising the production of non-traditional agricultural export (NTAE) crops over staple foods (Vorley and Keeney 1998). Extensive documentation indicates that the use of all pesticides (insecticides, herbicides, fungicides, nematocides) is much higher in most NTAE crops than in traditional crops. Over recent years, new input trait biotechnologies have also emerged as the manifestation of the continuing industrialisation of the agri-food systems in developing countries.

The WTO also indirectly promoted pesticide use by helping to create a context in which developing countries must reduce agricultural costs of production in order for exports to be competitive in the global market. In general, this has promoted the adoption of the industrial model of agriculture, which is dependent on chemical pesticides and the large-scale, capital-intensive farms specialising in single crops (GPC 2000). Low prices and high costs tended to force farmers to get bigger and to adopt whatever techniques and tools that promise to raise yields and/or reduce costs (Magdoff et al. 1998). At the same time, trade rules do not prevent subsidies for export and foreign investment, practices, which greatly foster larger-scale, highly pesticide-dependent agriculture (GPC 2000).

In addition to the enthusiastic endorsement of pesticide use by an assortment of UN and associated institutions, individual national governments in developing countries continued to support the use of pesticides in their agricultural development plans. The rational for the support of industrial agriculture centred on: increasing political and financial pressures to improve food self-sufficiency; the need to reduce import dependency; and conversely, the need to generate foreign income from agricultural exports.

Based on the economic importance of its rubber exports, Malaysia established its Department of Agriculture in 1905 (Boardman 1986). Pesticides were deemed as having a central role in the production of rubber. By the mid-late 1970s, pesticide production plants in Malaysia were operated by Bayer, ICI, Monsanto, Shell, Nioho Nohyako and Takeda, with involvement from Boots, Ciba-Geigy, Dow, Fisons, Hoechst and Rhône-Poulenc (Boardman 1986). By 1981, there were 113 companies registered as pesticide manufacturers or formulators. In 1974–5, approximately 135 AIs were being sold in Malaysia (Boardman 1986).

Furthermore, many governments continued to support the use of pesticides through pesticide subsidies. Repetto (1985) calculated subsidy rates for pesticide use that ranged from 20% to 90%, with an average of nearly 50%. For eight developing countries in Africa, Asia and Latin America, total pesticide subsidies were estimated at US$740 million. Pesticide support was widespread and comprehensive. Indeed, developing countries spent US$4 billion on pesticides in 1990. Applying Repetto's 50% average subsidy rate produces a tentative estimate of US$2 billion annually for pesticide subsidies in developing countries (E-council 2003). In 1989, Repetto calculated pesticide subsidies for a range of developing countries in the mid-1980s (see Table 5.1).

While the pesticide industry may not have been the instigators of the Green Revolution, it was not long before individual pesticide corporations and their industrial lobby recognised the significant market potential for pesticides in developing countries and increasingly began to act in a more proactive, organised and strategic manner.

Interestingly, renewed interest in the growth potential of developing country pesticide markets coincided with the slowdown in the growth and profitability of traditional pesticides markets in the USA and Western Europe. In 1982, IUPAC sponsored a number of international events, including CHEMRAWN II (Chemical Research Applied to World Needs) (Hough 1998) as a forum for crop protection chemicals. In a paper for the 1982 CHEMRAWN II Conference,

Table 5.1 Pesticide subsidies in selected developing countries (Repetto 1985)

Country	Subsidy (% of retailer cost)	Total value (US$ million)	Value per capita (US$)
China	19	285	0.3
Colombia	44	69	2.5
Ecuador	41	14	1.7
Egypt	83	207	4.7
Ghana	67	20	1.7
Honduras	29	12	3.0
Indonesia	82	148	0.8
Senegal	89	4	0.7

Table 5.2 Estimated losses in 1995 due to agricultural pests (US$ billion)

	Rice	Wheat	Maize
Pathogens	3.3	1.4	7.8
Insects	45.2	10.5	10.4
Weeds	43.2	14	9.3
Total	112.5	38.5	27.4

called 'Chemistry and World Food Supplies: The New Frontiers', co-hosted by IRRI, Pimentel demonstrated that 35% of the world's food crops per year were destroyed by pests and diseases prior to harvesting, plus a further 10-20% post-harvest losses (Pimentel 1983). See Table 5.2 for details. Pimentel's conclusion was that the continued use of both chemical and non-chemical pesticides was essential for food and fibre production in the world (Pimentel 1983).

As pesticide sales in the traditional markets of Western Europe, the USA, Japan, Canada and Australia began to stagnate and eventually decrease, sales opportunities for pesticides and emerging biotechnologies increased in the relatively new pesticide and biotechnology markets of Asia and South America (see Table 5.3 for details).

Pesticide consumption in developing countries increased dramatically throughout the 1970s, 1980s and most of the 1990s. By 1996, the global pesticide market was estimated to be US$30 billion, increasing to US$40 billion by 2010 (Popp 2011). The impacts of Green Revolution technologies in parts of the developing world have been substantial. The adoption of Green Revolutions technologies included significant quantities of pesticides, initially insecticides, and more recently herbicides. The fastest growing markets were in the developing world, particularly Latin America and Asia. Increased market orientation and industrialisation of agriculture in Asia led to a rapid growth in demand for pesticides in the late 1960s and early 1970s, particularly for Green Revolution cereal crops. Before 1960, pesticide use in Asia was not widespread (Pingali and Rogers 1995). By 1990, approximately 85% of agrochemicals used in Asia

Table 5.3 World pesticide consumption 1983-93 (US$ billion)

Region	1983	1993
North America	3991	7377
Latin America	1258	2307
Asia/Oceania	5571	6814
Africa/Mideast	942	1258
W. Europe	5647	7173
E. Europe	2898	2571

were imported from the UK, the USA, West Germany, France, Switzerland, Italy and Japan (UN 1990). In 1995, the production of organophosphorus insecticides in China totalled 177 900 tons, an increase of 28% over the previous year. Thailand, one of Asia's most rapidly expanding agrochemical markets, uses monocrotophos, methamidophos and methyl parathion as its main insecticides. Dramatic shifts towards the intensive use of pesticides were also starting to occur in Africa, particularly on export crops (FAO 1997). Developing country exports rapidly became an important part of the international pesticide industry (See Table 5.4 below for details).

However, it is important to note that, a significant proportion of pesticide imports into developing countries were either banned or severely restricted in their country of manufacture (Weir and Shapiro 1981).

5.2.6 Non-existent, or limited, regulation of pesticide production and use in developing countries

A predominant lack of meaningful mechanisms to ensure the protection of consumer health and the environment in developing countries created a range of accumulation opportunities for the pesticide industry. Cost minimisation, resulted in cost externalisation, as workers and ecosystems were exploited to bolster profits. First, not only did the pesticide corporations increase sales of pesticides and emerging biotechnologies in developing countries, but they were also able to do so without many of the costly restrictions imposed on

Table 5.4 Use of insecticides in selected developing countries 1986–95

Country	Insecticide use (mt)	Total ha (×1000)	Use per ha (kg)
Korea	9339	2160	4.6
Costa Rica	4655	2870	8.8
Honduras	2743	1650	1.6
Sri Lanka	394	1876	0.21
El Salvador	2060	755	2.73
Pakistan	5574	22 150	0.9
Papua New Guinea	60	60	1
Turkmenistan	1294	1450	0.9
Thailand	6010	17 370	0.35
Jordan	315	350	0.9
Colombia	3567	6200	0.5
Malaysia	1506	5800	0.3

them by the regulatory apparatus of developed country governments. This meant that pesticide corporations could benefit from the sale of: old, often hazardous, pesticides, which were either banned or restricted in their traditional markets, and; new pesticides and biotechnologies that had not yet been registered for sale in their traditional markets. Second, a general absence of human health and environmental protection regulatory mechanisms also allowed pesticide corporations to locate pesticide production plants close to their growing markets, reduce the cost of pesticide production and increase pesticide profits. Third, the absence, or weakness, of environmental, consumer and worker NGOs ensured little or no viable opposition to the production, sale or use of pesticides and biotechnologies in developing countries.

Although the situation is slowly beginning to improve, in comparison to their counterparts in the developed world, farmers and consumers in developing countries have suffered disproportionately from the negative effects of pesticide use because of the lack, or inadequacy, of pesticide regulation. In 2000, the global market value for pesticides was estimated at US$32 billion (FAO 2001), of which, around US$3 billion was destined for developing countries. However, because of the lack of adequate pesticide regulation, around 30% (FAO 2001) of pesticides marketed in developing countries, with an estimated market value of US$900 million annually, do not meet internationally accepted quality standards. These poor-quality pesticides frequently contain hazardous substances and impurities that have already been banned or severely restricted elsewhere (FAO 2001).

When the quality of labelling and packaging is also taken into account, the proportion of poor-quality pesticide products in developing countries is even higher. Labelling, often written in an improper language, fails to provide data on the AI, application and date of manufacture and safe handling of the chemical. For the consumer, the label is often the only source of product information that can guarantee a safe and effective use of the chemical. Falsely declared products continue to find their way to markets without quality control (FAO 2001). The problem of poor-quality pesticides is particularly widespread in Sub-Saharan Africa (SSA), where quality control is generally weak. The UN agencies urged governments, international and regional organisations to adopt worldwide-accepted FAO/WHO pesticide specifications to ensure the production and trade of good-quality products, and suggested that countries should make these voluntary standards legally binding. The FAO/WHO standards are especially important for developing countries that lack the infrastructure for proper evaluation of pesticide products. Pesticide industries, including producers of generic pesticides, should submit their products for quality assessment to FAO/WHO. In a memorandum of understanding signed between FAO and WHO, the two organisations agreed to cooperate in a joint programme to develop specifications for pesticides (FAO 2001).

According to Boardman (1986), pesticide regulations in developing countries lacked uniformity of approaches (Boardman 1986), and prior to the 1970s, many lacked pesticide regulations altogether (Pingali and Rogers 1995). However, even when pesticide legislation did exist, implementation could not be guaranteed. For example, pesticide legislation in the Philippines began in 1922, with supplementary legislation in 1963. Initial legislation was concerned with the assurance of adequate labelling of insecticide and fungicide products and the setting of pesticide residue limits on food. However, these measures were not implemented and existed only on paper (Boardman 1986). Indeed, it was not until 1978 that a pesticide regulatory institution was created to control the importation, manufacture, formulation, distribution, sale, transport, storage, labelling, use and disposal of pesticides (Pingali and Rogers 1995). By 1995, the pesticide regulatory institution's mandate had expanded to include pesticide registration, regulation of pesticides availability and use, handler licensing, the setting of residue limits on food and feed, the monitoring of compliance of pesticide policy, the monitoring of pesticide imports, and the design of pesticide training programmes (Pingali and Rogers 1995).

However, as with most developing countries, the capacity to execute their mandate was generally compromised. The failure of pesticide regulatory authorities to execute their mandates was often due to the complexity of pesticide legislation and substantial resource requirements (Pingali and Rogers 1995) and the corrupt nature of certain individuals within the pesticide policy development and implementation process. In India, pesticide regulations were established in 1968. However, corruption allowed, and encouraged, the widespread production and sale of some of the world's most potentially dangerous pesticides. In India, the Central Insecticide Board is empowered to deal with setting tolerance limits for residues, minimum harvesting intervals, labelling and packaging and, through the Registration Committee, applications for the registration and marketing of pesticides (Boardman 1986). Bribes to facilitate access to developing country markets have been estimated at 10% of the pesticide compound price (Boardman 1986).

5.3 Key characteristics of pesticide development processes and manufacturers

5.3.1 Corporate globalisation

Generally speaking, corporate globalisation embraces three elements: larger markets, capital mobility and greater specialisation. First, larger markets involve global rather than national or international markets. Second, capital mobility changes the emphasis of trade from production commodities to financial ones. Encouraged by fluctuating currency values, capital mobility has resulted in currency trading that surpasses trade in goods and services by more than twenty to one, with daily foreign currency transactions estimated at US$1.3 trillion). Such increased capital mobility has made economic well-being ephemeral, with poverty and prosperity alternating among different regions and localities, resulting in a kind of permanent insecurity (Daly 1999). This pandemic of insecurity is part of a mode of domination of a new type, based on the institution of a generalised and permanent condition of insecurity aiming to compel the workers to submission, to the acceptance of exploitation. Third, greater specialisation involves comparative advantage, which governs trade between nations. Although still promoted by global institutions like the IMF, in reality comparative advantage is being replaced by 'absolute advantage', by which global capitalists rather than nations compete with each other for labourers and natural resources, as well as markets, in all countries (see Fig. 5.1) (Daly 1999).

5.3.2 Increased economies of scale through global horizontal mergers and acquisitions within the pesticide industry

Whilst concentration and consolidation is not a recent phenomenon (Horstmeier 1996b), the sheer size of the companies involved, the geographic spread and the rate at which consolidation took place all greatly increased during the 1990s and more recently in 2016 and 2017. Consolidation is now a global phenomenon. During the 1990s, when world trade grew at more than twice the rate of outputs (Watson 2003), corporations were almost obliged to operate at the global level in order to remain competitive.

Subsequently, the value of global corporate M&A increased from US$462 billion in 1990 to US$2.4 trillion in 1998 (ETC 1999), and to over US$3.5 trillion in 2000 (Thompson Financial Services 2001), approximately 12% of total world economic output (ETC 2001a). Of the 100 largest economies in the world, 51 are corporations and 49 are nation states (Anderson and Cavanagh 2000). In the agro-food sector, growth of national, multinational and transnational corporations during this phase has been staggering (Watson 2003). In 1997, global M&A topped 1 trillion dollars, almost ten times the value of all takeovers at the start of the 1990s (ETC 1998); nearly three quarters of the M&A in 1997 involved US companies (PAN 1998b).

Whilst most M&A occurred between northern corporations, the number of north/south, south/north and south/south deals also increased (ETC 1998). According to the United Nations Conference on Trade and Development, 79% of all FDI to the south was in the form of corporate acquisitions. The world's top 100 TNCs accounted for approximately one-third of the combined outward FDI of their countries of origin (Greer and Singh 2000). Between 1988 and 1993, world FDI stock grew from US$1.1 to 2.1 trillion. However, even though FDI in developing countries rapidly increased during the early 1990s, northern

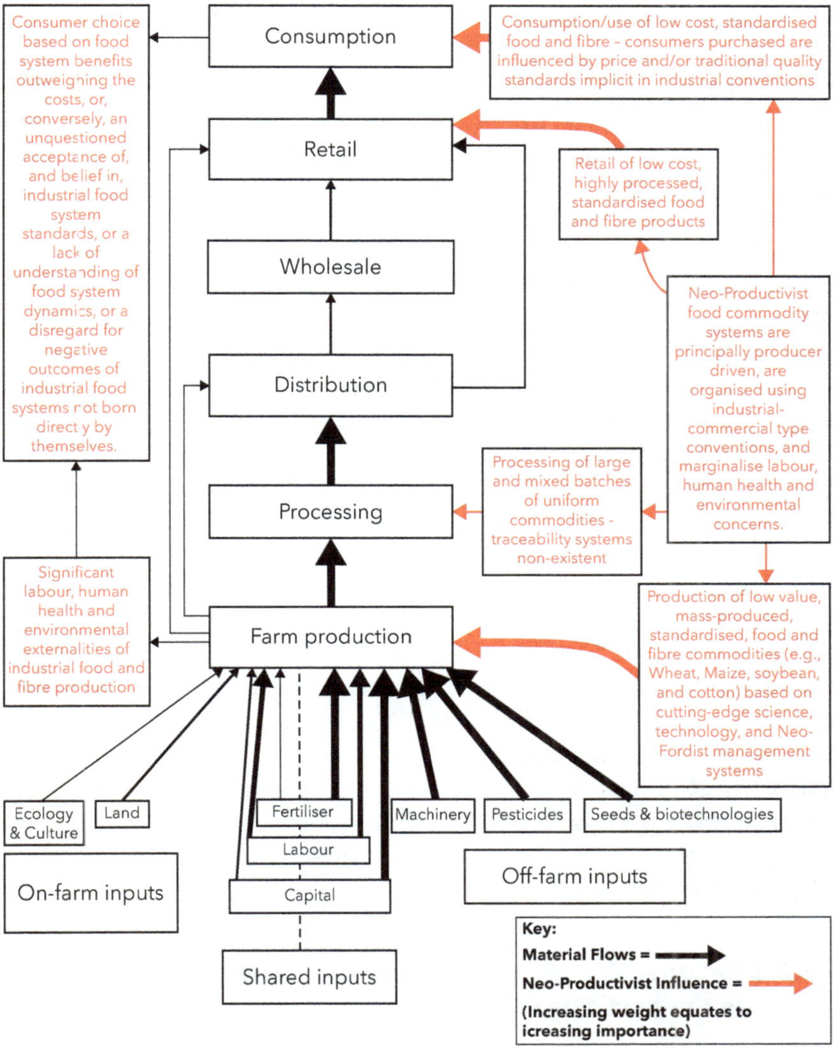

Figure 5.1 Neo-Productivist food system dynamics during the Divergent Period (1974–2017) of pesticide and pesticide industry development.

investments still dominated (UNCTAD 2003). In 1992, the world's top 100 firms had US$3.4 trillion in global assets, of which approximately US$1.3 trillion was held outside their home countries (Greer and Singh 2000). In 1994, the USA was both the largest source of outward investment (US$46 billion) and the largest inward flows (US$49 billion) (Raghavan 1996). Conversely, the Asian Pacific region, particularly China and Southeast Asia, accounted for 70% of the US$84 billion of FDI (Raghavan 1996).

According to UNCTAD's 1998 World Investment Report, FDI by TNCs reached a record US$430–440 billion in 1998 (ETC 1999). A parallel US government report suggested that 40–45% of all manufacturing sales are between subsidiary companies and parent multinationals (OTA 1994). During this time, all major pesticide companies engaged in global horizontal mergers and/or acquisitions (see Table 4.1). In 1996, the world's top ten agrochemicals corporations accounted for US$25.1 billion or 82% of the global agrochemicals market, with sales reaching US$30.5 billion (ETC 1998; PAN 1998b). In 1995, world pesticide consumption had reached 2.6 million tons of AIs, with a market value of US$38 billion (WRI 2002). By 2013, the six top pesticide companies alone had collective revenues of more than US$65 billion in agrochemicals/seeds and biotech traits (ETC 2015). Between 1984 and 1999, major US and European companies involved in the agrochemical discovery industry dropped from 43 to 17. In Japan, between 1989 and 1999, discovery companies dropped from 32 to 22 (Agranova 1999). However, Hartnell (1996) cautions interpretation of these figures suggesting that, by the early 1990s, there were only 10–15 companies with the resources to develop new AIs. Furthermore, this number was predicted to fall to as few as seven by 2010, a figure reached by the year 2000. The three other top ten companies are primarily generic pesticide producers (Bijman and Joly 2001a). In 1999, the top ten pesticide companies dominated 80% of global agrochemicals market. In 2000, 6 companies were responsible for 80% of pesticide sales, down from 12 in 1994. In 2001, the seven largest pesticide corporations account for 72% of the world market (Bijman and Joly 2001a). By 2013, the top six pesticide corporations dominated 75% of the global agrochemical market and 63% of the global commercial seed market. Table 5.5 illustrates major concentration in the pesticide industry and pesticide sales between 1994 and 2017.

5.3.2.1 Rationalising global horizontal concentration in the pesticide industry

In the early 1990s, a major structural crisis gripped the European chemical industry (Vorley and Keeney 1998). The industry reacted to this crisis with a series of structural adjustments and rationalisation measures, which resulted in considerable manpower reductions. In early 1995, the total number of personnel employed in the European chemical industry was approximately 255 000 less than in 1991, a decrease of 14%. European chemical corporations had severe cost disadvantages, namely, high prices for raw materials, labour and environmental protection. Chemical corporations in Europe have tended to push for either cost leadership or product differentiation (specialisation). Cost leadership is particularly prevalent in mass-produced plastics and base chemicals, but also in certain sectors of specialised chemistry, particularly the

Table 5.5 Top four agrochemicals companies (by sales in million US$) in 1994, 1998, 2006 and 2009, ranked by 2009 sales

Company name in 1994	Agrochemical sales 1994	Agrochemical sales 1998	Agrochemical sales 2006	Agrochemical sales 2009	Company name in 2017
Bayer	1962	2230	6120	7544	Bayer
AgrEvo	2054	2267			
Rhône-Poulenc	1796	2042			
Monsanto	2130	3480	3180	4427	
Syngenta	-	-	6030	8491	ChemChina
Zeneca	2078	2884			
Ciba-Geigy	2966				
Sandoz	1005				
Makhteshim Agan	-	-	-	2082	
BASF	1262	1946	4141	5007	BASF
American Cyanamid	1650	2141			
Dow AgroSciences	1737	2185	3368	3902	DowDuPont
Rohm & Haas	439				
DuPont	2140	2185	2211	2403	

production of generic pesticides (Chemische Industrie 1989). In recent years, there has been a further concentration in the market, and global companies such as Novartis, Merck, Monsanto, DuPont, Hoechst, Bayer, Rhône-Poulenc, Sumitomo and others have stratified their output, often by selling their chemical business and investing in biological sciences. Pesticide corporations and industry analysts have suggested that global pesticide industry concentration has been in search of: (1) Economies of scale; (2) Increased global market share; (3) Access to new pesticides and biotechnologies and important R&D pipelines; and (4) Increased control of pesticides in global agro-food systems.

5.3.2.1.1 Economies of scale

As in the Residual Productivist context, pesticide corporations could achieve economies of scale either by expanding their own sales or by merger with, or acquisition of, other pesticide corporations. However, many pesticide corporations were concerned that, even with the restructuring and consolidation of operations in Western Europe, the USA and Japan, they would still struggle to square high R&D costs with pressure on sales. By 1996, the average cost of developing a new pesticide was US$130 million (Hartnell 1996). Between 1971 and 1995 in the USA, the pesticide regulatory process was estimated to cost the pesticide industry US$7 billion, equivalent to 7.4% of sales value (Vorley and Keeney 1998). In addition, the pesticide industry, food industry and farmers spent US$10 billion in compliance with government pesticide regulations. By 2015, the cost of developing a new pesticide had skyrocketed to US$256 million, with regulatory compliance estimated at 35% of this cost (ETC 2015).

With the exception of two price spikes linked to two short-lived food crises, since the late 1970s, world crop prices have continued to fall in real terms and are currently at historic lows. Subsequently, this had a negative impact on both farming incomes and pesticide sales. Due to over-production and rock bottom commodity prices, global agrochemical sales fell by 0.6% in 2000, the second year of decline (ETC 2001a). Indeed, sales in the main pesticide markets in the USA, Western Europe and Japan remained virtually unchanged since the early 1980s. Growth in value between 1993 and 1998 was estimated to be 4% per year in the USA and 4.6% in Western Europe (Vorley and Keeney 1998). The search for greater economies of scale through global consolidation was a virtual necessity. M&A were expected to generate a competitive advantage and maintain, or even increase, profitability (Vorley and Keeney 1998), due to the attainment of lower costs, and to increase the relative disadvantage for those firms choosing not to expand (King et al. 2002; Shy 1996). While the search for economies of scale became commonplace, particularly by the end of the 1990s, as corporations differed in their specific context and associated room for manoeuvre, each corporation's search for economies of scale was also different. Throughout the past 40 years, the search for economies of scale intensified. Whilst M&A of relative small pesticide corporations, based in both the developed and developing world, have assisted what have almost become the elite of the pesticide industry to achieve cost-efficiencies, it has been the subsumption of some of the leading pesticide manufacturers that have made the most significant contributions to the attainment of economies of scale.

One of the first examples of the mega M&A occurred in 1986 when Du Pont (USA) acquired Shell Agrochemical Company's (UK/NL) US-based component, allowing the company to gain significant benefits through economies of scale. Analysts suggest that increased economies of scale helped to secure a higher rate of growth than its European competitors throughout the 1990s (AON 2001). Eight years later in 1994, Schering and Hoechst merged to form AgrEvo. According to Bijman (2001a), AgrEvo was specifically established to capture economies of scale in marketing and R&D. Indeed, Heissler (1995) suggested that this merger had become a necessity in order to survive in an increasingly competitive pesticide market (Heissler 1995). The merger facilitated a cost-saving reduction in its staff from more than 9000 to 8000 (Heissler 1995).

One of the most significant mergers in the 1990s was that between Ciba-Geigy and Sandoz in December 1996 to form Novartis (2003b). Savings were expected to come from shared promotional activities and reduced personnel. Ciba-Geigy and Sandoz CEOs predicted savings of US$1435 million in 3 years, including US$335 million in the agribusiness sector (Vorley and Keeney 1998). At its creation, Novartis was estimated as having approximately US$80 billion in assets (Ciba-Geigy 1996) larger than the 1997 GNP of 93 of the world's developing nations (World Bank 1999). In 1996, Novartis invested US$2.1 billion, US$675 million in agribusiness, and had 90 000 employees in more than a 100 countries (Syngenta 1997). In the year following the merger, Novartis reduced its costs by shedding approximately 3000 employees. In 1997, Novartis attempted to attain greater economies of scale by acquiring Merk & Co.'s (USA) Crop Protection business in May 1997 for US$910 million (Syngenta 1997). In 1996, Merck had sales of US$200 million (Syngenta 1997).

The penultimate year of the 1990s witnessed a flurry of key pesticide corporation M&A, namely those between Rhône-Poulenc and Hoechst, Dow AgroSciences and Union Carbide, and Monsanto and Pharmacia. The merger between Hoechst and Rhône-Poulenc in 1999 resulted in the formation of Aventis, which was expected to lead the world in health care and agriculture (Hoechst 1999). Savings were made through approximately 6300 staff reductions. However, according to Hoechst, these reductions were mainly due to early retirement and attrition (Hoechst 1999). In 2000, Aventis generated group sales of Euro 22.3 billion and employed around 92 500 people in its Pharmaceuticals and Agriculture businesses (Aventis 2001). Aventis CropScience employed approximately 15 300 people in more than 120 countries worldwide and generated approximately 4 billion euros sales in 2000. In the same year, Dow AgroSciences (formerly DowElanco) acquired Union Carbide. Both Dow and Union Carbide had already reduced staff numbers considerably during the 1990s. Dow reduced their workforce from 62 100 in 1990 to 39 000 in 1998, whilst Union Carbide reduced their workforce from 37 756 in 1990 to 11 627 in 1998. According to food system analysts, the acquisition of/merger with Monsanto had been expected for some time. Indeed, according to Reuters, there had been talk of a potential merger between Du Pont Co. and Monsanto Co. (Reuters 1999a). However, it was to be the giant pharmaceuticals corporation Pharmacia that eventually acquired Monsanto in 1999. By this time, Monsanto had already restructured and shed 5% of workforce (700 people) from its 14 600 employees (NGIN 2002).

The next key acquisition was BASF's purchase of American Cyanamid from American Home Products in 2000 for US$3.8 billion. BASF outbid Bayer, Dow AgroSciences and Sumitomo, at well above the rumoured asking price of US$3 billion (Moore 2000). According to Wield (2001), whilst BASF finished the 1980s in buoyant mood, the 1990s had been a difficult period. Plastics and fibres as well as agriculture were hard hit. As a result, BASF reduced its staff from 134 647 in 1990 to 105 954 in 1998. Analysts suggest that BASF began recovery from the European recession in 1994 (Wield 2001), and that the company's acquisition of American Cyanamid was undertaken in order that it could play a future lead role in the pesticide and biotechnology markets. Indeed, BASF planned to invest approximately US$700 million in plant biotechnology R&D over the ensuing decade (Successful Farming 2000). BASF's strategy can be characterised as 'Verbund'. Verbund has both a narrow position and broad path conceptualisation. The narrow focuses on economies of scale, while the broader focuses on network integration (Wield 2001). BASF said at its annual press conference that it booked integration costs with American Cyanamid at 344 million Euro in 2000. However, according to industry analysts, BASF's acquisition of American Cyanamid did not generate the expected synergies and concomitant savings. It is possible that because both BASF and Cyanamid had already streamlined their businesses that there was little scope for further significant economies of scale to be made. Cyanamid had eliminated 700 jobs, approximately 13% of its workforce during its last bout of restructuring (Moore 2000). Whilst relatively unchanged in 2017, according to pesticide industry analysts, BASF's pesticide and biotech components are becoming increasingly vulnerable to take over from other food system actors, especially machinery giants such as John Deere (ETC 2016).

One of the most important mergers occurred in 2000, when Zeneca Agrochemicals merged with Novartis' Agribusiness Division (Crop Protection and Seeds Sectors) to form Syngenta. This merger created the world's largest pesticide corporation, with annual sales of about US$8 billion (Chataway 2001). Syngenta was expected to benefit significantly from annual cost savings of US$525 million. Syngenta hoped that this significant saving, when combined with a broader range of innovative crop protection technologies and increased financial resources, would enable the company to survive the trans-Atlantic farming slump until a recovery arrived. However, the Zeneca and Novartis merger was overshadowed in April 2017, when ChemChina, which had previously blasted its way onto the global pesticide market through acquisition of Makhteshim Agan Industries (Israel), the world's seventh largest pesticide manufacturer in 2011 (ETC 2015), shocked the world by its acquisition of Syngenta for US$43 billion (Greenaway 2017).

While Dow AgroSciences' rate of growth throughout the 1990s was higher than their European competitors, concern grew that after recent consolidations in the European pesticide sector, their long-term future in the agrochemicals sector was in doubt (AON 2001). In order to maintain a critical mass in the pesticide and biotechnology sector (South West Farm Press 2003), Dow AgroSciences LLC acquired Rohm and Haas' agrochemicals business in 2001 for approximately US$1 billion (AON 2001). As with most pesticide corporations,

Rohm and Haas had been obliged to restructure operations, eliminating 1200 of its 18 000 strong staff, because of the poor sales performances in the 1990s (Chemical and Engineering News 2001). Annual sales were projected to reach approximately US$3 billion (AON 2001). Under terms of the agreement, Dow AgroSciences acquired Rohm and Haas' fungicide lines, insecticides, herbicides, other product lines, trademarks and license to all agricultural uses of its biotechnology assets (AON 2001). In 2000, Rohm and Haas' Agricultural products sales were US$531 million (AON 2001). According to Dow AgroSciences' CEO, this agreement continues to advance the company's strategy of value-added growth through mergers, acquisitions, alliances and internal R&D capabilities (AON 2001). Analysts feared that, after consolidations in the European pesticide sector, particularly the Zeneca-Novartis and Bayer-Aventis deals, both DuPont's and Dow's long-term future in pesticides and biotechnologies was in doubt (AON 2001). After growing faster than their European competitors in the late 1990s, Dow and DuPont were hit hard by the downturn in farming in 1999, and were expected to take a serious look at acquiring Monsanto, which had sales of US$5.5 billion. DuPont had already been obliged to shed over 12 000 staff from its 113 235 strong workforce in 1990 to just 101 000 in 1998, allegedly saving US$3 billion (Agrow 1995b). However, it was not until August 2017 that Dow and DuPont announced that the American plant science giants had agreed to merge (Agrow 2017).

However, probably the most significant strategic acquisition, in terms of perceived future industry profitability, has been the purchase of Aventis CropScience by Bayer in 2002. Bayer ended the 1980s in confident mood, with attention focused on health, nutrition, polymers and global expansion. Whilst the health sector continued to provide good returns, the early 1990s saw the chemicals sector falter. Its chemical businesses were only able to maintain market positions by making significant price concessions. Cost-cutting exercises were undertaken in plastics, fibres and other underproducing sectors. However, as with most other pesticide and chemical corporations, Bayer reduced its staff from 171 000 in 1990 to 145 000 in 1998.

Conversely, Aventis had suffered from falling sales since its inception, and faced problems with Balance, its corn herbicide, and the damaging StarLink affair, where GM corn not licensed for human use was found in a range of food products. Subsequently, in November 2000, Aventis announced its intention to divest Aventis CropScience by the end of 2001 in order to focus on its pharmaceutical business, deciding to sell rather than seek a separate listing for its unit. Aventis approached Monsanto Co., E.I. DuPont de Nemours Co., Dow Chemical Co., Bayer AG and BASF AG as potential buyers. Aventis S.A.'s plan was to focus on faster-growing pharmaceuticals and to reduce its debt (Financial Times 1998b). Aventis was expected to reap 3.7–3.8 billion euros from the deal, a source close to the transaction said (Financial Times 2001).

After reviewing different divestment options, Aventis continued exclusive negotiations with Bayer. This agreement represented a decisive step in the strategic shift of Aventis into a pure-play pharmaceutical company. For Aventis, the expected decrease in the net debt - through the proceeds from the divestment as well as the deconsolidation of debt associated with CropScience - would mean additional flexibility for the future development of its pharmaceutical business. Its drug business had one of the fastest rates of earnings growth in the sector, boosted by cost savings from the Hoechst-Rhône-Poulenc merger and increasing margins as it put more marketing muscle behind leading products. Aventis also intended to pay down debt from €11.5 to 5 billion. Schering, expected to earn €1.3 billion from the disposal of its 24% stake in the concern, was expected to use the funds to expand its presence in the USA and to acquire products to deepen its drug development pipeline. The company, best known for the contraceptive pill, was keen to expand its emerging cancer business.

Bayer's acquisition of Aventis was deemed as crucial in order to gain a critical mass and strong market position in the pesticide market, particularly in the USA (Financial Times 2001). With some 122 000 employees worldwide at the end of 2000, the Group recorded net income for the year of €2.2 billion on sales of €27.3 billion. Capital expenditures in 2000 amounted to €2.6 billion, R&D expenses to €2.4 billion. The acquisition of Aventis was expected to yield €500 million in annual synergies, requiring a one-time restructuring charge of the same amount. Substantial savings were expected to arise from an estimated 4000 redundancies, or 15–18% of the combined workforce (Financial Times 2001). Even though Bayer's debt burden was projected to increase from €7.8 to 15 billion in 2001, earnings were expected to increase post-2004. Indeed, Bayer predicted sales of the new Bayer CropScience unit to reach €8 billion, with an operating margin of 20%, by 2005. Spectacularly, further consolidation took place on 14 September 2016 when Bayer acquired Monsanto for US$66 billion (Greenaway 2017).

5.3.2.1.2 Increased global market share

Stagnant markets in developed countries encouraged pesticide corporations to rationalise industry capacity (Shy 1996), and look to developing export markets and sales in developing countries, particularly Latin America, Asia and the Middle East to increase returns on investments (Pan-UK 2003a; Vorley and Keeney 1998). According to the Bull (1982), in the period 1974–7, pesticide imports in developing countries grew by 15% (Bull 1982). At this time, developing countries accounted for approximately 15% of the global pesticide market (Bull 1982). As many developing countries did not have the capacity to develop and manufacture pesticides at that time, most pesticides were imported from Europe and the USA. For example, over the period 1975–9, UK pesticide exports to developing countries increased by 211%. In 1979, over 66 000 tonnes of pesticides, worth over £92 million, were exported from the UK to developing countries (Bull 1982). At this time, EC pesticide corporations were responsible for 61.5% of pesticide exports to developing countries, 4.5 times the value of pesticides exported by US pesticide corporations (Bull 1982). However, a global market for pesticides was not only desirable for pesticide corporations but, as time progressed, became increasingly necessary. As Bijman and Joly (2001a) suggest, the cost of developing new compounds had risen so much that a global market for pesticides was required in order to earn back R&D investments. Vorley and Keeney (1998) estimated that the agrochemical industry spent nearly US$3 billion per year on R&D and that a critical mass of around US$150 million (approximately 10% of sales) is required for a successful R&D programme. They flagged the Schering and Hoechst merger to form AgrEvo as a good example of pesticide corporations attempting to reach such a critical mass (Vorley and Keeney 1998).

Moreover, given such high costs, new compounds were primarily only developed for major global crops such as maize, wheat, soybeans and cotton grown in both developed and developing countries (Bijman and Joly 2001a). Alongside new biotechnologies, often reformulated older AIs remained a key part of their product range (Pan-UK 2003a). Since the late 1970s, corporate agrochemical companies in Western Europe and the USA sought to achieve additional and increased rates of sales outside their traditional markets, as well as securing a larger percentage of traditional markets. They competed for a significant share of the global pesticide market worth US$11.7 billion in 1980 (Vorley and Keeney 1998), US$32 952 billion in 2000 (Lerner 2001), US$44 billion in 2009, US$52 billion in 2014 (ETC 2011) and a predicted US$81.1 billion by 2021 (PR Newswire 2016).

Most pesticide corporations made substantial investments in the development of new production plants close to main trading areas in Asia and South America, and located R&D closer to consumers, and areas more favourable to R&D, such as the USA with gene technology (Lerner 2001). The availability of low-priced raw materials and the attainment of a high market share are vital to achieving cost advantages. More than 50% of firms pursued the goals of reducing their wage costs and material consumption. Export expansion was one way to deal with the problems of a maturing industry (Boardman 1986; Vorley and Keeney 1998). However, with the exception of Sumitomo and Nissan, most Japanese pesticide companies have remained focused on the national market. Kumiai, Sankyo, Nihon Nohyaku, Takeda and Hokko all sell between 83% and 97% of their products nationally (Pesticide News 2000). In 1998, Sumitomo, the largest Japanese pesticide corporation (ranked 12th globally) had sales of US$675 million. Nissan had sales of US$287 million. North America and Europe accounted for over 50% of their overseas sales (Pesticide News 2000). In 1984, regional agrochemical sales were: North America 30%, Latin America 13%; Western Europe 25%; Eastern Europe 2%; Asia 16%; Japan 12%; Africa 2% (WRI 2002). In 1999, regional agrochemical sales were: North America 29.4%, Latin America 15.3%, West Europe 22.4%, East Europe 2.9%, East Asia (including Japan) 25.3% and the rest of the world (Africa, Middle East, India, Turkey, Pakistan) 4.7% (Pesticide News 2000).

Top pesticide corporations were increasingly involved in the global expansion of R&D, production and sales. For example, Bayer led the development of cotton pesticides from the 1950s, with production facilities in Mexico, Guatemala, Colombia and Brazil (Søgaard 2001). Sales to North America and the Asia/Pacific region grew quickly during the 1970s, and by 1987, 78% of Bayer's customers were outside Germany (Bayer 2003e). According to Søgaard (2001), by the beginning of the twenty-first century, Bayer employed 120 000 people in 350 companies in 150 countries, and intended to continue its expansion into Asia, initially China and Thailand, and then to consolidate its position in Japan and Latin America. Bayer also expanded its hold on rubber by acquiring the Canadian Polysar Rubber Corporation in 1990, making it the world's biggest supplier of raw materials for the rubber industry (Bayer 1988). Bayer has had a pharmaceutical-related R&D institute in Kansai Science City, Japan, since 1995, and a crop protection research centre in Yuki (Søgaard 2001). The Japanese pharmaceutical R&D institute marked the basic completion of Bayer's Europe/North America/Japan pharmaceutical research triad (Bayer 1988). Bayer also owes much of its global coverage to its component subsidiaries such as AgrEvo. In 1998, AgrEvo's regional sales breakdown was: Europe 37%, North America 19%, Asia Pacific 18%, International 6% and others 6% (Hoechst 2003b). In 1994, AgrEvo held about 9% of the world market in agrochemicals; 93% of its sales were exports (Heissler 1995). European and North American markets were responsible for nearly 66% of its sales. However, sales in Latin America (mainly Brazil) and East Asia (Japan, China, Korea and Taiwan) were also important. Africa was of minor importance (Heissler 1995). Allegedly, the initial alliance between AgrEvo and Rhône-Poulenc (Aventis CropScience) resulted in cost and growth synergies, and combined distribution strength and greater market penetration for their key brands (Domain-B 2003). However, it also created the largest crop protection company in India with a combined market share of 14%, a field force of 496 (including 131 territory managers) promoting 30 brands through 2000 distributors (Domain-B 2003).

According to Novartis (Syngenta), one of the principal aims of its acquisition of Merck & Co., Inc.'s crop protection business was to achieve revenue enhancements and extend Novartis' geographical coverage (Syngenta 1997). In May 1998, Novartis Crop Protection AG acquired OCI in South Korea, which, according to Novartis, gave them 'direct access to the South Korean crop protection market, which is the 8th biggest national market in the world' (Syngenta

1998). Until recently, international R&D-based companies had developed and marketed their products through local formulators and distributors. Syngenta stated that acquisition of OCI offered them 'the best opportunity to access this market and to capture its dynamic growth potential' (Syngenta 1998). This included a production plant and distribution network.

By the 1980s, BASF had become a TNC focused primarily on Europe, the USA, Latin America and the Far East, with primary production in Europe. However, by the 1990s, BASF was extremely active in building production capacity in Korea, China and Eastern Europe (Wield 2001). Chemical sales quickly rose in the Asia Pacific region during the mid-1990s, and the acquisition of American Cyanamid was expected to give BASF a stronger presence in North and South America, Asia (Moore 2000) and strengthen its position in Europe (Successful Farming 2000), as well as strengthen its herbicide and insecticide components (Moore 2000). BASF acquired Sandoz US and Canada Corn in response to the growth in HT crops (Wield 2001).

US companies have also invested in developing country markets. For example, according to the Ground Up Organisation, by the turn of the twenty-first century, DuPont had major herbicide production plants in Australia, Indonesia, France and the USA; major fungicide production plants in France, Italy, Taiwan, Thailand and the USA; and major insecticide production plants in India, France, Italy, Mexico, Philippines, Taiwan, Thailand and the USA (Groundup 2003a).

5.3.2.1.3 Access to new pesticides and biotechnologies and important R&D pipelines

In addition to increasing economies of scale and the search for new markets, the pesticide and biotechnology industry has increasingly witnessed acquisitions and mergers in the search for more balanced product portfolios, and access to promising pesticide and biotechnology R&D pipelines (Vorley and Keeney 1998). According to Søgaard (2001), the acquisition of key products has been directly associated with attempts to exercise greater control over the global pesticide market. For example, Bayer's acquisition of Aventis CropScience was based partly on its broad and balanced portfolio, making it able to ride out regional volatility, as well as a strong pipeline of promising products. Importantly, Bayer acquired exclusive right to market products based on Novartis' AI cyproconazole in the EU. The FLINT line of cyproconazole fungicides are allegedly superior to currently available fungicides and the product covers a broad range of crop applications. FLINT has already been registered in 35 countries, including important markets like the USA and the UK; its global launch was expected to be completed by 2003. Bayer expected to exploit the full blockbuster potential of the FLINT® product line within 3 years. FLINT® will ideally complement Bayer's top-selling Folicur® fungicide line. In 2001, Bayer also acquired MIKADO, a maize herbicide, from Syngenta (Søgaard 2001). Finally, in November 2000, prior to its acquisition, Rohm and Haas acquired Zeneca's acetochlor herbicide (Lerner 2001).

5.3.2.1.4 Increased control of pesticides in the global agro-food sector

Another, and exceedingly influential, driver of horizontal integration is the benefits that can accrue through increasing control of pesticides in the global agro-food sector and the protection of valuable assets (Shy 1996). There are two important facets of corporate power in the global pesticide market.

First, due to the high level of sunk costs involved in pesticide R&D, significant amounts of capital are required to enter the search for new pesticide AIs. Thus, in effect, large sunk costs act as a barrier to entry (King et al. 2002). Historically, new entrants to the pesticide industry have acquired their stake through the acquisition of, or merger with, an existing pesticide corporation or pesticide division of a large diversified company. Alternatively, it has been possible for corporations to enter the pesticide market through the production and sale of generic (off-patent) products. Over the past 30 years, several companies, such as Nufarm, Platform Speciality Products and Makhteshim Agan (now part of ChemChina), have utilised this strategy to break into the pesticide market (ETC 2011).

Second, alongside significant barriers to entry, concentration in the pesticide industry has facilitated the evolution of giant global pesticide corporations with significant political and economic power. Since the late 1970s, the pesticide corporations of Western Europe and the USA increasingly operated at a global level, in an attempt to maximise revenues to corporate shareholders and high-level corporate managers. In order to accomplish this, TNCs roamed the world searching for opportunities that would yield the highest returns on capital investments. Not only were pesticide corporations keen to exploit the poor labour and environmental regulations in developing countries, they were also prepared to use their political and economic power to influence future pesticide and biotechnology regulation in ways that did not impede negatively on their profits or, indeed, to promote regulation that increased their revenues.

As Daly (1999) suggests, the globalisation of private enterprise resulted in a greater proportion of world product being produced under conditions that are most successful in externalising costs. 'It is this combination of ever higher profits with ever lower costs that creates the incubus that is corporate globalization. Under its agenda, environmental damage, child labour abuses or hazardous workplaces are externalities that must not affect the constant search for the lowest costs possible'. Globalisation increasingly facilitates corporations in their attempts to incorporate whole societies and regions as low-cost sites of production. According to the ETC Group, by 2001, the top ten agrochemical corporations accounted for 84% of US$30 billion world market (ETC 2001a). Due to recent mega M&A, by 2017, just three agrochemical corporations controlled 70% of the global pesticide market (Greenaway 2017).

However, academic, NGO and government analysts have recognised concentration within the pesticide industry as potentially worrisome. Indeed, the creation of Novartis initiated concerns in the US Federal Trade Commission (FTC) that the merger would substantially reduce competition in the US$1.4 billion US market for maize herbicides. Because of this concern, Sandoz was required to divest its US maize herbicide business to BASF for approximately US$780 million. This divesture is one of the largest ever requested by the US FTC (Ciba-Geigy 1996). At the time of the merger, Ciba had 35% of the maize herbicide market and Sandoz had 10% (Ciba-Geigy 1996). In addition, Novartis was also required to divest its FLINT® line of fungicides for US$782 million (Lerner 2001), including the rights to the sale of straight cyproconazole. The FLINT® purchase included global ownership of all associated IPRs, registrations and trademarks as well as production and formulation know-how. With the acquisition of Aventis CropScience, regulators were also expected to look closely at the implications of Bayer's position in insecticides and sugar beet herbicides (Financial Times 2001).

5.3.3 Increased control over the agro-food sector through global vertical mergers and acquisitions

In line with rapid global horizontal integration in the pesticide industry during the 1970s, 1980s and 1990s, global vertical integration of pesticide, seed and biotechnology corporations also took place at a rapid rate (Vorley 2002). See Table 5.6 for details. See Table 5.7 for global seed sales in selected years between 1995 and 2009.

5.3.3.1 Rationalising acquisitions, mergers and developments in global seed and biotechnology industries

In parallel to the explanation of horizontal integration in the pesticide industry, vertical integration through acquisitions, mergers and new developments in the global seed and biotechnology industries was again primarily driven by the growing internal demands of pesticide company shareholders for high rates of return on investments. Pesticide companies invested in global seed and biotechnology companies based on five key assumptions. The first assumption is that higher profits can be expected from synergies and cost savings between seed, biotechnology and pesticide businesses. This assumption is in accordance with that used in Chapter 3 to justify vertical integration in the

Table 5.6 Key global acquisitions and mergers in the pesticide, seed and biotechnology industries 1974-2017

Company name	Date of merger/acquisition and company/companies involved
Bayer	Bayer, as AgrEvo acquired: Agro Stefes in 1995; the rice breeding programme and associated assets of Granja 4 Irmaos (Brazil); ProAgro Group (comprising ProAgro Seed Company Ltd., ProAgro Seeds-PGS India Ltd., Hybrid Rice International, and MISR Hytech); Sementes Ribeiral (Brazil); Sementes Fartura (Brazil) and Mitla Pesquisa Agricola in 1999; Aventis in 2002; Gustafson 2004. AgrEvo acquired Misung (Korea) in 1994. Hoechst acquired a 51% stake in Bilag India in 1999. Chunjin (Korea) was acquired by Rhône-Poulenc in 1998.
Dow AgroSciences LLC	Dow AgroSciences acquired Morgan Seeds (Argentina), and Dinamilho Carol Productos Agricolas (Brazil) in 1998; Hibridos Colorado (Brazil); FT Biogenetica de milho (Brazil), and established Advanced AgriTraits.
DuPont (Pioneer)	DuPont acquired Sementes Dois Marcos (Brazil) in 1999. DuPont acquired Sentrachem (South Africa) and AO Khimpron (Russia) in 1997 and Nocil Crop Protection (India) in 2000.
Monsanto	Monsanto acquired Agracetus, and Monsoy (Brazil) in 1996; Sementes Agroceres (Brazil) in 1997, and; First Line Seeds (Canada), and Mahyco (India) in 1998; and Delta Pine Land Co., and subsidiaries in 1999.

seed, biotechnology and pesticide sectors in Western Europe, the USA and Japan. Major biotechnology companies increasingly purchased seed companies as a source of seed material in which to insert the genes for herbicide or insect resistance (Kalaitzandonakes and Hayenga 2000). Second, and most importantly, the pesticide industry assumed that the subsumption of globally dispersed biotechnology and seed companies would provide controlled, and

Table 5.7 Top nine seed companies (by sales in US$ 000 000) in 1995, 2000, 2006 and 2009 ranked by 2009 sales

Company name	Seed sales 1995	Seed sales 2000	Seed sales 2006	Seed sales 2009
Monsanto (Delta & Pine Land, Dekalb Plant Genetics and Seminis)	-	1600	4028	7297
DuPont (Pioneer)	1500	1938	2781	4641
Syngenta (Novartis and Advanta)	900	958	1743	2564
Limagrain	525	622	1035	1252
Land O'Lakes	-	-	756	1100
KWS	315	332	615	997
Bayer (Aventis)	-	267	430	700
Dow AgroSciences	-	-	-	635
Sakata	300	-	401	491

highly profitable, access to extensive seed production and distribution networks through which their new biotechnologies, and pesticide/biotechnology packages, could reach growing markets in the rest of the world (Goodman and Redclift 1991). Third, as opposition to pesticides increased, particularly in Western Europe, new markets in developing countries offered less resistance to the production, and use, of pesticides and new biotechnologies. The fourth assumption was that increased subsumption of biotechnology and seed companies would facilitate greater control over food systems, particularly in developing countries, and, therefore, higher profits. The final assumption was that in order to achieve high returns on investment, IPRs associated with the development and sale of new technologies must be protected.

There were several reasons for pesticide corporations to invest in biotechnologies (Bijman and Joly 2001a). First, the slowdown in market growth, at least in developed countries, caused firms to reconsider their strategies. This slowdown was due to the end of production growth in agriculture and to low prices for agricultural products forcing farmers to economise on plant protection. Second, societal concerns about the negative environmental impact of chemical plant protection products, expressed by strong environmental NGOs, led to more stringent environmental policies. Third, the potential of biotechnology opened new opportunities for integrated crop protection methods. Thus, market, regulatory and technology developments provided incentives for the pesticide industry to look beyond the traditional focus on AI development (Bijman and Joly 2001a).

Whilst seed companies have also invested in biotechnology, the extremely high cost of R&D, concerns over public acceptance, patent concerns and impacts of regulation have severely limited their involvement (Bijman and Joly 2001a; Ollinger and Fernandez-Cornejo 1998). Between 1999 and 2000, the number of publicly traded biotech companies decreased due to the loss of 33 biotech companies through consolidation (Lahteenmaki and Fletcher 2001). As early as the early 1980s, companies like Sandoz, Ciba-Geigy, Shell, ICI/Zeneca, Hoechst and Rhône-Poulenc acknowledged the potential of biotechnologies for their innovation strategies (Bijman and Joly 2001b). Some pesticide manufacturers, however, such as Shell, divested their agricultural and biotechnology interests, while others, such as BASF and Bayer, have only recently invested in the biotechnology sector.

5.3.3.1.1 Access to new markets through extensive seed production and distribution networks

In addition to the perceived benefits of national vertical integration of pesticide, biotechnology and seed companies, international M&A have been used as a vehicle to attain perceived benefits across the globe. Indeed, in 2000, cross-border sales accounted for 35% of all M&A (ETC 2001a). Most of the top ten pesticide and biotechnology corporations initiated international M&A throughout the 1990s. While companies on both sides of the Atlantic have invested considerable amounts of capital into biotechnology research, when it comes to the global development and sale of new biotechnologies, Monsanto, to date, is still the dominant force. However, much of Monsanto's success has been achieved through the subsumption of biotechnology and seed companies within the Monsanto group.

In the mid-1990s, Monsanto began making significant investments in biotechnology and seed companies in non-traditional markets. In 1996, Monsanto acquired Cargill's (USA) international seed division, Agracetus (Brazil) and Monsoy (Brazil). This gave Monsanto easy access to existing seed marketing networks in Asia, Africa, Central and South America and Europe (Groundup 2003j). In 1997, Monsanto strengthened its biotechnology capacity through the acquisition of Sementes Agroceres (Brazil). This acquisition gave the company access an estimated 30% share of the Brazilian maize market, (ETC 1998) and, with its purchase of Cargill's international seed division, it controlled over 50% of Argentina's maize seed market (Fulton and Giannakas 2001). In 1998, Monsanto purchased First Line Seeds (Canada). In addition to its important access to the US cotton seed market, its purchase of Delta and Pine land and subsidiaries in 1999 (Groundup 2003j) gave Monsanto access to the important Asian cotton markets (RAFI 2000a). In 2000, Monsanto acquired Mahyco, Maharashtra Hybrid Company, E.I.D. Parry and Rallis (The Hindustan Times 2000). Monsanto stated that they proposed to penetrate the Indian Agricultural sector in a big way. Monsanto believed that the new acquisitions would foster synergies and economies of scale (New York Times 2003). In 2001, the Canadian division of Monsanto Co. acquired Limagrain and Canada Seeds Inc., a small Saskatchewan-based canola seed research, production and marketing company (Reuters 2001), giving Monsanto a leadership position in a North American canola business with links to cutting-edge canola technology (Reuters 2001). In 2002, Monsanto received approval for the commercialisation of insect-protected hybrid cottonseed in India (Monsanto 2003a). There are approximately 8.8 million ha of cotton grown in India annually, and about half of those are suitable for the Monsanto's cottonseed (Associated Press 2002). This approval opened more doors for Monsanto to sell its GM seeds in Asia (James 2015).

Monsanto's commitment to biotechnology is reflected in its R&D investment. In 2000, Monsanto invested approximately US$588 million in R&D for new and improved products (Monsanto 2001). Monsanto's seed and genomics business was expected to increase its portion of overall revenue from 31% in 2001 to 55% in 2007 due to the release of new biotechnologies (NGIN 2002). Monsanto's pipeline GM products include rootworm resistance maize.

According to Associated Press (2002), the introduction of rootworm-resistant maize could eliminate or reduce the need for the rootworm insecticides applications on 26 million ha of maize. Monsanto also hoped to introduce high-yielding maize sometime after 2006 (Associated Press 2002). The new varieties are suggested to produce up to 10% higher yields compared to conventional hybrid maize varieties. Monsanto also invested in the development of high-yielding hybrid wheat varieties (ETC 1998).

However, despite Monsanto's run of high-profile acquisitions, and its position as undisputed leader (Reuters 1999a) in input trait biotechnologies, breaking into the market as early as 1995, profits, and, in turn, share prices, have been deflated for several years (New York Times 2003). Analysts blamed Monsanto's poor performance on a range of problems, including weakness in Latin America, where inventory and management problems have taken a toll and the unexpected resistance to biotechnologies, particularly in Western Europe (New York Times 2003). Investors say Monsanto has also been weighed down by its heavy cost structure. According to *The New York Times*, return on equity is weak because of the US$10 billion, or so, that the company invested in seed company acquisitions. Indeed, according to the New York Times (2003), they do not even make their cost of capital. It has been suggested that part of the problem was that Monsanto's infrastructure is for a global market (New York Times 2003). If the world market had accepted biotech 2 years ago, or even now, they would be much more profitable (New York Times 2003). Many believe that whilst Monsanto dominates the biotechnology market through the sale of approximately a dozen varieties of GM crops with access to an estimated 140 million acres, Monsanto timeframe for the release and uptake of biotechnologies was too ambitious, and based on unrealistic assumptions (New York Times 2003).

Bayer has also been incredibly active in the search for the access to new markets through the vertical integration of pesticide, biotechnology and seed companies through its acquisition of Aventis CropScience and, latterly, through the acquisition of Monsanto. Bayer (as AgrEvo) acquired the rice breeding programme and associated assets of Granja 4 Irmaos (Brazil) in 1998, ProAgro Group (comprising ProAgro Seed Company Ltd., ProAgro Seeds-PGS India Ltd., Hybrid Rice International and MISR Hytech), Sementes Ribeiral (Brazil), Sementes Fartura (Brazil) and Mitla Pesquisa Agricola in 1999. In addition, Bayer subsequently acquired Genex (Australia). In India, Bayer controls 50.1% of AgrEvo India, 100% of Rhône-Poulenc Agrochemicals India Ltd., 51% of Bilag India Ltd. and 98% of ProAgro Seeds (98%), a seed breeding, seed production and distribution company (Domain-B 2003). The ProAgro Group was the second largest seed company in India, ranking number one in the hybrid maize, millet and forage sorghum markets, and second in the sunflower and sorghum grain markets. The ProAgro Group was developing hybrid rice, cotton,

OSR and vegetables for India and other markets in Asia, the Middle East and Africa (Bijman 2001a). In India, Sunseeds India was integrated with ProAgro-PGS to create a single venture called Nunhems-ProAgro. With the acquisition of ProAgro by AgrEvo in 1999, the company was fully integrated into Nunza (Bijman 2001a).

In 1998, DuPont acquired Sementes Dois Marcos (Brazil) and the remaining shares in Pioneer Hi-Bred for US$7.7 billion in 1999 (Lipin et al. 1999). DuPont chose to acquire Pioneer for its germplasm/seed bank base, global seed sales force and Pioneer conventional seed breeding expertise and farmer markets (Groundup 2003c). For Pioneer, DuPont offered substantial new financial resources to invest in competitiveness enhancing R&D (Groundup 2003c). In 1998, Pioneer Hi-Bred had seed production facilities in Argentina, Australia, Austria, Brazil, Canada, Chile, Egypt, Ethiopia, France, Germany, Hungary, India, Indonesia, Italy, Mexico, Pakistan, Thailand, Turkey and Zimbabwe (Groundup 2003d). In addition, Pioneer has 110 research facilities: Argentina (2); Australia (2); Austria; Brazil (3); Canada (6); Chile (2); Egypt; France (7); Germany (4); India (4); Italy; Mexico; New Zealand; Philippines; Puerto Rico; South Africa; Spain; Thailand; Turkey; UK; US (43); Venezuela; Jamaica; West Indies; and Zimbabwe (Groundup 2003d). In 1998, DuPont-Seed sales for Pioneer were US$1.835 billion; 40% were for GM crops (Genewatch 2003b). In 2001, DuPont acquired the Yun Meng Protein Co. (China) (Corporate Watch UK 2003c).

Dow AgroSciences has also been active. In 1998, Dow AgroSciences acquired Morgan Seeds, Argentina's second largest maize seed company, and Dinamilho Carol Productos Agricolas (Brazil), another key South American maize seed company (Fulton and Giannakas 2001). Dow AgroSciences also acquired Hibridos Colorado (Brazil) and FT Biogenetica de milho (Brazil).

5.3.3.1.2 Higher levels of appropriation could be achieved through greater control over biotechnology and seed production and distribution networks

M&A in the seed, biotechnology and pesticide industries could be justified by the following: (1) A wish by some firms to pre-empt other firms from taking over valuable assets. (2) A result of the need to consolidate costs and rationalise industry capacity. (3) A desire by the management of the firms involved to extend their sphere of influence, and, in doing so, appropriate higher levels of value added (profit). From a value chain perspective, input manufacturers, as suppliers to the least profitable sector of the agri-food system, namely farming, are in a strategically weak position. The level of concentration in the business is in part a desperate drive to maintain profitability against declining strategic value of chemicals, seeds and biotechnology (Vorley 2002).

However, as the global market for pesticides and biotechnologies expanded, so did the capacity of pesticide corporations to affect industry capacity, and more importantly, the capacity to, at least in some small way, determine product/technology prices and therefore the level of profitability.

Throughout the 1990s, pricing behaviour of vertically integrated pesticide/seed/biotechnology TNCs was highly strategic in nature. While market concentration does not necessarily confer market power, incontestable markets do (Baumol et al. 1992). Most pesticide/biotechnology markets are incontestable due to the presence of high sunk costs associated with pesticide and biotechnology development. Hence, it is impossible for new entrants to use a hit and run strategy, particularly when IPRs are well protected (Fulton and Giannakas 2001). Subsequently, pesticide and biotechnology corporations are able to keep prices above average costs of production (Fulton and Giannakas 2001). In practice, pricing tends to be based on the level of competition from other products, particularly the impact that prices might have on competition, and the value created by the products being sold. Strategic pricing means that prices are actively chosen by firms rather than being determined by the market. As mentioned above, strategic pricing is closely linked to sunk costs, which are a key feature of the pesticide/seed/biotechnology industry.

Sunk costs have two important consequences for the pricing of a product: (1) pricing will not be at marginal cost and (2) pricing will typically not reflect average cost. The existence of substantial sunk costs and relatively low marginal costs mean that firms must have some ability to price above marginal cost if they are to have an incentive to undertake R&D expenditures. In addition, investments in biotechnology R&D capacity have anti-competitiveness effects because of large sunk costs that constrain entry (King et al. 2002) and result in economic rents that are not dissipated by competition and entry (Werden and Froeb 1998).

There is substantial evidence of price discrimination, whether it is in the form of (1) two-part pricing, (2) differential pricing or (3) tied sales. An example of a two-part pricing scheme is the Technology Use Agreement (TUA). Although TUAs are being phased out, they were an important pricing feature in the industry. Under a TUA, farmers paid a set fee for the right to use the seed, as well as the cost of the chemical to which the seed is resistant each time they applied the chemical. TUAs are an example of second-degree price discrimination – farmers with greater weed control problems pay a smaller per-unit price (the TUA is spread over a greater amount of chemical) than those with a smaller weed control problem.

An example of differential pricing occurred in the USA in both 1997 and 1998, when the introduction of herbicide-tolerant soybean varieties led to a 50% discounting of the prices of post-emergence soybean herbicides by DuPont and American Cyanamid. This price reduction substantially limited the flexibility of Monsanto to competitively price Roundup Ready® soybeans and its related herbicide. In another well-publicised case, the US General Accounting Office provided evidence of how Roundup Ready® soybeans are priced significantly lower in Argentina than in the USA. The lower price in Argentina appears to be linked to the weaker IPR regimes in place in that country. The key impact of this strategic pricing is not on the total economic surplus created as a result of R&D, but rather on the distribution of this surplus (Sexton 2000). Indeed, differential pricing also occurs in the pesticide market. According to Vorley and Keeney (1998), the price of AIs varied by a factor of two in Europe, and prices in Europe can be ten times higher than those in developing countries (Vorley and Keeney 1998).

Throughout the 1990s, the market for biotechnologies expanded rapidly. In 1996, the global GM crop area stood at 2 million ha. This quickly expanded to 39.4, 53 and 63 million ha in 1999, 2000 and 2001, respectively (NGIN 2002). By 2013, 27 countries, including 19 developing countries, had approved and adopted insect-resistant and herbicide-resistant GM crops. Insect resistance and herbicide resistance were often stacked together in the same seed (James 2015). By 2013, the adoption of GM crops had reached a near-optimal level of around 90% in mature markets in both the developed and developing countries that had wholeheartedly embraced GM crops. These were the USA, Canada, Australia, China, India, the Philippines and South Africa (James 2015). By 2016, approximately 185 million ha were planted with GM seed (ISAAA 2016). Of this area, the USA accounted for approximately 72.9 million ha; other major countries include Brazil (49.1 million ha), Argentina (23.8 million ha), Canada (11.6 million ha) and India (10.8 million ha). More than 54% of the farmers using GM crops are from developing countries (ISAAA 2016). According to James (2015), approximately 18 million farmers are planting GM seed. The most important GM crops are maize, soybean, canola and cotton in that order. By 2016, 64% of all maize planted was GM (ISAAA 2016).

In 2000, industry analysts predicted that by 2004, global GM sales could reach US$2.9 billion (Agrow 2000). However, this figure was only a fraction of the value placed on biotechnologies by the largest pesticide corporations. For example, in 2000, Zeneca (Syngenta) (Agrow 2000) and Rhône-Poulenc (Assouline and Pierre-Benoit 2001) predicted that the GM market could reach US$5 billion by 2005, and around US$75 billion by 2020 (Reuters 1999a). Monsanto, whose germplasm in 2000 accounted for 94% of the total sown to GM (Monsanto 2001), was even more optimistic, suggesting that, by 2015, the biotech industry could be worth US$100 billion (Reuters 1999a). However, even this figure was dwarfed by DuPont's prediction of a staggering US$500 billion a year by 2020 (Reuters 1999a; Agrow 2000). Ultimately, by 2016, the actual global market value of GM crops was a relatively modest US$15.8 billion (ISAAA 2016).

Rightly or wrongly, most pesticide corporations still expect significant returns on their biotechnology investments, particularly when concentration in the seed and biotechnology sectors has created opportunities to extract even greater proportions of value added in food systems.

The result of global M&A meant that by 1999, the top five vegetable seed companies controlled 75% of the global vegetable seed market (Grooms 1999), while just four companies (Monsanto, Pioneer, Novartis and Dow) controlled 47% of the commercial soybean market in 1998 (Hayenga 1998). Due to mega concentration, by 2017, only three corporations (Bayer, DowDuPont and ChemChina) controlled almost 60% of the global seed market and 70% of the global pesticide market (Greenaway 2017).

However, food system consolidation is not limited to pesticide, biotechnology and seed corporations. The 32 leading grocery retailers accounted for 34% of the total global food retail market, estimated at US$2.8 trillion in 2000, and the top ten grocery retailers accounted for US$513.7 billion, or 54% of total sales for top 32 retailers (ETC 2001a). According to Angus (2008), 60% of global food stocks are in corporate hands, with just six corporations controlling 80% of the global trade in wheat and rice. Processing industries are also rapidly consolidating their economic and market power. According to Vorley (2002), the economic power of the top eight food multinationals is equal to half that of Africa. In 2000, food industry mergers, acquisitions and associated agreements were valued at US$87 billion. Nestlé, Philip Morris and Unilever now dominate the food-making sector. Processors justify the substantial accumulation of market power in order to compete and remain profitable in a market place increasingly dominated by the retail sector. Vertical concentration is also rapidly taking place, as cross-ownership between food processing and the seed sector increases. Farmers are increasingly being placed within a 'contractual sandwich', where the farmer is exposed to most of the risks associated with production while receiving an increasingly smaller proportion of the value added (Vorley 2002). Indeed, the Foundation on Economic Trends and the US-based National Family Farm Coalition as well as individual farmers stated their intent to sue companies such as Novartis, DuPont and Monsanto in up to 30 countries, claiming that they are exploiting bioengineering techniques to gain a stranglehold on agricultural markets (Reuters 1999d).

5.3.3.1.2.1 USA

Since the mid-1990s, farmers in the USA responded favourably to aggressive GM marketing strategies from companies such as Monsanto and Novartis AG (Washington Post 1999). By 1998, 20.5 million ha of GM crops were being grown in the USA (74% of the world total), with significant areas also planted in Argentina and Canada. The market value for GM crops grew to an estimated

£94 million in 1998 from only £47 million in 1995 (James 2000). By 1999, 55% of the soybeans and 35% of maize grown in the USA were GM (Washington Post 1999). However, this radical transformation of the agricultural economy to incorporate emerging biotechnologies was planned and executed behind closed academic and corporate doors. There was very little public input (Deutsche Bank 1999).

In 1992, the FDA unilaterally decided that as long as a GM food is no more toxic, allergenic, or any less 'substantially equivalent' than its standard counterpart, it need not be labelled to show the process that created it (Deutsche Bank 1999). This action meant that GM technology was invisible to the US consumer. In turn, the USDA rubber-stamped applications for GMO field tests. On average, the USDA rejected only 4% of all applications (Belsie 2001). In the period 1995–2000, there were US$15 billion worth of amalgamations in the US seed industry (Vorley 2002). Just four companies (DuPont/Pioneer, Monsanto, Novartis and Dow) controlled 69% of the North American seed corn market (Hayenga 1998). In 1998, Monsanto controlled 15% of the US maize market and 24% of the soybean market. Delta & Pine Land controlled 71% of the cotton seed market in 1998 (Kalaitzandonakes and Hayenga 2000). According to Pechlaner and Otero (2010), the rapid expansion of GM crops in the USA and across much of the world demonstrates that the USA, pursuing a neo-productivist/neoregulatory agenda, is 'determined to capture this comparative advantage and make biotechnology an American industry'. This is not surprising given that the USA's two principal agricultural exports (soybeans and maize) were predominantly GM. Having bet heavily on biotechnologies, the USA is strongly committed to see the formal acceptance of GMOs across the globe. The USA constantly challenges the EU regarding its stance towards GM foods (Pechlaner and Otero 2010).

5.3.4 Increased political and economic power of transnational pesticide corporations

Despite declining agro-food industry growth, CEOs of leading agri-corporations have been expected to respond to increasingly ambitious targets for growth in market share and high levels of returns on capital investments. Whilst agriculture is one of the world's largest industries, employing 1.3 billion people and producing US$1.3 trillion worth of goods, competition for an increasing share of value added in a relatively stagnant market has become incredibly fierce and innovative. CEOs of major agribusinesses have adopted a wide range of strategies with which to capture value added, including the aforementioned horizontal and vertical integration, the development of new and improved technologies, and the manipulation of the political and economic environments in which they operate (see Fig. 5.2). According to Vorley (2002),

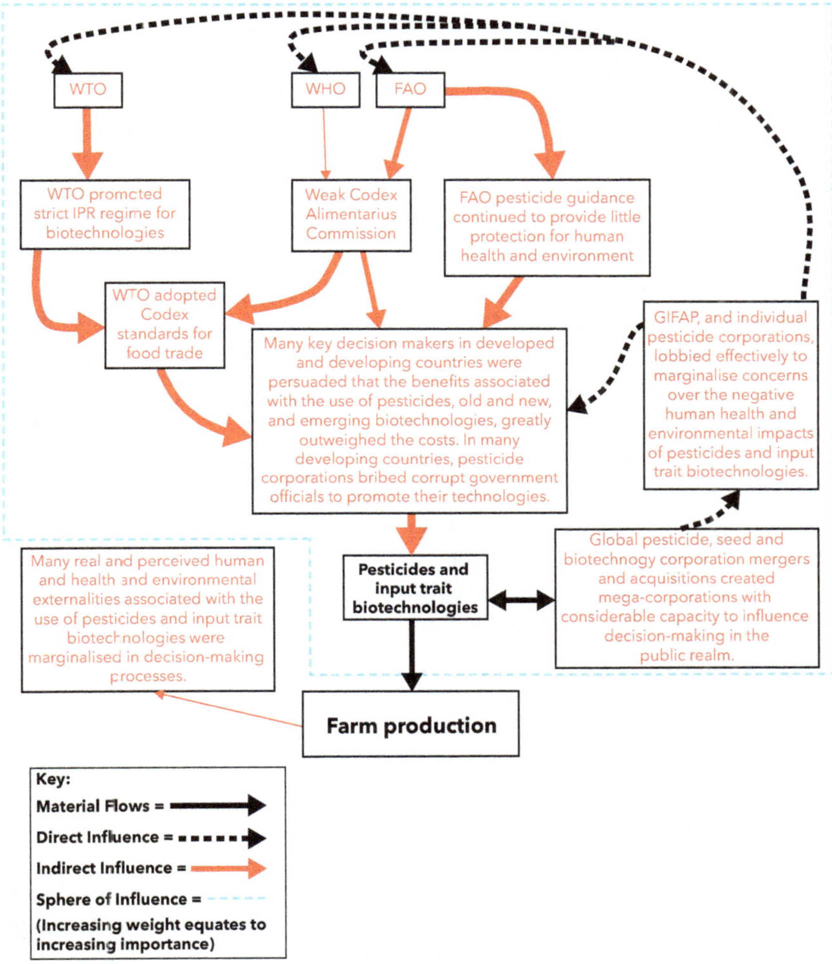

Figure 5.2 Pesticide manufacturers and the Neo-Productivist model of agriculture in the Divergent Period.

'Size confers logistical control, reduced transaction costs, economies of scale, improved market and meteorological intelligence, and access to and control of the most valuable intellectual property and the most comprehensive distribution network. Size confers "absolute cost advantage" – the ability to outbid smaller farmers or companies for resources and ideas, to invest more heavily in R&D and patent protection (for instance, to obtain critical mass in genomics), to set predatory prices, to externalise risk, to raise external capital, and to mount lavish promotional campaigns. Size also confers access to information related to the workings of government that selectively benefit the company and the ability to remould the social and political environment to an individual's or company's own benefit' (Vorley 2002).

In evolving food systems, only the most powerful, generally large, corporations have been able to satisfy shareholder demands. Food system actors with decreasing economic and political power have either been absorbed by their more aggressive and/or successful counterparts or have been exposed to declining returns on capital and, in many cases, have chosen to exit the system. Despite the considerable financial support that many farmers receive, their share of value added has fallen rapidly over the past 30 years, and is predicted to fall further. For example, in 1950, agribusiness was worth US$420 billion, and farmers added more than 33% of the value. By 2028, the market could be worth US$10 trillion, and the farmers share could fall to as little as 10%. As one of the principal suppliers of inputs to agricultural production process, pesticide producers have tended to suffer when farmers suffered. Subsequently, the pesticide industry has become increasingly innovative to capture higher and perhaps more predictable returns on investments. However, the transfer of power from national and supranational state apparatus to international state apparatus and MNCs/TNCs, and the subsequent, and highly unpredictable, distribution of the costs and benefits associated with this transfer of power is cause for alarm.

5.3.4.1 IPR

In accordance with the discussion in Chapter 3, because of the high levels of investment needed to produce emergent biotechnologies, pesticide corporations assumed that there would be a growing need to ensure increased IPR protection for their new biotechnologies. The way in which organisations and contractual arrangements are structured is also influenced by IPRs. Concerns over IPR protection can also create pressures for greater vertical integration. Vertical integration, through M&A, is more likely if IPRs are not well defined. Concern over IPRs, for instance, can help explain why multinationals have bought out local seed companies in developing countries, where IPRs are not well protected, rather than license genes to them. Furthermore, the buyout decision is also generally dependent on the degree of appropriability of the technology and the relative strength of complementary assets.

Vertical integration also becomes attractive when IPRs create opportunities for exploitation. If IPRs give a company the ability to exert market power vis-à-vis the companies with which it trades, the up, or down, stream companies may be deterred from investing in new technologies or developing new products that would be traded with the dominant firm. The firms are deterred because of a concern that the dominant firm will appropriate the benefits from their innovations. To remedy the situation, companies with market power may decide to vertically integrate and take over R&D and market development. The decision by multinationals to integrate the sales and distribution systems of the local seed companies they purchased into their own businesses may be linked, in part, to a concern that the local seed companies will not have the appropriate incentives to maintain and strengthen their retail operations.

Indeed, in most developing countries, M&A are the only sure way to protect IPR. It is ironic that a lack of pesticide and biotechnology regulation encouraged pesticide corporations to develop markets and establish research and production facilities in developing countries. Conversely, in those same developing countries, the obvious lack of IPR regulation virtually obliged corporations to protect IPR through acquisitions and mergers (ETC 1998).

As IPR protection on crop varieties, and latterly biotechnologies, became increasingly important to the success of seed companies. Corporations have lobbied, at both national and international levels, for robust regulatory mechanisms. At a national level, the US Administration passed the 1970 Plant Variety Protection Act (USDA-ARS 2004a), which aimed to provide patent-like protection for new non-hybrid seed varieties. The ultimate goal was to create an incentive for public and private research on new commercial plant varieties by making it possible for scientists to benefit financially from developing them. In 1980, the US Supreme Court determined that, whilst patent protection could not be extended to 'naturally occurring living things, it should apply to living organisms that had been altered by human intervention'. Furthermore, later interpretations extended this coverage to new processes, which may or may not give rise to a new product (World Bank 2003).

IPR protection at an international level was established by the International Convention for the Protection of New Varieties of Plants (UPOV). The Convention was adopted in Paris in 1961 and it was revised in 1972, 1978 and 1991. The objective of the Convention is the protection of new varieties of plants through IPRs (Etc 1998). In 1978, the UPOV Act allowed farmers to save seed of a proprietary variety for their own use. However, under the 1991 UPOV Act, replacing the 1978 Act, the right to save and re-plant seed is recognised only if member countries make special provision for it in national legislation (Etc 1998).

As discoveries in the field of molecular genetics exploded in the late 1980s, the US Patent and Trademark Office became inundated with pending applications on Expressed Sequence Tag (EST) sequences. In 1991, 4000 ESTs were pending. This compared to approximately 350 000 EST sequences in 1996 and over 500 000 EST sequences in 1998 (ETC 1999). In attempts to standardise biotechnology-based IPR, amendments to the Plant Variety Protection Act (PVPA) in 1994 made US law consistent with the UPOV of 19 March 1991, to which the USA is a signatory. In February 1999, the 1994 PVPA amendments were formally accepted by UPOV as being in conformance with the International Convention. USDA, rather than the Patent and Trademark Office, administer the law (Webref

2003). It should be noted here that the USA is a staunch supporter of national IPR and leads global efforts to develop and enforce a globally standardised system of IPR protection (Pechlaner and Otero 2010).

5.3.4.2 WTO

While IPR is also regulated through bilateral and regional trade deals (McAfee 2003), the WTO has become the principal regulatory institution for biotechnology-related IPR (see Fig. 5.2). Attempts to increase the scope of industrial patents first surfaced in the 1986 Uruguay Round of the Trade-Related Aspects of Intellectual Property Rights (TRIPS) in the GATT. This discussion soon became one of the fiercest arenas for North–South patent controversy. The USA, Japan and the EC advocated the principle that all countries should provide and respect IPR protection in all technical fields – including biotechnology. Disregarding this principle would constitute a contravention of GATT regulations, making the offending country liable to economic sanctions. Third World governments were strongly opposed to these proposals, maintaining that patents benefit those states that are already technologically and economically strong. This is a point hard to refute, as Third World countries hold no more than 1–3% of all patents worldwide. By 1994, and initiation of the next round of GATT, now WTO, talks in Marrakesh, TRIPS had significant support from a US-fronted and TNC-based coalition.

Subsequently, in 1995, the TRIPS Agreement arose as the principal regulatory mechanism of the WTO through which IPR is controlled (World Bank 2003). The TRIPS Agreement stipulates that WTO member states must adopt and enforce patent laws 'in all fields of technology'. TRIPS treats the genetic components of organisms, as well as genetically altered varieties of living organisms, as ordinary commodities subject to private ownership and standardised rules of transnational commerce (McAfee 2003). In its present form, TRIPS makes it illegal under most circumstances for citizens, businesses or government agencies to commercialise or distribute brand-name plant varieties and privatised gene sequences, proprietary medicines, research technologies and databases (McAfee 2003).

Article 27.3 (b) of TRIPS calls for WTO members to protect IPR over plant varieties either by patent or by an effective 'sui generis' system or by a combination of the two (FAO 2017b). Developing countries had until January 2000 to comply with TRIPS, and were pressured to adopt the 1991 UPOV as the mechanism for their obligations. The USA was keen to recommend their own patent-based model to other members, stressing that other mechanisms would need to be looked at on a case-by-case basis (BWTND 2003).

Resistance by developing countries to this version of TRIPS grew, and still continues in the WTO TRIPS Council (McAfee 2003). Concern over the appropriation of common property and indigenous people's rights culminated in the Indigenous People's statement on TRIPS: 'We, indigenous peoples from around the world, believe that nobody can own what exists in nature except nature herself. A human being cannot own its own mother. Humankind is part of Mother Nature, we have created nothing and so we can in no way claim to be owners of what does not belong to us. But time and again, western legal property regimes have been imposed on us, contradicting our own cosmologies and values' (Tebtebba Foundation 1999).

The WTO is not the only arena in which IPR struggles are articulated. A moratorium on IPR claims on germplasm held in trust (CGIAR system) stopped the growing trend of trait patenting by private plant breeding companies; something many observers quickly coined as bio-piracy (RAFI 1998a). The story first broke when two Western Australian research institutes improperly claimed monopoly rights on two chickpea varieties from the International Crops Research Institute for Semi-Arid Tropics (ICRISAT) in Hyderabad, India. One of the two varieties was part of ICRISAT's trust agreement with the FAO (RAFI 1998a).

Biotechnology is also a significant factor in the controversial US effort to strengthen the WTO Agreement on TRIPS. The USA wants to require countries to recognise and enforce property rights to drugs, food products, crop varieties and the chemicals they depend upon, as well as biotechnology tools, techniques and expertise. According to McAfee (2003), these trade rules foster the expansion of established biotechnology industries, which are strongest in the USA, and raise barriers to entry against newer private or public sector biotechnology enterprises. US firms and trade organisations have had a direct role in determining US international biotechnology policies, for instance, through their close collaboration with the Bush/Quale and Clinton administrations (McAfee 2003; Pechlaner and Otero 2010).

In effect, the WTO's neo-liberal paradigm produces universally applicable categories and a discourse that bifurcates nature and society (McAfee 2003). 'Globalized' property regimes and techno-scientific standards, and universal measurements of resource values - such as the dollar prices of genetic resources in international markets - lend themselves to regulation under the WTO because they produce 'global' criteria for managing the contents of the biophysical world (McAfee 2003).

Subsequently, the WTO promotes industrial agriculture by requiring member nation states to adopt an essentially US-styled system of copyright, trademarks and patents. Such a global system of IPR increases big agribusiness's ability to control the range and prices of agricultural inputs and technologies (RAFI 1998b). IPR on GM seed through the WTO has spurred investment in biotechnologies in recent years. IPR allows corporations to recover R&D costs and make potentially high rates of return on invested capital. GM plants are also quicker to market, taking on average 6 years instead of 10 years for conventional breeding. This gives the patent owner longer to recoup R&D costs and generate high levels of profit (Clapp 2002). In addition, getting regulatory clearance for GM crops is considerably cheaper (US$5–30 million) than the parallel process for pesticides (New Scientist 2002).

Furthermore, under the WTO, a decision to reject genetically engineered imports for socio-economic reasons or on environmental grounds not unequivocally justified by existing science might be deemed an 'unfair trade practice' punishable by trade sanctions (McAfee 2003).

The level of protection afforded under the TRIPS Agreement of the WTO has stimulated substantial biotechnology R&D efforts. By the end of 1998, the US Patent and Trademark Office had granted 1370 agricultural biotechnology patents to the top 30 patent assignees; 74% of the agricultural biotechnology patents awarded to top 30 assignees were held by six Gene Giants: Pharmacia (Monsanto) 287; DuPont 279; Syngenta 173; Dow 157; Aventis 77 and Grupo Pulsar 38 (RAFI 2000b; Clapp 2002).

Terminator technology has been one of the most controversial agricultural biotechnologies in recent years. Terminator technology refers to GM crops engineered to allow the successful completion of one life cycle of the crop plant but that does not permit its progeny to successfully complete its life cycle. This is usually achieved through the production of sterile seeds. In effect, in order to benefit from the enhanced attributes of terminator GM seeds, farmers must purchase new seed from the GM seed company each year. Terminator genes were first targeted for use in maize and cotton, but could potentially be placed into any crop. Wheat and rice were potentially next on the list in countries such as India, China and Pakistan. According to RAFI (2000a), all major pesticide corporations, with the possible exception of Dow, have terminator gene patents. Delta & Pineland stated that the terminator technology could be applied to 400 million ha of cropland. According to Collins (2000), the US Government and Delta & Pine Seed Co. were heavily involved in promoting the speedy commercialisation of terminator technologies; the USDA and Delta & Pine Seed Co., are co-owners of the terminator patent (Collins 2000).

The second generation of terminator technology, known as Genetic Use Restriction Technologies (GURTs), first emerged in the late 1990s. GURTs are much more sophisticated than terminator technologies as they generally do not result in sterile seed. However, whilst still viable, farmer-saved seed from GURTs do not demonstrate the key genetic improvements associated with the GURT parent plant. Syngenta held the largest number of GURTs (36 of 71 patents). Among these patents are techniques that control the following characteristics in staple food crops: crop disease susceptibility, crop fertility, crop flowering, crop sprouting and crop ageing (Corporate Watch UK 2003d). According to RAFI (2000a), GURTs were being field-tested in the UK as early as 1999.

5.3.4.3 Codex Alimentarius Commission

Since its establishment in 1963, the Codex Commission has endeavoured to develop and apply global food safety standards. Government acceptance of Codex MRLs has been encouraged by allowing two tiers of compliance, 'full' and 'free'. Full compliance requires governments to ensure that imported and domestically produced food conforms to Codex specifications and that it will not enact legislation to restrict distribution of food conforming to the limit. On the other hand, free compliance does not require a country to ensure that domestically produced food conforms to Codex specifications but prevents the country enacting any form of legislation that hinders the distribution of imported foods conforming to the limit (CODEX 1989). In addition, governments that do not accept Codex MRLs are requested to state their reasons (CODEX 1989).

In October 1989, the US Bush Administration presented the double zero plan at the 1986 Uruguay Round of GATT talks, which aimed to curb domestic agricultural subsidies and the right of states to limit the importation of food on non-scientific grounds. The criteria deemed not to represent scientific evidence included social, economic, cultural and religious grounds. In an attempt to harmonise the scientific basis of food trade, the USA advocated the adoption of Codex as the WTO standard (see Fig. 5.2) (Hough 1998). By 1992, over 2000 MRLs had been set for 80 pesticides (Hough 1998). While not compulsory, Codex MRLs acted as an international point of reference for pesticide residues and are often quoted by governments, NGOs and the private sector publications on pesticide residues (Hough 1998). In 1993, the WTO/GATT Accord made the Codex Alimentarius Commission the reference point against which national food safety standards are assessed and the basis of international food standards for trade between member countries. The 'Agreement on the Application of Sanitary and Phytosanitary Measures', known as the SPS Agreement, makes Codex recommendations the benchmark against which national food standards are judged. If a country opts for a stricter standard, it is open to challenge unless it can prove that these are the least trade restrictive and scientifically justifiable (Macfarlane 2002).

For most agri-food interests, the marriage of the WTO and Codex was a match made in heaven. In many respects, the convergence of WTO and Codex can almost be considered as an arranged marriage; as agri-food interests had, over several decades, fiercely lobbied both bodies and, indeed, had been instrumental in designing and developing Codex activities and standards and WTO mechanisms. Indeed, Codex meetings quickly grew to be dominated by organised and well-informed negotiators from industrialised countries. This left many developing countries disadvantaged in the decision-making process as they did not possess the institutional or financial resources to generate scientific evidence in support of their positions.

In 1993, the UK-based pressure group National Food Alliance published a report based on 2-year monitoring of the Codex Commission Meetings. The report found that, of the 2578 participants in the various meeting, only 26 came from consumer organisations with none coming from environmental groups. In comparison, 660 came from the food industry (Avery et al. 1993). Indeed, GCPF is represented at most FAO and WHO Expert Committee meetings dealing with pesticides and has observer status on the Codex Alimentarius Commission (Hough 1998). Furthermore, many developing countries are unable to send delegations to Codex. If delegations are sent, they tend to be small; many African countries can send only one. In contrast, OECD countries generally send much larger delegations of up to a dozen, including representatives of agrochemical companies. In 1997, the GCPF sent 45 observers to the Codex meeting. Because of the dominance of developed countries and the agrochemical sector, Codex has a decidedly non-precautionary approach to regulation, in which practices and chemicals are effectively presumed safe until proven otherwise. Indeed, scientists from the US Environmental Working Group concluded that of 3285 pesticide/crop combinations for which Codex has set standards, 1539 are illegal in the USA (Wallach 1999). In addition, Codex compliance issues have also stifled implementation of the EPA's ban on probable carcinogenic fungicide Folpet and have led to the amendment of federal food safety regulatory acts. The EPA has also delayed action on another probable carcinogenic pesticide, procymidone, residues of which were found in wine imported from Europe (GPC 2000).

While the WHO and Codex Secretariat provide information and training programmes to assist developing countries, representations of these countries remain decidedly weaker than those of the developed countries. Furthermore, many developing countries lack capacity to develop national food residue standards, as well as the institutional infrastructure to support residue compliance and reduce chemical inputs. According to Longfield (1993), because of the power asymmetry, many consumer and environmental groups increasingly came to view the international legitimisation of Codex standards via the WTO as a 'global conspiracy riding roughshod over the interests and rights of consumers' (Longfield 1993). They claimed that the food standards setting process should be transparent and representation should reflect the consumer and environmental concerns of less powerful actors and not just those of dominant developed country governments and the private sector (Macfarlane 2002). As Longfield (1993) suggested, 'individual nations should not be obliged to sacrifice high standards in order to facilitate world trade and the movement of global capital'.

5.3.5 New markets for old off-patent pesticides

Since the late 1980s, exclusive patents on many of the most widely used pesticide AIs expired. In turn, this has enticed both existing and emergent pesticide corporations to produce them. However, as the sale of older and generally more hazardous off-patent pesticides became increasingly restricted or prohibited in developed countries, pesticide corporations, both old and new, began to exploit new markets, particularly 'Green Revolution' developing countries.

In 1976, approximately 30% of US pesticide exports were banned in the USA (Peterson and Harris 1981). By 1991, The Foundation for Advancements in Science and Education estimated that in 1990, nearly 60 tons of extremely toxic US-banned pesticides were being exported per day, increasing to 80 tons/day in 1991 (Smith 2001). In the years 1992–4, at least 25 million (Smith 2001) pounds of domestically banned pesticides were being exported, falling to 21 million pounds in 1995–6 (Smith and Root 1999). According to US customs records, nearly 3.2 billion pounds of pesticide products were exported between 1997 and 2000, an increase of 15% over the period 1992–6 (Smith and Root 1999); glyphosate accounted for between 22% and 27% of all pesticide exports, linked to use on glyphosate-tolerant crops (Smith 2001).

Other important off-patent pesticides banned or restricted in developed countries but exported to developing countries included: aldicarb, heptachlor, chlordane, leptophos, endrin, dieldrin and parathion (Bull 1982). In 1997, exports of aldicarb from the USA totalled 4.7 million pounds, increasing to 8.9 million pounds in 2000 (Smith 2001), of which 70% were destined for developing countries. Although banned in the USA, Velsicol continued to produce chlordane in the USA for export to developing countries into the 1990s (Fairly 1996).

According to Smith (2001), the FAO determined that pesticide traders advertised restricted use pesticides to the public in over 60% of Latin American countries. Between 1972 and 1976, even aid-funded exports to developing countries included banned or restricted heptachlor, chlordane, leptophos, endrin, dieldrin and parathion (Bull 1982). Between 1969 and 1974, aid-funded exports of US pesticides to developing countries totalled £7 million/year or 12.5% of total exports (Bull 1982).

While the USA at least attempted to regulate the export of banned pesticides, European pesticide manufacturers wanting to export banned or restricted pesticides to developing countries were allowed to do so freely (Bull 1982). However, whilst the USA demonstrated a willingness to respond to NGO pressures to regulate pesticide exports to developing countries, measures were, at best, minimal. All that was required was a statement of acknowledgement from the importing agent recognising the legal status of the pesticide in the USA.

The exporting agent was then required to send a copy of the statement to the EPA who then forwarded it to the importing country's administration with the first shipment of the year (Bull 1982). The Carter administration made further attempts to tighten up export restriction by insisting on better labelling and improved export procedures (Boardman 1986). Indeed, the FIFRA amendment in 1978 insisted that pesticide exports to developing countries must be labelled in English and the language of the importing country (Bull 1982). This measure reflected some of the recently published International Labour Organization's (ILO) Guidelines on the Safe Use of Pesticides (Bull 1982), and echoed the FAO's attempts in 1977 for the establishment of Guidelines on Good Labelling Practice. However, much of this progress was abandoned when the incoming Reagan administration in 1981 attempted to remove unnecessary trade-limiting restrictions (Boardman 1986).

5.3.5.1 Growth in generic pesticides, and generic pesticide manufacturers

While the bulk of generic (off patent) pesticides are still produced and sold by GCPF companies, since the mid-1980s, there has been a rapid expansion in the number of newly established formulators and producers of generic pesticides (Dinham 1994). Many of the emerging generic pesticide producers came from developed OECD countries including: Israel (Makhteshim Agan), New Zealand/Australia (Nufarm), Denmark (Cheminova), the USA (Griffin), France (Elf) and Italy (Sostra) (Dinham 1994; Pesticide News 2000). However, in addition, many generic pesticide manufacturers and formulators have been established in developing countries, particularly in China and India, but also in other Asian and some Latin American countries. Most of the pesticides produced by emergent generic manufacturers are generally based on older pesticide technology, specifically organochlorines and organophosphates, and are destined for developing countries, particularly China and India (Dinham 1994).

In the year 2000, five generic pesticide producers were in the top 25 pesticide companies, based on world pesticide sales, namely – Makhteshim Agan, Nufarm, Cheminova, Griffin and Elf (Pesticide News 2000). By 2013, Nufarm and Makhteshim Agan Industries ranked eighth and ninth in the world based on pesticide sales. Interestingly, in 2011, Makhteshim Agan Industries was ranked seventh in the world based on pesticide sales. This was, in part, due to Makhteshim Agan Industries' acquisition of Albaugh, which was the largest off-patent pesticide firm operating in the Americas. Albaugh sales of generic pesticide in 2009 amounted to almost US$1 billion (ETC 2011). In 2011, Makhteshim Agan Industries was acquired by The China National Agrochemical Corporation, a subsidiary of ChemChina, and renamed ADAMA. In 2013, ADAMA generated over $3 billion in sales of generic pesticides in over 120 countries (ETC 2015).

Whilst Europe (37%) and Latin America (25%) are currently ADAMA's largest markets, according to ETC (2015), 'the company's strategic goal is to capture China's domestic agrochemical market, which is currently fragmented and ripe for consolidation'.

5.3.5.1.1 Price leadership

One of the most important characteristics of generic pesticides is their price (Pesticide News 2000); generic brands are usually 25% cheaper than brand names (Hartnell 1996). Because of this, generic pesticides are extremely competitive when priced against patent-protected products. In 1994, there were approximately 64 generic pesticide producers, whose products accounted for 12–15% of the world market (Dinham 1994). In 1993, 14% of Western Europe's pesticide market was taken by generics from generic producers (Farm Chemicals 1996). Due to the competitiveness of generic pesticides, AgrEvo expected this figure to rise to 25% by 2000, and off-patent pesticides to 50% of the market from 33%. In 2002, generic pesticides accounted for 53% of global pesticide sales. This figure was expected to rise to 69% by 2005 (Kuyek 2000).

In order to capitalise on growing markets for generic pesticides in developing countries, the liberalisation of agricultural trade under GATT/WTO agreements, and less stringent labour and environmental protection legislation, producers of generic pesticides have increasingly relocated manufacturing and formulation sites close to some of the fastest growing and largest generic pesticide markets (Dinham 1994, 1996). The major target countries include India, China, Taiwan, Malaysia and Thailand, with growing interest in Argentina, Brazil, Colombia and Mexico (Dinham 1994). Whilst Eastern Europe is recognised as having investment potential, it seemingly has not yet reached a critical size or stability. Likewise, the Sub-Saharan market is still too small and cash poor to attract investment (Dinham 1994). According to Dinham (1994), market size and growth potential are both key to investment in pesticide production facilities. In the early 1990s, the Brazilian market for pesticides was around US$900 million, India US$740 million, China US$485 million, Argentina US$336 million and Mexico US$270 million. National pesticide use in China was estimated at 230 000 tonnes in 1993 and rising (Dinham 1994).

5.3.5.1.2 Market proximity

Established pesticide corporations have responded to this growing demand by investing in generic pesticide production plants close to new markets rather than producing generic pesticides in the USA or Western Europe and then exporting them to developing country markets. For example, Novartis/Zeneca (to become Syngenta and now ChemChina) established production plants in China for monocrotophos and paraquat, which came online in the late 1990s/early twenty-first century with capacities of 3000–6000 tonnes/year (Vorley and Keeney 1998). Syngenta's US$85 million paraquat production plant in China was a joint venture between Jiangshan Agrochemical and Chemical Co. and Nantong Petrochemicals Corporation, and was expected to supply China and other Asian Pacific countries. Syngenta's existing paraquat production plants are in the UK (1), the USA (1), Brazil (1) (Dinham 1994) and, until March 2001(PAN-UK 2003b), India (1). Syngenta also owns 50% of a paraquat production plant in Japan (PAN-UK 2003b). Ciba-Geigy (to become Syngenta and now ChemChina) was also in the process of building a formulation plant near Shanghai, ready for use in 1996. There was also strong interest in China from BASF, Bayer and AgrEvo. DuPont was also establishing a new pesticide plant (Dinham 1994). India is another target for expansion because foreign investment is now encouraged. The biggest foreign producer is Bayer. Others include ICI India (Syngenta/ChemChina), producing cypermethrin, Shell (American Cyanamid-now BASF) and Ciba-Geigy (now Syngenta/ChemChina). Japanese companies were also actively looking to invest (Dinham 1994). Makhteshim Agan Industries, Dow and DuPont were also looking for Indian partners (Dinham 1994). Recent expansion in India has been for the production of some of the most hazardous pesticides, including: AgrEvo (now Bayer) manufacturing the organophosphate anilophos (300 tonnes/year), sourcing its worldwide requirements from this plant for export to the growing Asian Pacific region (Dinham 1994). Bayer opened a new plant in 1991 to formulate and package 24 000 tonnes of methyl parathion a year (Dinham 1994). American Cyanamid's (now BASF) also planned to expand phorate technical production; it already produces pendimethalin and mephosfolan for export (Dinham 1994). DowElanco exports technical-grade AIs to several Indian formulation companies and set up a joint venture in 1993 with Nocil (Dinham 1994) that produced monocrotophos (1127 tonnes), cypermethrin (103 tonnes), phosphamidon (415 tonnes) and dichlorvos (169 tonnes) (Dinham 1994). During the period 1992–4, India produced 70 000 tonnes of technical-grade pesticides in 125 manufacturing plants and 500 formulation plants. Past laws stated that pesticides used in India must have been manufactured there. Government policy now encourages exports. Between March and December 1992, exports totalled US$52.7 million, a rise of 45% on the previous year. Exports included DDT, BHC, Lindane and endosulfan. Pesticides still legal in developed countries are also exported (Dinham 1994). Two of the largest Indian pesticide manufacturers, Hindustan Insecticides and Gharda Chemicals, produce a range of pesticides for domestic use and export. Exports of Hindustan Insecticides increased 500% to US$1.4 million in 1992–3. The main markets are the USA, Mexico, Argentina, Australia and Europe. Indeed, in 1993, Gharda Chemicals held 25% of the

European market for isoproturon (Dinham 1994), and Bilag India specialised in synthetic pyrethroids for use in cotton, vegetables and fruits (Domain-B 2003).

Whilst there is no indication that major corporations are transferring the production of banned organochlorines to the south, some pesticide producers remain committed to producing and exporting pesticides to developing countries that are banned or restricted in developed countries. For example, Hoechst resisted attempts to eliminate the production and export of endosulfan, and, in 1993, Velsicol still made chlordane and heptachlor for export. In 1994, there was still widespread production of methyl parathion, monocrotophos and phosphamidon. In addition, DDT was still in use in China and India, and toxaphene was in use in Nicaragua (Dinham 1994). While the pesticide cost rationale for the use of many generic pesticides is obvious, the human health and environmental impacts of many of these pesticide products are often greater than those associated with their early production and use because of the amounts of often unregulated contaminants contained within them (Pesticide News 2000).

5.3.5.1.3 The emergent relationship between major innovation-based pesticide producers and emergent generic producers

What is interesting is the emerging relationship between the major innovation-based pesticide producers and the emergent generic producers. In general, innovation-based pesticide producers have taken several approaches to safeguard their market share and profitability in light of a growing proportion of their pesticide products that are no longer afforded patent protection. First, major innovation-based pesticide producers have attempted to restrict entry into markets for their off-patent pesticides by attempting to achieve cost leadership on their products by increasing production efficiency, reducing unit costs of production and reducing the profit margins on their products. There are two good examples of this strategy. The first is the case of Zeneca/Syngenta and its production of paraquat. Although Zeneca/Syngenta's patent expired in the late 1970s (Agrow 1995a), and the fact that, in Europe alone, another 11 companies requested registration of paraquat (PAN-UK 2003b), Syngenta still produces 50% of the world's paraquat (PAN-UK 2003b). Paraquat generated £1 billion annually in 2002. In 2001, Syngenta produced approximately 25 000 tonnes of paraquat (PAN-UK 2003b). Indeed, such was Zeneca's/Syngenta's faith in its ability to retain market share and profitability, after it had lost its patent protection, it decided to build the aforementioned paraquat production plant in China. This was because of strong market growth for paraquat in China and the perceived barriers to entry of generic producers (Tait 2001). The second case is that of Monsanto and its production of glyphosate. According to Vorley and

Keeney (1998), Monsanto, which still dominates the glyphosate market (90% of world production), dramatically reduced the price of Roundup in order to achieve cost leadership and protect its market share for glyphosate.

Second, major innovations-based pesticide producers have improved the formulations of many of their generic pesticides in an attempt to maintain both market share and profitability. Syngenta and Monsanto have improved the formulation of paraquat and glyphosate, respectively, in order to do this (Agrow 1995a).

Third, innovations-based pesticide producers have also attempted to lock in sales of their generic pesticide products with the production of GM crops. The principal example of this strategy is demonstrated by the development of Roundup Ready crops by Monsanto.

Fourth, many innovations-based pesticide producers have either acquired, or are collaborating with, generic pesticide producers. For example, DuPont formed a joint venture with Griffin in 1998 (Groundup 2003e) to produce and market generic crop products worldwide. BASF bought Micro Flo to channel all its commodity pesticides. Sumitomo of Japan acquired an equity stake in Hikal Chemicals (Financial Daily 2000), and, Dow AgroSciences acquired Sentrachem (South Africa), an equity stake in Nocil Crop Protection, and formed a joint venture with Cheminova (Chemical Week 1998; Pesticide News 2000). In addition, many innovation-based pesticide producers license AIs to generic companies (Chemical Week 1998). In turn, many OECD-based generic pesticide producers are buying into smaller generic producers in developing countries. For example, Cheminova acquired an equity stake in Lupin Agrochemicals (India). In addition, some of the larger specialist generic pesticide producers are acquiring proprietary pesticides from innovation-based pesticide producers that have been required to divest pesticides as part their M&A activities. For example, Cheminova acquired the fungicide Flutriafol from Syngenta in 2001, as part of an EU ruling on Syngenta for the fungicides divestment (Cheminova 2002).

5.3.5.1.4 India

India is one of the largest producers of pesticides in Asia. Indian companies have developed process technologies for more than 30 pesticides. Indeed, many Indian companies have opened offices and stock depots in Europe and Australia to ensure quick delivery, and exports have increased. In India, pesticides are regulated under the Insecticides Act of 1968 and Insecticides Rules of 1971 (Beyond Pesticides 2003). However, despite several amendments, the Insecticide Act of 1968 provides little protection for farmers, consumers and environment. For example, there is no provision in the Act to deal with counterfeit pesticides, and the reputable companies, in whose name they are sold, have to prove to the satisfaction of the court that they have not manufactured any substandard materials (Beyond Pesticides 2003).

In addition, the power of both domestic and foreign pesticide industry interests have corrupted the system and choreographed the drafting of pesticide legislation and amendments that affords them considerable room for manoeuvre. Pesticide companies have allegedly corrupted top officials in pesticide registration bodies. Pesticide companies allegedly employ brokers to enter dialogue with the Central Insecticides Board (CIB), who are authorised to encourage junior and senior officials at CIB to take a favourable official position regarding the company's pesticide product (Beyond Pesticides 2003). In May 2000, the Insecticide (Amendment) Bill 2000 was passed, making the punishment for adulterated pesticides more stringent, but it did little to clean up the regulations, register pesticides and monitor the pesticide industry (Beyond Pesticides 2003). The Insecticide Act regulates the import, manufacture, sale, transport, distribution and the use of insecticides to prevent any risk to people and animals. There are basically two types of registration under Section 9 of the May 2000 Insecticides Act (primary and secondary). When a new molecule is registered in India, it gets a primary registration. Subsequent applications for the same molecule get the secondary registration (Beyond Pesticides 2003). The Registration Committee, constituted under Section 5 of the Act, registers an insecticide after verifying its efficacy and safety to human beings, animals and the environment. The CIB based in Faridabad, Haryana, advises the Union and state governments on technical matters (Beyond Pesticides 2003). In 2001, a total of 2718 applications were received for registration, of which, according to the government, 1439 were approved. In India, registration of a new molecule requires the generation of environmental studies, conducted under Indian agroclimatic conditions. Once a primary registrant exists, others applying for fresh registration are given a secondary registration. Environment independent data is generally borrowed from existing data elsewhere in the world. The generation of environment-dependent data usually takes around 4–5 years, and includes toxicology and phyto-toxicology tests. However, while these rules exist, they are rarely followed (Beyond Pesticides 2003). In the initial data generation stage, companies have two options; they can either go to a government laboratory or a government-approved private laboratory. According to Beyond Pesticides (2003), companies usually prefer government-approved commercial laboratories as they can quickly generate the desired results, and are less bureaucratic. According to Beyond Pesticides (2003), 'industry pays bribes in order to register a new molecule or to attain data for an existing registered molecule or even bribes the registration committee members to influence decision-making at the top level. But the bribe is not always money. According to Beyond Pesticides (2003), many top CIB officials, including the registration committee members, regularly receive junket invitations. The company pays for all the expenses - air tickets, five-star hotel accommodation and even shopping expenses. Again, there is no direct involvement of industry'. According to Beyond Pesticides (2003), Indian registration procedures are useless, claiming the existence of government-approved laboratories that can generate toxicological data for a product within a day, a process that usually takes up to 1.5 years.

5.3.6 New markets for input trait biotechnologies

Crop biotechnologies have primarily emerged from the USA and Western Europe, with the first commercial GMO released in 1996 (Deutsche Bank 1999). Because the cost of developing new biotechnologies, between US$200 and 400 million over 7–10 years (Le Monde diplomatique 2001), used to be even higher than those associated with traditional pesticide products, most biotechnologies were: (1) Based on relatively simple single gene input traits for major temperate commodities such as maize, cotton, soya and wheat. (2) Exclusively produced and marketed by the largest pesticide corporations (Horstmeier 1996b).

In 1997, companies began marketing crops with stacked gene traits. The first included a maize hybrid with tolerance for two different classes of chemical herbicides, plus herbicide and insect resistance, grey leaf spot resistance and high soil pH tolerance (Horstmeier 1997). These releases were based on approximately 25 000 transgenic crop field trials, involving more than 60 crops and ten traits, which were conducted in 45 countries between 1986 and 1997 (ETC 1999). Unlike pesticides, the lack of a strong body of scientific literature on the environmental and health impacts of GM seeds and crops has benefited the agricultural biotechnology sector in terms of costs to bring a product to market. In addition to the new scientific advances, improved laboratory equipment generated significant efficiencies in nucleotide sequencing. For example, in the mid-1970s, it took a laboratory two months to sequence 150 nucleotides. By 1998, one commercial genome firm could sequence 11 million nucleotides/day (Moukheiber 1998). The cost of DNA sequencing also dramatically decreased. In 1980, the cost of sequencing a base pair was approximately US$100. By 1998, the cost of sequencing a base pair had fallen to less than US$1, and experts predicted that it would be less than a cent by 2002 (Sansom 1998). By 1982, researchers had sequenced about 680 000 base pairs. By the end of 1999, the number was expected to stand at more than 2.6 billion (DuPont 1999).

Between 1996 and 2001, the area of GM crops expanded more than 30 fold, from 1.7 to 52.6 million ha (James 2001). Four countries accounted for 99% of this GM area: the USA, Argentina, Canada and China (Clapp 2002). However, since 2000, developing countries have become increasingly important markets

Table 5.8 Area of transgenic crops planted (millions of ha)

Year	1997	1998	2000	2013
USA	8.1	20.5	30.3	70.1
Argentina	1.4	4.3	10	24.4
Canada	1.3	2.8	3	10.8
Australia	0.1	0.1	1.5	0.6
Mexico	0.1	0.1	0.1	0.1
Spain	0.0	0.1	0.1	0.1
France	0.0	0.1	0.0	0.0
South Africa	0.0	0.1	0.1	2.9

Source: James (1998, 2000).

for GM seed. The area of GM crops in developing countries increased from 7.1 million ha in 1999 to a staggering 100 million ha in 2016 (ISAAA 2016). In 2002, it was estimated that China, the world's largest cotton producer, planted 2.2 million ha, 53% of the total cotton area, with insect-resistant GM cotton (Planet Ark 2003a). According to the New Scientist (2002), the use of Bt cotton has resulted in an 80% reduction in pesticide applications. In addition, only 5% of farmers growing Bt reported pesticide-related illness compared to 22% growing conventional cotton. Table 5.8 illustrates growth in the area of GM crops sown across a cross section of countries. It can be seen that growth in the use of GM has been variable. Some countries such as the USA, Argentina and Canada wholeheartedly embraced biotechnology and exponentially increased the area of GM crops. Other countries, such as Australia, Mexico, Spain and France, have failed to fully embrace GM crops. Indeed, in Australia and France, the area sown with GM crops actually decreased.

All major pesticide producers are involved in the production of input trait biotechnologies. Table 5.9 highlights some of the key input trait biotechnologies that were being developed by the major pesticide corporations back in 2003.

5.4 Key outcomes of pesticides and biotechnologies

5.4.1 National food surpluses and national food security

One thing that cannot be denied is that, while population growth continues apace, the global spread of modern intensive agriculture throughout the past 40 years has, to date, been able to match aggregate food production with aggregate food consumption. Not surprisingly, GCPF (1997) claims that, with the aid of pesticides, farmers have been able to double production of world food calories since 1960; triple the output of intensive foods such as cooking

Table 5.9 Key input trait biotechnologies under development in 2003

Corporation name	Input trait	Countries where trials were conducted
Monsanto	Glyphosate-tolerant cotton	Greece and Spain
	Bt insect resistance in cotton	Greece and Spain
	Glyphosate-tolerant and Bt insect resistance in cotton	Greece and Spain
	Glyphosate-tolerant maize	Belgium, France, Germany, Greece, Italy, Bulgaria, Czech Republic, Hungary, Romania, Ukraine and Spain
	Bt-insect-resistant maize	France, Italy and Spain
	Glyphosate-tolerant and Bt insect-resistant maize	Belgium, Italy, France and Spain
	Glyphosate-tolerant OSR	Belgium, France, Germany and UK
	Glyphosate-tolerant OSR, plus male sterility/fertility restorer system	UK
	Glyphosate-tolerant soybean	France, Italy, Bulgaria, Romania and Spain
	Glyphosate-tolerant sugar/fodder beet	Belgium, France, Ireland, Netherlands, Germany, Italy, Spain, Sweden, Czech Republic, Hungary and UK
	Glufosinate-ammonium-tolerant maize	France and Italy
Syngenta	Glyphosate-tolerant fodder/sugar beet	Belgium, Finland, France, Germany, Greece, Netherlands, Italy, Spain, Sweden and UK
	Fungi-resistant potato gene switch technology OSR	UK
	Glufosinate-ammonium-tolerant fodder/sugar beet and maize	France
	Bt-insect-resistant maize	Spain and Italy
	Bt-insect-resistant and glufosinate-ammonium-tolerant maize	France, Italy, Portugal and Spain
	Bt-resistant and glyphosate-tolerant maize	France
	Fungi-resistant and glufosinate-tolerant spring wheat	UK
DuPont	Bt-insect-resistant and glyphosate-tolerant cotton and Bt maize	Spain
	Glyphosate-tolerant OSR	France, Germany and UK

(Continued)

Table 5.9 (*Continued*)

Corporation name	Input trait	Countries where trials were conducted
	Bt-insect-resistant maize	France, Greece, Italy, Portugal, Spain and Japan
	Bt-insect-resistant and glufosinate-ammonium-tolerant (liberty) Maize	France, Italy and Spain
	Bt-insect-resistant, glufosinate-ammonium-tolerant (liberty) and glyphosate-tolerant maize	France
	Glufosinate-ammonium-tolerant maize	France, Greece, Italy, Spain and Portugal

Source: Corporate Watch UK (2002a), Groundup (2003k), Genewatch (2003a,b,c) and BIO (2004b).

oils, meat, fruit and vegetables; and increase per capita food supplies in developing countries by 25% (GCPF 1997).

As the global population increases, and arable land declines, a 2–3% growth in agricultural productivity is required to meet growing food demand. Indeed, many within and outside the pesticide industry contend that intensification of existing land through increased inputs remains the only way to ensure adequate food production for the world's still growing population (Pesticide News 2000). Indeed, Bayer continually stresses the important business opportunity and mission to help feed the world's growing population (Søgaard 2001).

Their position has been supported by a string of academic studies, which developed models to assess the impact on crop yields with and without the use of pesticides. In the 1960s, estimates of crop losses in India without the use of pesticides totalled 25% for vegetables, 20% for potatoes, 20% for cotton and 20% for rice (Boardman 1986). In 1978, National Academy of Sciences (NAS) estimated that post-harvest losses ranged from 10% to 20%. When combined with pre-harvest losses, this figure increased to 45% (Boardman 1986). Indeed, according to Boardman (1986) and Oerke (2005), whilst significant variation exists between crops, on average, pesticides prevented crop losses of up to 70% or more.

Fear of significant crop losses, particularly in developing countries, who are dependent on maintaining yields of economically essential cash crops, has been, in many cases, enough to maintain the intensive use of pesticide inputs (Boardman 1986).

5.4.2 Improved farmer profitability?

The impact on farmer productivity of pesticide use throughout the past 40 years has generally been positive, with the exception of their often excessive and prophylactic usage, particularly in relation to the use of insecticides in developing countries. Traditional benefit/cost ratios for the use of pesticides ranged from 3:1 to 13:1, and up to 15:1 for herbicide use (Vorley and Keeney 1998). However, a growing number of food system analysts question the validity of traditional pesticide-based cost/benefit studies. Their concern is that inadequate provision is made for the human health and environmental externalities of pesticide use (Rola and Pingali 1993). For example, Rola and Pingali (1993) suggested that pesticide use in the Philippines actually cuts rice productivity instead of improving it when associated health costs are counted as production costs, and that during normal years, returns are low on high levels of pesticide use due to the presence of natural predators (Rola and Pingali 1993).

The impacts of the introduction of biotechnologies on farmer productivity have also been varied. For example, while Fernandez-Cornejo and McBride (2002) suggest that the adoption of HT maize has improved net returns on specialised production units, another study by Ferrell and Witt (2002) suggested that there is no cost advantage of using HT varieties compared to the best alternatives. In the case of cotton, the USDA-ERS determined that adoption of HT cotton was associated with increases in yields and variable profits, but with no significant change in the amount of herbicide use (Monsanto 2001).

Conversely, the adoption of HT soybean was associated with small increases in yields and variable profits, and a significant decrease in herbicide use (Monsanto 2001). In another study, the National Centre for Food and Agricultural Policy (NCFAP) indicated that the introduction of Roundup Ready soybeans had saved American growers hundreds of millions of dollars since 1996. In 1998 alone, the NCFAP report noted that Roundup Ready soybeans saved US soybean growers more than US$220 million (Monsanto 2001). Monsanto's Roundup Ready soybean system is estimated to have eliminated approximately 19 million acre treatments of herbicides during the 2000 crop year, according to the USDA Economic Research Service (Monsanto 2001). However, it was acknowledged that the impacts on net returns for HT soybean varied from farm to farm and from farming systems to farming systems and are significantly affected by weed pressures.

In addition, Fernandez-Cornejo and McBride (2002) were keen to point out that other less tangible benefits of using a herbicide such as Roundup (glyphosate), such as management ease, speed of operation and ability to keep fields weed free for landlords, are not reflected in standard calculations of net returns to farming (Fernandez-Cornejo and McBride 2002).

The adoption of insect-resistant crops has also received mixed reviews. According to Fernandez-Cornejo and McBride (2002), adoption of Bt cotton was associated with significant increases in yields and profits and decreased insecticide use. According to Purcell and Perlak (2004), Bt cotton increased US farmers' incomes by US$20/acre and the total net value of US cotton production by US$103 million in 2001. In China, income increased due to Bt cotton by between US$357/ha and US$549/ha (Pray et al. 2002). In South Africa, incomes increased due to Bt cotton by US$25.51/ha (Ismael et al. 2001). This was due to higher yields and reduced production costs.

Conversely, Fernandez-Cornejo and McBride (2002) suggested that, while adoption of Bt maize acts as a good insurance against severe insect losses, when stem borer pressure is low, Bt corn premiums are not matched by higher returns through yields. The best Bt hybrids had a yield increase of 5% compared to non-Bt hybrids (Deutsche Bank 1999). According to Fernandez-Cornejo and McBride (2002), BT maize generates positive returns only 1 year in 3 and has generally had a negative impact among specialist maize farmers. Ultimately, according to Deutsche Bank (1999), it was the higher yields and associated cost savings that assured wide acceptance of the Bt maize.

However, while economic studies have demonstrated that both farmers and consumers have benefited from the introduction of new biotechnologies (Duffy 2001), as a rule, the gene developer and germplasm supplier receive between 26% (Falck-Zepeda et al. 2000) and 45% of the total benefits.

5.4.3 Human health

While the drafting and implementation of pesticide and food safety legislation reduced the negative human health impacts of pesticides in developed countries, the use of pesticides remains a major risk in many developing countries. According to the FAO (2002), while more than 80% of pesticides are applied in developed countries, 99% of all poisoning cases occur in developing countries where regulatory, health and education systems are weakest. Indeed, according to FAO (2001), 30% of pesticides marketed in developing countries do not meet internationally accepted quality standards. According to the FAO (2001), these pesticides pose a serious threat to human health and the environment. Indeed, the FAO and WHO warned that 'these poor-quality pesticides frequently contain hazardous substances and impurities that have already been banned or severely restricted elsewhere, and that poor-quality pesticides are particularly widespread in SSA, where quality control is generally weak' (FAO 2001).

Estimations of both acute and chronic pesticide poisoning vary considerably. After a revision in 1977 by Copplestone (1977), a regular member of the WHO Expert Committee, the WHO upgraded its estimates of global unintentional deaths from pesticides from 5000, in 1973, to as many as 20 000 in 1977, primarily in developing countries (FAO 2002). In addition, due to the lack of official figures in many countries, these figures could be even higher (Pesticide News 1998b). In 1985, the WHO confirmed an annual death toll of around 20 000, and the number of unintentional poisonings at 1 000 000 annually (WHO 1986). By the end of the 1980s, acute pesticide poisoning reached 2–3 million annually, mostly in the developing world (WHO 1990; Pesticide News 1998b). An estimated 40 000 poisonings were fatal (ILO 1999). The World Resources Institute – WRI (2002) – suggested that the WHO's figure of 3 000 000 acute poisonings per year could be matched by a greater number of unreported but mild intoxications and acute conditions. For example, Vorley and Keeney (1998) suggested that there were 25 million cases of acute pesticide poisonings per year among agricultural workers in developing countries. The Helsinki Institute of Occupational Health reported 11 million acute pesticide poisonings in Africa alone (Vorley and Keeney 1998). Another study in the 1990s estimated that 50–100 million people in developing countries may receive intensive pesticide exposure, and another 500 million receive lower exposures. These exposures may result in some 3.5–5 million acute poisonings per year, with a much larger number of people suffering from subacute effects (WRI 2002).

In country-level reports of pesticide poisoning, China reported 48 377 poisoning cases in 1995, including 3204 fatalities. More than 7500 of these cases were mostly attributed to normal agricultural use of parathion and methamidophos (FAO 1997). In Malaysia and Sri Lanka, 7–15% of all farmers experience poisoning at least once in their lives. In Thailand, a survey of 250 government hospitals found that 5500 people were admitted for pesticide poisoning in 1985 alone, of who 384 died. In the Philippines, 50% of rice farmers have suffered from sickness due to pesticide use (Vorley and Keeney 1998). According to Knight (1997), pesticides affect at least 280 000 Brazilians, 2% of the population, each year (Knight 1997). However, the worst example of pesticide-related poisoning occurred on the night of 2 December 1984, when 42 tonnes of lethal methyl isocyanate (Vorley and Keeney 1998), approximately twice the amount the factory was authorised to store (Le Monde diplomatique 2004), leaked from Union Carbide's pesticide factory in Bhopal, India. It was the worst chemical disaster in history. By the third day of the disaster, between 2000 and 20 000 (Hough 1998) people were dead from direct exposure to the gases, and an estimated 17 000 permanently disabled (Le Monde diplomatique 2004). Union Carbide was manufacturing Carbaryl, AKA Sevin, which is not a particularly hazardous pesticide (WHO category II). However, the intermediate chemical methyl isocyanate is extremely toxic, approximately five times as powerful as mustard gas (Le Monde diplomatique 2004). Several accidents had already occurred at the plant (established in 1981) because of allegedly lax safety precautions. In 1982, an internal report by Union Carbide described the site as possessing 'serious potential for sizeable releases of toxic material' (Weir 1987). Articles were even printed about the danger of chemical leaks in the local press (Hough 1998).

Conversely, there are others who believe that figures related to pesticide poisoning are exaggerated and that, where pesticide poisoning has occurred, the cause has been the improper use of pesticides or the implementation of inadequate safety procedures and/or equipment. The Ecologist (1980) published an article by Professor Keith Mellanby insisting that the WHO's pesticide poisoning figures in 1977 were probably grossly overexaggerated and that the true figure was likely to be in the hundreds and not hundreds of thousands (Ecologist 1980). Not surprisingly, the WHO's figures are also hotly contested by the pesticide industry. They claim that the WHO's estimates are based on unrepresentative sampling strategies. For example, the 500 000 poisonings estimated in 1972 were based on a survey of only 19 countries, and the 1977 study based on only 9 countries and governmental notifications (Hough 1998). US EPA estimates of between 20 000 and 300 000 acute pesticide poisonings per annum among agricultural workers are also much lower than the WHO (ILO 1999).

While the chronic effects of pesticides are much less well understood, regular exposure to pesticides has been associated with a growing list of medical conditions, including: dermatitis, immune system suppression, cancer, reproductive damage, eye problems, pulmonary disease, pancreatitis, hyperamylasaemia, kidney and neurological effects (Pingali and Rogers 1995; WRI 2002; FAO 2002). Recent studies have linked 2,4-D with a heightened risk of birth defects and non-Hodgkin's lymphoma (WRI 2002), but the probable effects of 2,4-D, as with many other pesticides, are hard to prove (The Ottawa Citizen 2003; Pesticide News Brief 2003). For example, it has been suggested that 2,4-D may be harmful to the male reproductive system but as it is obviously difficult to test the toxic effects of 2,4-D and other pesticides directly on humans, virtually all studies on their effects on human fertility are either case reports or epidemiological (i.e. population-based, case control or cohort studies). As such, it is difficult to obtain clear 'proof' of their adverse effects, as might be done with a prospective placebo control study of a new medication for the treatment of a disease. However, available studies do provide very compelling data showing that herbicides and pesticides are likely to have profound adverse effects on human reproduction. Exposure may reduce sperm counts, affect egg quality, increase miscarriage risk and increase the risk of birth defects. Adverse effects of some herbicides and pesticides have been reported in animal studies, lending biological plausibility to the argument that exposure to these substances may lead to reproductive problems in people (Pesticide News Brief 2003). Glyphosate has also been linked to cancer (PAN

1999; Pesticides and You 2015). Organophosphate and carbamate insecticides have also been linked to anxiety, confusion, depression (including suicidal thoughts in severe cases), headache, fatigue, impaired concentration, incoordination, increased sensitivity to repeated exposure to organophosphates and a higher incidence of leukopenia (very low white blood count) (Davignon et al. 1965; The Lancet 1981).

Whilst most agree that the human suffering and fatalities attributed to pesticide use is unacceptable, pesticide poisoning is not always the result of the inherent properties of the pesticides themselves. While it is possible to infer a causal link between the use of older and generally more toxic pesticides in developing countries and the predominance of pesticide-related human health problems, most pesticide corporations have worsened the situation by actively promoting potentially hazardous pesticides as safe (Bull 1982).

Indeed, American Cyanamid's Cythion (Malathion) was sold as 'the safe insecticide', which enables you to 'be sure of a bumper coffee crop' (Bull 1982). Sumitomo also advertised Sumithion (fenitrothion), another organophosphate, as 'safe', whilst it was classified as moderately hazardous by the WHO. Sumitomo admitted that the advert was not sophisticated but would try to improve in the future (Bull 1982). In India, the pesticide lobby's response to a ban on aerial spraying of endosulfan allegedly consisted of a barrage of disinformation and the lobbying of government agencies (Downtoearth 2011). Articles, interviews and advertisements began appearing in the media suggesting that endosulfan was a safe pesticide, including a study undertaken by the Fredrick Institute of Plant Protection and Toxicology that reconfirmed endosulfan's safety. Activists opposing endosulfan were threatened with legal action, and the extensive lobbying of scientists and government officials eventually resulted in a lifting of the ban on endosulfan (CSE 2011).

On the other hand, however, a significant proportion of pesticide poisonings appear to result from the inappropriate use of, or inadequate protection from, moderately to extremely hazardous pesticides. A study by Jeyaratnam et al. (1982) demonstrated that 73.1% of the 13 000 people admitted to hospitals in Sri Lanka because of acute pesticide poisoning, of which 1000 died, were suicide attempts (Jeyaratnam et al. 1982). A later survey led by Jeyaratnam in 1987 found that Indonesia, Malaysia and Thailand all had suicide levels between 60% and 70% of total pesticide poisoning cases (Jeyaratnam 1987). In Papua New Guinea, during the late 1980s, paraquat killed 100 people annually (Boardman 1986). Based on Jeyaratnam's work, the WHO estimated that there were approximately 2 000 000 suicide attempts, and 200 000 were successful (WHO 1990). In addition, many cases of pesticide poisoning result from lack of appropriate safety precautions (Rola and Pingali 1993). These include the absence or inadequacy of protective clothing when handling pesticides (ZIPR

1997), lack of information about the dangers of pesticides, lack of understanding of the dangers of pesticides, and poor personal hygiene.

Pesticide poisoning is worse in developing countries due to the neglect of safeguarding employee-working conditions by management (Hough 1998). On the one hand, employers may not provide protective clothing because of its high cost. On the other hand, even if provided, many employees resist using of protective clothing (which is often rubber-based) in hot and humid environments (FAO 2002). In addition, older pesticides, which are cheaper and easily available, are generally preferred by poor farmers, and cost-conscious employers. However, as mentioned earlier, these pesticides are often the most dangerous to both health and the environment (Pesticide News 1998b). In addition, many pesticide products manufactured and/or sold in developing countries are often of poor quality and may be contaminated with toxic substances or impurities (FAO 2001).

A study by the Zimbabwe Institute of Permaculture (ZIPR 1997) found that despite the dangers of dimethoate (an organophosphate pesticide), farmers did not use any protective clothing except a piece of cloth to cover the mouth or nose. All application techniques involved direct contact with pesticides through handling, breathing and sometimes orally (ZIPR 1997). In Pakistan, in 1976, 2800 workers spraying Malathion during a malaria eradication programme were poisoned, and five died due to insufficient safety precautions (Baker 1978). The major determinant of the poisoning was identified as isomalathion present as an impurity in the Malathion (Aldridge et al. 1979). A study conducted by Knight (1997) suggested that, in most cases, hired farm workers often knew little, if anything, about the chemicals they were using or the precautions they need to take, and that they often worked with faulty or outdated equipment (Knight 1997).

In many developing countries, health and education systems are weak or virtually non-existent (FAO 2002). When combined with the advertising strategies employed by many pesticide corporations, suggesting that many potentially hazardous pesticides are indeed safe, pesticides are often seen as miracle cures for agricultural problems and, with the exception of paraquat and other pesticides used to aid suicide attempts, are not regarded threats to farmer health. In turn, this leads to the institutionalisation of a lack of respect for pesticides and, ultimately, their abuse both in their handling and storage (ANOHS 1997; FAO 2002). A 1991–5 survey of 2500 farmers and agricultural workers (mostly women) from Indonesia, Malaysia, Korea, India, Sri Lanka, Pakistan and the Philippines, by the Pesticide Action Network, recorded women's frequent, direct exposure to pesticides during application and again when entering fields to hoe, weed, thin plants and harvest. Additionally, women were exposed when mixing pesticides, washing spray tanks, washing pesticide-soaked clothes and disposing of pesticide containers. Illiteracy and access to information and health care compound the risk to farmers (particularly women) as they have limited ability to read or understand the written warning labels on pesticide containers, let alone the instructions for use and disposal of unused pesticides. Despite this, government-sponsored, often internationally financed, training programmes which teach men about the hazards of pesticides rarely reach women (not only in Asia, but throughout Africa as well). Such programmes mistakenly assume that only men need this information, or that men will pass their new knowledge on to the women who are both applying pesticides and storing food in old pesticide containers. Problems relating to poor or inadequate information are also widespread in Africa and Central and Southern America, where information regarding the safe use of pesticides is often technical and complex. In addition, labels are often in the wrong language and 'fail to provide data on the AI, application, and date of manufacture and safe handling of the chemical' (FAO 2001). Indeed, labels from pesticide containers may be missing or damaged.

5.5 Summary

Chapter 5 focuses on the impact of the global Neo-Productivist developmental context on the strategies employed by pesticide corporations from the mid-1970s onwards. It suggests that, in contrast to the USA, Western Europe and Japan, where Keynesian-styled support for Fordist-styled industrial agro-food systems remained pretty well entrenched, the agricultural sectors of many other countries, in both the developed and developing world, became increasingly exposed to relatively unregulated market forces, brought about by the implementation of neo-liberal economic policies. In parallel with the Residual Productivist context, the Neo-Productivist context of pesticide evolution arose during the early to mid-1970s. Indeed, it shared many of the same drivers of Residual Productivism, namely: falling real unit prices for staple commodities and stagnant world markets; resistance to a growing range of pesticide AIs; expiry of pesticide patents; increasing pressure to reduce negative human health impacts of pesticide in Western Europe, the USA and Japan; and scientific and technological advances. However, the Neo-Productivist context of pesticide industry evolution differs from the Residual Productivist context in several important respects: (1) Concern about the continued durability of political and financial support for high-input/-output industrial farming systems in Western Europe, the USA and Japan. (2) A reduction in production-based (and by default pesticide-based) farm subsidies. (3) Introduction of neo-liberal economic and trade policies. (4) Increased globalisation. (5) Industrialisation of farming systems in developing countries. (6) Non-existent, or limited, regulation of pesticide production and use in developing countries.

This chapter focuses on the strategic responses of the pesticide industry to declining profitability in the core pesticide markets in Western Europe, the USA and Japan, particularly global strategies that attempted to increase profitability and market domination by creating and capturing pesticide markets in developing countries. Increased profitability and market domination was primarily achieved through price competition on generic pesticides, and, latterly, through the application of new plant biotechnologies. These strategies include: (1) Attaining increasing economies of scale. (2) Increased global market share. (3) Access to new pesticides and biotechnologies and important R&D pipelines. (4) Increasing control of pesticides in the global agro-food sector through global horizontal M&A. (5) Extending control over extensive seed production and distribution networks. (6) Increasing levels of appropriation from new biotechnologies. (7) Increasing political and economic power through global vertical M&A of biotechnology and seed companies. (8) The production and sale of off-patent generic pesticides in low-cost locations, specifically areas where weak regulations fail to protect workers, consumers and the environment.

Private sector TNCs, including agro-biotechnology corporations, have been particularly successful in helping to persuade international regulatory institutions, such as the World Bank and the WTO, to open up new and profitable markets in developing countries. Overall, whilst pesticide corporation strategies generated significant negative outcomes, on a positive note, agricultural output, productivity and national food security rapidly increased in the majority of developed and developing countries throughout this Period.

Chapter 6

The evolution of the pesticide industry under Post-Productivism (1974–2017)

6.1 Key characteristics of urban-centred industrial agro-food systems and broader economic, social and political contexts

Chapter 6 outlines strategies employed by the pesticide industry in response to a combination of key drivers that created a Post-Productivist or Post-Fordist context of development. The term 'Post-Fordism' has been used to describe a range of key social, political, economic, organisational and technological characteristics that define, what, in line with Neo-Fordism, many have proclaimed to be, the beginnings of a new and stable regime of accumulation and mode of social regulation. Indeed, in contrast to McMichael's proclamation of an emergent corporatist food regime, Friedmann (2005) has hinted at the possible emergence of a corporate-environmental food regime.

According to Friedmann (2005), the corporate-environmental food regime represents a 'convergence of environmental politics and corporate repositioning, especially through retail-led reorganization of food supply chains (the "supermarket revolution") aimed – in very different ways – at increasingly transnational classes of rich and poor consumers'.

Just as Fordism was production-led, Post-Fordism is consumer-led. Whereas, Neo-Fordism represents an adjustment to the problems of Fordism, a way forward that extends the phase of Fordism, Post-Fordism is suggested to represent a qualitatively new economic direction, a step beyond Fordism. Post-Fordism and the Third Food Regime have also been described as a distinctive movement from mass production to flexible specialisation; Fordism to

Post-Fordism or flexible accumulation; organised to disorganised capitalism; and industrial capitalism to the information age or the network society. These labels have been useful in raising awareness of a whole series of changes in capital and labour markets, corporate strategies, industrial technologies, industrial geographies, financial regulation, labour processes, managerial hierarchies and the restructuring of the (welfare) state.

Having said this, Post-Fordism embodies several characteristics in common with the Fordist Regime of accumulation, namely, continued: urbanisation; separation of ownership and control; and the growth of the professional, managerial and administrative (service class). However, as the prefix 'Post' implies, Post-Fordism embodies a range of characteristics that allegedly endow it with the capacity to succeed where Fordism could not. These subtle differences include; (1) Supply-side restructuring of production; (2) Supply-side restructuring of labour; (3) Demand-side adjustments (market fragmentation and information flow); (4) Usurpation of the power of national governments in economic management; and (5) Enhanced accumulation through increased commoditisation.

6.1.1 Supply-side restructuring of production

Post-Fordism signifies a qualitative shift in the organisation of production and consumption, as well as a break in the mode of regulation. While Post-Fordism remains committed to the 'scientific management' of labour and assembly line production, it does so in such a manner as to overcome the slowdown in productivity and declining profitability that allegedly signalled the end of the Fordist Regime. In line with Neo-Fordism, Post-Fordist principles are suggested to enhance labour productivity and therefore reduce unit costs in order to remain both competitive and profitable. To this end, the scientific management of labour and assembly line production is refined in such a way as to minimise costs and maximise productivity. Indeed, proponents of Post-Fordism have also laid claim to 'lean production', 'Toyotism' 'Just-in-Time' (JIT) systems and Total Quality Management. However, whilst the elimination of unnecessary actions of people or equipment that do not contribute to added value is also a key part of production, Post-Fordism embodies a range of additional productivity-enhancing characteristics. These include the flexibility to undertake short production runs and product life, greater product innovation and the production of a wider range of products on offer in response to changing consumer demand (Piore and Sabel 1984). Whilst both Neo-Fordist and Post-Fordist manufacturers share the ability to switch from economies of scale (mass) to economies of scope (batch), and develop new ways of organising work to improve product quality, Neo-Fordism does not embody the degree of flexibility and fluidity and demand-side responsiveness as Post-Fordism.

6.1.2 Supply-side restructuring of labour

In the Post-Fordist context, bureaucratised unions no longer represent the interest of the new, highly differentiated labour force. Decentralised collective, even individual, bargaining replaces centralised negotiations. However, contrary to Neo-Fordism, which associates the adoption of new technologies with job deskilling and increasingly centralised managerial control, Post-Fordism insists that productivity and profitability can be assured through the hard work of a highly skilled (often reskilled and multi-skilled) and an often higher paid (due to the possession of specialist and idiosyncratic skills) labour force operating in a less hierarchical work environment (Hodgson 1999).

Proponents of Post-Fordism herald a post-division of labour between workers and management, with concomitant flatter management structures (de-layered firms, more lateral communication, organic rather than bureaucratic, professionalism and entrepreneurism). Worker empowerment and involvement is said to greatly improve product quality and timeliness of delivery and even foster continuous equipment maintenance. It is empowered individuals, with greater responsibility and autonomy, which are more likely to become more intimately involved and concerned with the production process. Indeed, the concept of 'quality circles' is based on the premise that workers with higher motivation and capacity that perform the same tasks will meet on a voluntary basis to discuss improvements in the product and the way they work. Whilst reducing unit costs of production is of paramount importance in both the Post-Fordist and Neo-Fordist context, it is suggested that higher levels of productivity and profitability can be achieved through the utilisation of flexible technologies in conjunction with flexible labour in the production/delivery of increasingly complex, sophisticated and knowledge-based products and services (Hodgson 1999). Learning, learning processes, knowledge management and innovation seem to become more and more important in the economic world. The interaction between knowledge, learning and economics becomes so intense that this gives rise to terms as the knowledge-based economy and the learning economy. After all, many countries, which cannot, or do not want to, provide even cheaper labour (race to the bottom), choose the phrase that who cannot be cheap (Neo-Fordism), has to be smart (Post-Fordism).

A dynamic process of continuous improvement and learning is suggested to replace the rather static price competition. TNCs now rely upon factories located around the world to produce parts that are later assembled in one locale. Post-Fordism is suggested to create a new international division of labour in which manufacturing activity is increasingly located in the Third World, where labour costs are low and potential profits are high, whilst R&D takes place in places like Silicon Fen (Cambridge) and Silicon Valley (California), which can attract top-quality scientists and engineers. TNCs increasingly rely upon subcontractors, part-time labour, JIT inventory systems and other information technologies to minimise costs. Indeed, advanced knowledge and key skills are being required in many processes of production to cope with growing difficulty and complexity (Hodgson 1999). In this Post-Fordist context, individual workers would display heterogeneity not only in relation to the terms and conditions of employment, and the variety of skills that they possess, but also in their demands as consumers.

6.1.3 Demand-side adjustments (market fragmentation and information flow)

Breakthroughs in communication technologies, the increased mobility of people and the deregulation of exchanges have induced significant changes in the pattern of consumption, characterised by a broader, more diverse, and fluctuating demand.

The Post-Fordist market place is therefore segmented, niche and one that changes as quickly as consumer tastes (Amin 1994). In this context, goods and services increasingly become specialised, customised, high quality and high value added. The Post-Fordist regime of accumulation is driven by new information technologies and organisational networks. Knowledge and information is now essential for the speed and astuteness of market trading and in the coordination of responses to volatile global demand. In order to further enhance flexibility, Post-Fordist strategies embody the vertical disintegration of production. New relations of production are based on the decentralisation of tasks and responsibilities, to loosely interconnected units, and increased cooperation through joint ventures and strategic alliances. Indeed, the Post-Fordist period has been characterised as an era of 'disorganised capitalism', where corporations coordinate activities between themselves and focus on continuous improvement of processes of the entire value chain. In this way, suppliers and buyers are treated as long-term partners; they often must be trained in ways to reduce set-up times, inventories, defects, machine breakdowns and so forth in order to enable them to take responsibility for delivering the best possible parts/services to the manufacturer in a timely manner. The introduction of computers has meant that distribution systems have enabled retailers to avoid overstocking and has allowed mass markets to be split up into the targeting of specific groups. Targeting specific groups of consumers has meant that design has become a major selling point. Commodities are no longer bought simply for the use value they have, but also for the lifestyle that goes with their design.

Five developments have enhanced the importance attached to knowledge production, use and circulation in economic life today. (1) Processes of production and their products are becoming more complex and sophisticated.

(2) Increasingly, advanced knowledge and skills are being required in many processes of production to cope with growing difficulty and complexity. (3) There is an increasing reliance on specialist and idiosyncratic skills. (4) The use and transfer of information is becoming more extensive and important for economic activities. (5) There is increasing uncertainty in contemporary economic life (Atkins and Bowler 2001).

As all manner of knowledge is being commodified in contemporary capitalism, it is clear that there are wide variations in the capacities of agents to participate in the processes of knowledge creation, circulation and management (Atkins and Bowler 2001). Yet, under informational capitalism, such capacities are critical to the futures of cities, regions and their inhabitants. The production and transformation of knowledge thus becomes another element in the experience of uneven development (Atkins and Bowler 2001).

6.1.4 Reconditioning of national governments in order to develop new areas for accumulation

On a macro-level, Post-Fordism involves a reconditioning of the state in line with Post-Fordist goals. In line with Neo-Fordism, Post-Fordism embodies the mobilisation of key social groups to displace Keynesianism and replace it with a neo-liberal economic strategy, that is, one based upon the economic rationality of the private market. Many see the liberalisation of the banking and financial system as central to the rise of Post-Fordism since the 1970s. Indeed, the banking and the financial system has achieved a substantial degree of autonomy, enabling it create havoc with the stability of material production as well as to overcome the rigidities of Fordist-type production and consumption. The formation of new financial markets, the introduction of new financial instruments, and the opening up of new systems of global coordination between financial centres, have, according to Harvey, carried capitalism into a new era in which the rapidity and scale of capital flows make it more difficult for nation states to secure stable accumulation strategies. At the same time, innovation within financial systems has enabled companies, governments and consumers to adopt more flexible strategies towards the 'blockages' of Fordism (Harvey 1996).

Whereas both Neo- and Post-Fordists stress the importance of greater competitiveness, only Post-Fordists stress the need for greater morality in the marketplace. Whereas Neo-Fordism sought the almost total withdrawal of the state, except for vital areas over which corporations had no direct control, Post-Fordism envisions a key role for the state through the development of new, generally post-industrial, opportunities for accumulation. These opportunities were expected to arise in the fields of public services, welfare and education.

6.1.5 Enhanced accumulation through increased commoditisation

On many occasions, Post-Fordism has been likened to post-industrial and even post-modern society, and has often been described as a period of Post-Productivism and de-industrialisation, characterised by growth in the service and leisure sectors, and a growing concern for quality of life and the environment. In the rural domain, the Post-Productivist climate is said to be creating the opportunity for new rural accumulation, associated with the production and consumption of a wide range of value-added commodities and aspects of what Bourdieu (1984) calls 'culture capital'. Capitalism itself is said to be moving into a phase in which the cultural forms and meanings of its outputs become critical if not dominating elements of productive strategy, and in which the realm of human culture as a whole is increasingly subject to commodification, that is supplied through profit-making institutions in decentralised markets. In other words, an ever-widening range of economic activity is concerned with producing and marketing goods and services that are infused in one way or another with broadly aesthetic or semiotic attributes. Increasingly, local and national governments have recognised this connection between the cultural and economic, and have sought to reap the benefits by deliberately formulating and implementing policies that harness the linkage.

6.2 Key external drivers of pesticide evolution

6.2.1 Growing intergovernmental pressure to reduce negative human health and environmental impacts of pesticide use

Pressure to reduce the negative human health impacts of pesticides continued to originate from the international scientific community, emerging grassroots environment and consumer organisations, and UN organisations, particularly FAO and the WHO (see Fig. 6.3). In many ways, Earth Day, held on 22 April 1970, in the USA, where over 20 million concerned citizens voiced their anger over the impact of modern industrial economic activity on human health, was the release of years of public concern. Their actions, in conjunction with the resonance and sympathy within/from the American public, acted to raise the political profile of public health issues, and, once again, placed public health issues in the political arena. Following quickly on the heels of Earth Day public demonstration, the Stockholm Conference of 1972 continued this momentum by stimulating a great deal of scientific interest in the effects of human economic activity, on human health. This had the effect of spurring a spate of scientific analyses, including studies analysing the effects of pesticides on human health, specifically the impacts of organochlorines, organophosphates and carbamates.

In 1975, the WHO Standard protocol was published in the guise of the 'Survey of Exposure to Organophosphorous Pesticides in Agriculture'. The Standard Protocol could be used to determine the levels of organophosphorus exposure deemed hazardous (Hough 1998). In addition, because of the growing controversy surrounding the unwanted effects of DDT and concerns over growing mosquito resistance to DDT, the WHO formally abandoned its global eradication programme in 1975 (Hough 1998). Maintaining the momentum generated by its publication of a 'Global Assessment of Pesticide Poisoning' in 1973, the WHO raised the pesticide stakes by releasing its very first 'Recommended Classification of Pesticides by Hazard' publication in 1978 (Hough 1998). This publication classified pesticides in accordance with their toxicity to humans. By default, WHO singled out what they perceived to be the most hazardous pesticides on the market. This has been, and still is, one of the most influential and widely used publications (WHO 2003a). Indeed, it has been adopted as standard pesticide classification mechanism across the world, with the notable exception of the USA, Bulgaria, Brazil, Canada and Japan (Hough 1998). Also, in 1978, the WHO's Expert Committee on the Safe Use of Pesticides recommended the extension of the worker exposure safety protocol to include other types of pesticides.

Subsequently, in 1980, and because of an acknowledgement of significant overlap in their areas of activity, the WHO, in collaboration with the United Nations Environment Programme (UNEP), and the ILO, incorporated its activities concerned with the safety assessment of food chemicals into the International Programme on Chemical Safety (IPCS) (IPCS 2003) (see Fig. 6.3). This was undertaken as a follow-up commitment to the Stockholm Convention of 1972, in which the WHO was expected to establish mechanisms to provide 'early warning and prevention of harmful effects of chemicals to which humans were being increasingly exposed, and for the assessment of the potential risks to human health' (WHO 2003b). The two main roles of the IPCS were to establish the scientific health risk assessment basis for safe use of chemicals (normative functions) and to strengthen national capabilities for chemical safety (technical cooperation) (WHO 2003b). IPCS areas of activity include the evaluation of chemical risks to human health and environment (including a major initiative on endocrine disruptors). Between 1976 and 2000, approximately 220 risk evaluations of priority chemicals were undertaken. This effort culminated in the production of Environmental Health Criteria Documents. Priority chemicals included individual chemicals, groups of chemicals, forms of radiation and biological agents. In addition, the IPCS also produced Health and Safety Guidelines and 20 Concise International Chemical Assessment Documents (DEFRA 2001). The IPCS has also been actively involved in the development and validation of methods; and harmonisation of approaches to risk assessment of chemical exposure, with the objective of harmonising generic and technical terms used in chemical hazard/risk assessment (DEFRA 2001).

A major harmonisation effort occurred in October 1977 (Boardman 1986), when the IPCS attempted to address harmonisation concerns regarding the chemical/physical properties of pesticides; assessment of biological activity; efficacy and crop safety; toxicology; determination of pesticide residues in food and feed; environmental impacts; and problems with labelling (Boardman 1986). Coincidentally, GIFAP was also in favour of the harmonisation of pesticide registration (Hough 1998). Industry stressed that harmonisation was a prerequisite for solving the world's food problem, and that it was in everybody's interests (Boardman 1986). However, their interpretation of harmonisation of pesticide registration schemes rested on assumptions of their financial costs and benefits rather than hazards. IUPAC also regularly provided technical assistance to IPCS with regard to pesticides and human health risks (Hough 1998).

From 1977 onwards, industry forums attacked issues of environmental criteria for registration because it left them vulnerable to environmental groups and regulatory authorities. However, confidentiality was still the main concern (Boardman 1986). In 1979, the FAO's specification group met again to discuss guidelines for confidentiality of company pesticide data (Boardman 1986). FAO expert consultations with GIFAP were convened to discuss environmental criteria for registration of pesticides in 1979 and 1981 (Boardman 1986). The 1979 consultation adopted the principle that developing countries 'should be encouraged and supported in the initiation and/or development of pesticide registration schemes, which include attention to potential environmental effects' (Boardman 1986).

In 1981, the Director General of FAO suggested that a code of conduct on the distribution and use of pesticides could help to overcome a number of difficulties associated with pesticides (see subsections 6.3.1.1.2.2 and 6.4.1.2.1). Later that year, the FAO Second Government Consultation on International Harmonisation of Pesticide Registration Requirements endorsed the recommendations of PEPSRRAS and authorised drafting of a code after consultation with relevant UN bodies and non-UN international organisations (UN 2003a). This action represented a response by FAO to NGO pressure for action to be taken on the issue of human poisoning by pesticides. David Bull of Oxfam was even given the task of drawing up an initial list of proposals for the code (Paarlberg 1993). GIFAP supported FAO's role in the Code of Conduct development, if there was going to be one (Boardman 1986). In 1982, a US delegation stressed the need for protection for IPR on proprietary data, plus coverage of the Code of Conduct for pesticides in developing countries. This meeting was attended by IOCU (Boardman 1986). However, whilst FAO and Oxfam's desire was to raise standards across the world, GIFAP's motivation was principally to reduce industry costs incurred under heterogeneous registration systems

(Hough 1998). In the same year, the WHO published a revised and updated standard worker exposure (Hough 1998).

6.2.2 Increasing pressure from NGOs for national governments and intergovernmental institutions to impose tighter regulations on pesticides and biotechnologies in order to reduce negative human health and environmental impacts of pesticide use in both developed and developing countries

As the evidence of negative human health and environmental effects of pesticides became overwhelming, public concern coalesced through the establishment of new environmental pressure groups and a rapid expansion of many existing ones throughout the mid-1970s and 1980s (Winter 1996). For example, Friends of the Earth formed in 1970 was followed by Greenpeace in 1971 (Winter 1996). The People Party formed in 1973 was re-launched as the Ecology Party in 1974 and then renamed the Green Party in 1985 (Winter 1996). In the UK, the Royal Society for the Protection of Birds (RSPB) membership expanded from 12 000 in 1960 to nearly 900 000 by 1995, and to over 1 000 000 by the turn of the twenty-first century.

Increasingly, pressures to reduce negative human health and environmental impacts of pesticide use have arisen from a growing number of environmental and non-governmental organisations and civil society groups in developing countries (see Fig. 6.3). Although global data on environmental NGOs is scarce, one estimate suggests that by 1990 there were more than 100 000 environmental protection groups worldwide, most of them founded in the previous decade (WRI 2003). In the developing world, many environmental and civil society organisations arose in response to increased incidences of environmental pollution and health problems, including those involving pesticides. Environmental and consumer health-based NGOs have expanded in line with the rapid explosion of NGOs across the world, particularly since the early 1990s.

According to the Union of International Associations, the number of recorded NGOs more than doubled between 1985 and 2002 to over 47 000 (WRI 2003). By the mid-1990s, approximately 1 000 000 NGOs were operating in India, 210 000 in Brazil, 96 000 in the Philippines, 27 000 in Chile, 20 000 in Egypt and 11 000 in Thailand (WRI 2003). In conjunction with higher incomes, better education and more information, The Rio Earth Summit in 1992, and its emphasis on community/grassroots participation, initiated a significant increase in the number of NGOs and civil society organisations and dramatically enhanced their powers to set and influence agendas at national and international levels (WRI 2003).

Accordingly, in the UN system, 2143 NGOs held consultative status in 2003 (WRI 2003) compared to 928 in 1992 and just 222 in 1952 (WRI 2003).

Civil society organisations have been increasingly effective in both the national and international policy arenas. Since 1982, NGOs and consumer organisations have been putting pressure on the World Bank to improve its pest management activities and reduce pesticide use. In addition, contacts between European environmental organisations and developing countries expanded by the 1980s, as well as a considerable body of literature on pesticides problems (Bull 1982). Consumer protection groups actively forged transnational links through such bodies as the International Organisation of Consumers Unions (IOCU) (Boardman 1986).

As more scientific information on the potential impacts of pesticides became available during the 1970s and 1980s, concern over the use of pesticides increasingly spread from the international scientific and UN communities into the public domain. Much of this information, in the form of scientific and industrial reports, was released into the public domain by national governments in an attempt to increase transparency and allay criticisms that public health was continually put at risk in order that the pesticide industry could maintain its profits.

A growing tide of concern eventually led to the creation of a string of single-issue anti-pesticide organisations. The two most prominent examples are the National Coalition Against the Misuse of Pesticides (1981) (Beyond Pesticides 2003) and Pesticide Action Network International (PAN) in 1982 (PANNA 1995). PAN emerged as a result of a conference in Malaysia in 1982, where international and national NGOs agreed to coordinate efforts to tackle the growing pesticide problem. PAN's strategy was to influence the regulation of pesticide use and trade through the FAO (Hough 1998). Indeed, in its founding year, PAN proclaimed that it 'recognised the pivotal role that FAO plays in formulating and promoting major advances in agricultural policies and practices around the world' (PAN 1987). At the same time, Oxfam's publication 'A Growing Problem: Pesticides and the Third World' explicitly set out the safety standards it recommended governments, international organisations and the agrochemical industry to observe (Bull 1982). In 1984, PAN International compiled its own list of 12 of the world's most hazardous pesticides, many of which corresponded to those identified by the WHO in 1975. The so-called 'dirty dozen' list was comprised of; aldicarb, campechlor, chlordane, heptachlor, chlordimeform, DBCP, DDT, aldrin, endrin, EDB, HCH/BHC, Lindane, paraquat, parathion, methyl parathion, PCP and 2, 4, 5-T (PANNA 1995). A year later, on World Environment Day, PAN International launched its 'Dirty Dozen' campaign, aimed at putting pressure on the pesticide industry, the UN and national governments to do the following: (1) Ensure that human safety is considered foremost in all pesticide policy decisions. (2) End the use of dirty dozen pesticides whenever their safe use cannot be assured. (3) Generate support for research and implementation of safe pest control methods (PANNA 1995).

PAN International and other anti-pesticide groups have ardently strived to catalogue the growing list of human tragedies, caused by both appropriate and inappropriate use of pesticides, and document an expanding inventory of pesticide-based problems, in order to further their aims. For those closest to the pesticide industry, it was no surprise that organochlorine pesticides were classified by the WHO as extremely hazardous, and were conspicuously present in PAN's dirty dozen. The most hazardous pesticides were the 'drins'. This group included popular chlorinated hydrocarbons insecticides such as aldrin, dieldrin, endrin, chlordane, DDT, HCB, Heptachlor, mirex and toxaphene. According to PAN, exposure of people to drins has led to numerous poisonings and deaths. Many incidents have been reported. For example, the consumption of bread made from endrin-contaminated flour that affected at least 936 people and caused 26 deaths (Greenpeace 2002).

In addition to organochlorine pesticides, many other popular agricultural pesticides at the time also contained chlorine. Indeed, worldwide, half of the ten top-selling herbicides were chlorinated (alachlor, metolachlor, 2, 4-D, mecoprop, cyanazine and atrazine). Four of the top ten insecticides (chlorpyrifos, fenvalerate, endosulfan and cypermethrin) and fungicides (propiconazole, chlorothalonil, prochloraz and triadimenol) were chlorinated. In the USA, the chlorine industry reported that all herbicides, 87% of insecticides, 96% of fungicides and 85% of other crop protection chemicals depended on chlorine chemistry in some way. In the UK, approximately 25% of pesticides used were chlorinated. In the arable sector, five of the top 12 pesticides used were chlorinated pesticides. In decreasing order, they were atrazine, simazine, diuron, 2, 4-D and mecoprop (PAN 1994).

According to the WRI, a high percentage of pesticide poisonings worldwide are attributable to two particularly toxic classes of pesticides: organophosphates and carbamates, many of which are banned or restricted in developed countries. The toxicity of these chemicals stems from their ability to inhibit the action of cholinesterase (WRI 2002). Although large-scale poisonings by pesticide-contaminated food are unusual, they do occur.

Organophosphates have been implicated in human health scares, but as a pesticide class, they vary considerably in their toxicity to humans. Two of the most potentially harmful examples are parathion and Malathion (particularly impurities contained in Malathion-based pesticides). Between 1984 and 1990, parathion was ranked second as the cause of reported systemic respiratory pesticide-related illnesses in California (William et al. 1993). Poisoning from parathion occurs even in industrialised countries despite stringent protection. According to the FAO (1997), there are more reported cases of pesticide poisoning with parathion than with any other pesticide currently in use. Malathion is also suggested to be linked to the onset of overt renal insufficiency with massive proteinuria (Albright et al. 1983). Reports have shown that consumers have also been at risk of acute poisoning following the consumption of green leafy vegetables that had been treated shortly before harvest with organophosphate pesticides (FAO 1997).

Carbamates have also been linked to human health problems. In 1985, almost 2000 consumers on the west coast of the USA and Canada were poisoned by aldicarb-contaminated watermelons (Schonfield et al. 1995); six deaths and two still births were reported (WRI 2002). In 1989, US EPA toxicologists estimated that tens of thousands of infants and children a day were exposed to enough aldicarb residues in bananas and potatoes alone to pose a risk of illness (Siedenburg 1991).

Increasingly, efforts throughout the 1980s and 1990s aimed to secure universally accepted standards and conventions for the production, distribution, use and indirect consumption of pesticides, particularly the most hazardous examples. Momentum towards the global regulation of pesticides, initially generated during the Stockholm Conference of 1972, continued throughout this period. While many consumer and environmental groups demonstrated vehement opposition to the manufacture and use of pesticides, some NGOs developed a more balanced and less confrontational approach to pesticides. Indeed, Oxfam openly acknowledged the importance of pesticides in integrated agriculture and public health programmes, particularly in developing countries (Bull 1982; Boardman 1986).

6.2.3 Weakened legitimacy of pesticide use in developed countries

Since the humble beginnings of Rachel Carson's 'Silent Spring', the Earth Day demonstration in 1970, the Stockholm Conference of 1972, the publication of WHO's Global Assessment of Pesticide Poisoning in 1973 and the JMPR's establishment of ADIs and MRLs, anti-pesticide rhetoric in Western Europe and the USA, has steadily grown. Key (often single-issue) environmental and consumer-based organisations have arisen to argue the need for greater pesticide regulation on the grounds of unacceptable human health and environmental impacts. Indeed, public concerns became so intense that many governments in developed countries began to lose their credibility as custodians of safe and environmentally benign food systems. Conversely, with the aid of positive media coverage, anti-pesticide organisations have increasingly enhanced their credibility as champions of consumer health and the environment and gained critical public support (Ishii-Eiteman 2000).

The loss of credibility in nation state governments has primarily materialised through an erosion of trust. Ironically, the erosion of trust had arisen due to one of the central tenets of the Second Food Regime, that is, the existence, and persistence, of exceedingly close ties between government agencies involved in agricultural production, food safety and environmental protection and private sector corporations/organisations involved in the food system. Claims continued to be made that government was too closely allied with powerful elements in the agro-food systems, including agrochemical companies and corporate farming interests. Increasingly, many attributed the skewed distribution of costs and benefits associated with the agro-food systems to corporatist alliances. Corporate interests were seen to be the principal beneficiaries, while many of the costs are borne by farm workers and consumers (Macfarlane 2002). Indeed, in the case of the Alar scare, the US government was suggested to be more concerned about the negative publicity attached to the apple industry, particularly as the apple was used as an icon of healthy food, then they were for the safety of the American public. In this case, as with the use of most pesticides, the principal beneficiaries of Alar use were not seen as the consumers but growers and the food industry (Erlich and Erlich 1996; Irwin et al. 1997). This, and many other similar incidents, generated both a distrust of urban-centred industrial agro-food systems and of reassurances of pesticide safety emanating from governmental sources. In addition, there is a growing belief in many developed countries that there is a definite tendency for decision makers (agro-policy communities) to adjust science to meet their policy goals, and that this leads to a credibility gap. Indeed, according to a British parliamentary report, the way that the UK Government handled the outbreak of Bovine spongiform encephalopathy (BSE), better known as Mad Cow Disease, created a 'crises of confidence' in both science and government (World Bank 2003).

Both the perceived and actual adjustment of science increasingly became an Achilles heel of government agricultural, food safety and environmental policymaking processes. One of the key accusations levied at governments has been their over-reliance on industry-supplied data when making decisions concerning pesticide safety. Another problem relates to the interpretation of pesticide safety data. For example, using the same data, the risk of the pesticide alachlor as calculated by Monsanto was six orders of magnitude smaller than the one estimated by Health and Welfare Canada. Macfarlane (2002) suggest that this was due to variations in judgements made, with one side guided by potential loss in benefits and the other erring on the side of caution. Having established that, aside from one or two notable exceptions, there is rarely scientific consensus regarding the costs and benefits of using a particular pesticide; it is in the economic interest of pesticide manufacturers to use disagreements among experts to minimise concerns about risks. Given that risks are more acceptable if the source of information is perceived as trustworthy and credible, reassurances of safety from those with self-interest in certain technologies or activities are viewed with suspicion.

A more fundamental disagreement also arose over the objectivity of science itself.

The anti-pesticide lobby increasingly criticised developed country governments on two counts. First, in the 1960s, there was a growing consensus that science was not value free and therefore, by default, the process and products of science, should be open to public debate, and, if necessary, publicly censorship, and that this debate, and indeed, censorship, should be retrospective. Secondly, and more importantly, governments, allied with pesticide corporations, were criticised for their continued promotion of science as value free throughout the 1970s and 1980s (Irwin et al. 1997). Indeed, science is not just a simple 'body of facts' or a given 'method' but rather a diffuse collection of institutions, areas of knowledge and theoretical interpretations (Irwin et al. 1997); this means that decisions must be made while there is scientific uncertainty (Macfarlane 2002). However, many governments continued to present their views as if there were a consensus in the evidence, even though this was/is generally not the case.

Retaining the institutionalised positivist science/modernist paradigm also perpetuated the status quo by placing more importance on avoiding false positives than on false negatives. The benefit of doubt falls on concluding no effects: it is more likely that scientists will conclude that there are no adverse effects when there is one than the other way round. This can be thought of as the burden of proof required – 'beyond all reasonable doubt'. Consumers, however, are deemed to favour precaution as a form of insurance. This would suggest that if too high a burden of proof is required before preventive action is taken, consumers will view science as a barrier to a more environmentally sound society. Therefore, a decision to wait until the uncertainty is clarified is not a neutral one as it supports the status quo (Macfarlane 2002). For example, the US Government's rebuke by the American public over the Alar scare was accentuated because it had known about the potential problems associated with Alar for a long time, but had not acted. Indeed, in this case, as in the earlier case of EDB, if people overreacted, it was probably as much due to the reluctance of the US regulatory system to act on scientific evidence as to public misunderstanding of the risks (Krimsky and Plough 1988).

Whilst acknowledging that progress has been slow, the anti-pesticide movement made considerable political inroads with regard to pesticide regulation. Consumer and environmental groups organised themselves into formidable and legitimate mouthpieces for the publics' growing disillusionment with industrial agro-food systems. With this new found legitimacy and constituency, the anti-pesticide movement began to capture and preserve significant political space (Ishii-Eiteman 2000).

The changing twists and turns of pesticide and agro-food systems debates have significantly altered the dynamics of agro-food system regulation. Stakeholder dynamics within the agro-industrial policy community institutions changed dramatically, and with them, so did the outcomes. In time, Carson's work, in conjunction with a growing body of supportive literature, and heightened public concern was enough to fracture the previously impenetrable barriers of agro-food policy communities and open up food system development processes to the influence of alternative agricultural and natural resource development paradigms based on environmentalism (Sauer 1990). The growing political authority of environmental and consumer NGOs enabled them, albeit slowly, to adjust the discovery, development, production, distribution and use of pesticides. Partly as a result of the growing importance of consumer and environmental interests and the growing concerns regarding food safety and nutrition, the food retail sector has become the most influential actor in agro-food system regulation. Since the 1970s, food safety and environmental concerns, along with their anti-pesticide champions, have thus entered the political arena.

The culmination of both the public mistrust of governments and of corporate agribusiness is that, the public does not trust the government to set or to enforce safety standards, nor corporate agribusinesses to ensure the foods they sell are safe. A number of important food safety scares, including Kyushu rice contamination with PCBs in Japan, 2-4-D/2, 45-T in Vietnam, the Bhopal case in India, and the 'Alar' scare in the USA, have accentuated this mistrust. Lately, technical risk information has become far less important than other factors, such as equity, trust or the degree to which the exposure is voluntary, in resolving conflict over the use of pesticides. A decision-making process that ignores these wider facts is not going to satisfy the consumer (Macfarlane 2002).

One of the major changes in agriculture in recent decades is the rapid increase in the importance of consumers. Whereas 40 years ago, most ministries of agriculture would consider farmers as their main clients, today, and especially in Western countries, consumers are equally as, or even more, important. The rise in consumer power is the result of a variety of factors. In part, they relate to the variety of food scares. They also relate to better awareness, and information exchange among consumers. Consumers are more vocal, and demand that policymakers focus on food safety. Environmental interest groups also believe that there are viable alternatives to productivist model of agriculture, which relies too heavily on off-farm inputs, particularly pesticides (Macfarlane 2002).

Many governments began to sway because of increasing public pressure. Although consumer and environmental groups have seen this as a success, it has also led to claims from both industry and scientists that government priorities are too strongly influenced by public perception rather than scientific evidence and that the public does not accurately perceive risk. Governments and corporate agribusinesses are often frustrated by what they consider too extensive a worry by the public about small or unfounded risks and their amplification. Many believe that the solution is risk communication, defined as conveying technical risk information to citizens in order to lead them to the same risk management conclusions as the experts. This aside, governments, particularly in Western

Europe, have introduced a significant number of policies and regulations that reflect public concern, whether or not these concerns have a sound scientific basis. Indeed, particularly during the late 1980s and throughout the 1990s, many individuals within governments and their legal apparatus have evolved from official positions of almost total unanimity in the acceptance and wholehearted support for pesticide use to one that promotes the use of more integrated, and often organic, forms of agricultural production. Whilst acknowledging the difficulty of establishing direct links between pesticides and negative effects on human health and the environment, governmental decisions in Western Europe, the USA and Japan have often endorsed the 'precautionary principle' and acted to protect both human health and the environment in the face of probable threats. In many cases, this led to the establishment of legally binding policies. In turn, this provided the legal leverage for environmental and community action that, until this period, had essentially been missing (Ishii-Eiteman 2000).

In many developed countries, and Western Europe specifically, conservative agro-food policy communities began to fracture, allowing the penetration of new stakeholders with overt environmental and consumer safety agendas. However, penetration of agro-food policy communities has been uneven and, where it has occurred, the conditions of penetration and the agency of new stakeholders have varied considerably. In some instances, governments (international, national, regional and local) committed to real and rapid change in policy direction have openly invited environmental and consumer interest actors into agro-food policy communities in order to promote change. Secondly, governments (international, national, regional and local) have invited environmental and consumer interest actors to join agro-food policy communities in order to create the image that they are dedicated to the initiation of real change when in fact they were not. Many environmental and consumer protection actors have been literally shoe-horned into agro-food policy communities. The extreme expression of this has been where governments have attempted to 'de-claw' active individuals and groups by bogging them down in slow and incredibly bureaucratic processes where policy changes, or lack of, are developed by consensus and fail to make any real environmental or consumer protection progress. Because of the highly political nature, actors belonging to the most radical environmental and consumer safety groups continued to be excluded from agro-food policy communities.

6.2.4 The Post-Productivist transition

Critical observers of agriculture in Western Europe (Lowe et al. 1993; Ilbery and Bowler 1998) have been writing about a 'post-productivist transition' (PPT) for some years (Ward 1993; Ilbery and Bowler 1998). Such a transition conforms to the concept of a Third Food Regime in which farmers no longer struggle to maximise their productivity under support from state subsidies and protected domestic markets. Rather, alternative objectives are sought, such as higher food quality, the production of environmental goods and a more diversified rural economy, with the market for mass-produced food more open to global competition. But the PPT contains within it two parallel, some would argue contradictory, developmental processes: one through which elements of an 'industrialised', productivist agriculture are continued and enhanced; the other through which post-productivist objectives are developed. Outside Western Europe, the balance between these two processes remains skewed towards productivist development, particularly in rural economies that historically have been oriented to the mass production of agricultural raw materials for export onto global markets (e.g. the USA, Australia, New Zealand).

6.2.4.1 Promotion of Post-Productivist agriculture

Growing concerns and changing public demands, and the subsequent shift in production methods, in both developed and developing market economies (Barrett and Browne 1996; Watts 1995) are part of what has been labelled the Post-Productivist Transition (see Fig. 6.1) (Shucksmith 1993; Ilbery and Bowler 1998). It is a transition that embraces the whole food chain from production and processing to consumer choices and the marketing systems that have evolved to link them (Shucksmith 1993; Ilbery and Bowler 1998). Within Western Europe, tangible evidence of post-productivist agriculture can be observed, albeit with regional variations. However, even in Western Europe, many question whether or not these changes are as profound and widespread as suggested.

While short-term fluctuations in agro-food system regulation and practice have little effect on pesticide and seed development strategies, pesticide/biotechnology corporations do respond to long-term trends in food system development. For example, in Europe, reform of the EU's CAP was predicted to lead to a more competitive agriculture with reduced production-based support, and greater protection for the environment. Policy instruments such as cross-compliance and agri-environmental programmes have already had an impact on pesticide use because they provide governments with indirect means to influence the intensity, and consequent environmental impact, of farming practices (Bijman and Joly 2001a).

6.2.4.1.1 Agri-environmental policies, precision agriculture, integrated crop management and organic agriculture

During the 1980s, European Directives and Regulations and EC member states legislation began to attribute a greater value to the environment. For example,

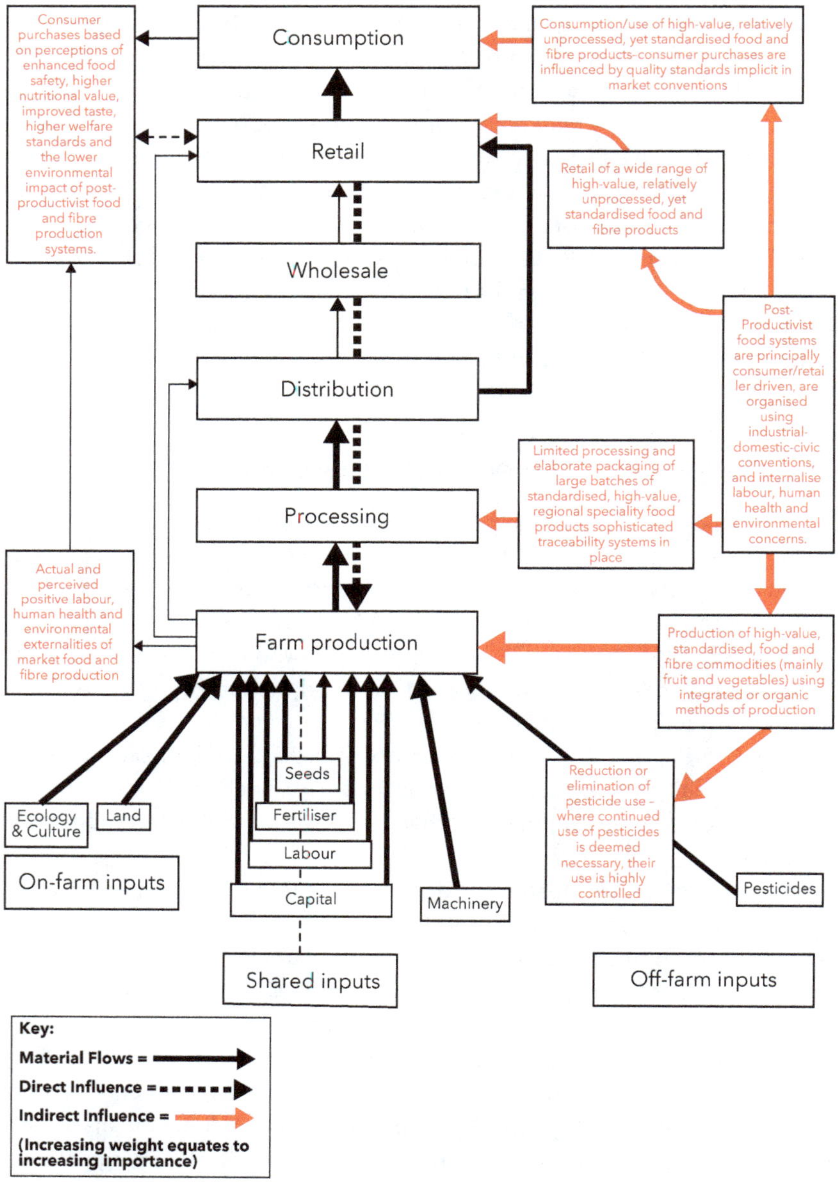

Figure 6.1 Post-Productivist food system dynamics during the Divergent Period (1974–2003) of pesticide and pesticide industry development.

the UK's 1985 Green Paper recommended the cessation of agricultural activity adjacent to freshwater and ecological corridors (Ilbery 1992). In 1988, the UK Government launched its Nitrate Sensitive Areas (NSA) scheme in an attempt to reduce the pollution of streams and groundwater in a number of designated locations that were prone to nitrate contamination. Farmers were encouraged to join the scheme via financial incentives. NSAs have since been superseded by the establishment of Nitrate Vulnerable Zones (NVZs), which now cover most of the UK (MAFF 1994).

In the UK, the establishment of Environmentally Sensitive Areas (ESAs) was another prominent and successful attempt to protect the environment. The ESA scheme is designed to help conserve agroecological areas of high landscape and or wildlife value, which are vulnerable to changes in farming practices (MAFF 1989). Payments are received for the adoption of more traditional and environmentally balanced farming operations. Agri-environmental schemes such as ESAs have gained wide acclaim from academics and landowners, and have served as a template for the development of similar schemes across the rest of Europe (Gasson and Potter 1988). ESAs are just one element of what has come to be known as 'Green' or 'agri-environmental' payments, and form part of the UK's agri-environmental package. Green payments offer farmers financial incentives for the positive management of important landscapes or habitats and enhanced countryside access. Land diversion and the extensification of production have also been advocated (Swinbank 1992; Turner 1993; Gasson and Potter 1988). Other national and supranational initiatives such as Less Intensive Farming and the Environment, Linking Environment and Farming (LEAF) and the FWAG are currently promoting integrated systems of agricultural production that combine profitable farming and care for the environment. Organic agriculture, which avoids the use of modern synthetic agricultural inputs, such as synthetic pesticides and fertilisers, and biotechnologies, and enhances employment and safeguards the environment, also grew in importance in both Western Europe and the USA. Organic farming was seen as an attractive alternative to modern intensive farming techniques (Bowler 1988). In the USA, introduction of the Farm Subsidy Law's 'green payment' programme, called the Conservation Security Programme, rewards growers who practice land, water and wildlife stewardship. Payments of up to US$50 000 a year could be claimed (Planet ark 2003d).

Precision farming is another recent development that attempts to manage crop variability by tailoring inputs to crop needs. Applied effectively, it increases profitability by reducing unit costs of production. It is also suggested to benefit the environment by enabling more efficient input use. Integrated Crop Management (ICM) on the other hand stresses the importance of integrating approaches in crop production in order to optimise output and profitability while significantly reducing negative impacts on the environment. Integrated approaches include chemical, biological, mechanical and cultural methods of crop production. IPM is a constituent part of ICM. IPM is a pest control system that utilises all available pest control techniques (chemical, biological and cultural) in an environmentally compatible manner in order to prevent economic damage (Metcalfe et al. 2002).

6.2.4.2 Evolution of rapidly changing and highly differentiated globalised food markets

During the 1980s and 1990s, trade in fresh horticultural products became increasingly global, a trend that led to claims that the fresh fruit and vegetables sectors had become truly transnational (Friedland 1994). For many, vertical integration was seen as a necessary prerequisite to globalisation. Whereas Neo-Fordist-styled appropriation is based on increasing levels of controls through the ownership of food system components, Post-Fordist appropriation in this newly emerging trade is via control exerted through vertically integrated contracts. This trend was clearly encouraged by a liberalising international and national regulatory framework, associated with GATT/WTO, IMF and World Bank policies, and further facilitated by considerable improvements in communications and packing technologies. For an overview of the role of neo-liberal economic and trade policies, see Chapter 4.

In 1990, it was estimated that trade in fresh fruit, vegetables and cut flowers was equivalent to 5% of global commodity trade, roughly equivalent to that of crude petroleum (Watts 1996; Watts and Goodman 1997). Horticultural exports from the developing world have become a major growth sector in international trade (Friedland 1994; Thrupp 1995). By the mid-1990s, developing countries contributed a third by value to this trade, twice the value of their traditional agricultural exports. In 1990, 24 low- and middle-income countries, mainly in Asia and Latin America, exported high-value fresh horticultural products to the value of more than US$500 million. Argentina, Brazil, China, Kenya and Mexico account for over 40% of world trade in these highly perishable items from developing countries (Barrett et al. 1999). North America's all year-round demand for fresh vegetables, fruits and flowers was responsible for catalysing much of the booming export trade from Latin America and the Caribbean, while the same demands originating from Western Europe are primarily responsible for the escalation of fruit and vegetable exports from Kenya (WRI 1998).

6.2.4.3 Consumer and retailer-initiated food chain restructuring

Throughout the past 40 years, food has acquired a high profile in many developed market economies. Concerns over food safety and, increasingly, food quality have raised the political and economic profile of food (Ilbery and Kneafsey 2000). Increasing, public awareness of the ethical and environmental implications of intensive farming systems and the vogue towards healthy eating, the body beautiful and fitness in general indicate a growing public interest in, and knowledge of, food (Ilbery and Kneafsey 2000).

Furthermore, among the relatively affluent, well-educated consumers, the preparation and consumption of some types of food have become a statement of fashion and an indication of the possession of cultural capital, good taste and sophistication (Ilbery and Kneafsey 2000). It is within such a climate of change that the issue of food quality has become increasingly important (see Fig. 6.1). However, quality is a complex notion, the meaning of which may vary for specific products and between individuals, regions and countries (Ilbery and Kneafsey 2000). It is socially constructed through the interplay of different actors who may seek, for various reasons, to interpret, represent and regulate quality in particular ways (Ilbery and Kneafsey 2000). Furthermore, quality in relation to regional speciality food products (SFPs) is defined by making references to other socially constructed and value-laden concepts such as authentic, healthy and traditional. Of particular interest to the study of emerging markets for regional SFPs is the idea that food quality may be signified through associations with particular places or regions (Ilbery and Kneafsey 1998). Consumers, for example, are increasingly concerned to know where products come from and how they are produced, not only for health and safety reasons, but also in terms of satisfying a current nostalgia which harks back to a perceived time of real and wholesome foods (see Fig. 6.1) (Ilbery and Kneafsey 2000).

Glocalisation is a term used to describe how global pressures and demands are made to conform to local conditions. Whereas powerful companies might 'customise' their products for local markets, glocalisation operates in the opposite direction. Local actors select and modify elements from an array of global possibilities, thereby initiating some democratic and creative engagement between the local and the global (Cohen and Kennedy 2000). The concept of glocalisation has obliged sociologists, political scientists and economists to re-acknowledge the importance of geography to the way the world works (Harvey 1996; Storper and Salais 1997). Ultimately, globalising forces are paradoxically being played out at the regional level. It is perhaps also worth pointing out that one effect of this academic outpouring has been a tremendous political appeal to the regional scale in territorially configuring economic prosperity (Jones and MacLeod 1999). Indeed, John Lovering (1999) labels this a 'new regionalism' in academic and political discourse.

Even though short-term consumer preferences during the 1990s remained focused on lower priced products, longer-term trends became increasingly focused on higher-quality, more diverse, convenient, healthy and environmentally friendly products (OECD 1994; Ilbery and Kneafsey 2000; Morris and Young 2000). Whilst this may not have represented a coherent and sustained transformation of fundamental values away from production towards consumption (Marsden 1999), the market in Western European, and increasingly in the USA, indicated that consumers increasingly wanted to purchase pesticide and GMO-free foods (Vorley and Keeney 1998) (see Figs. 6.1 and 6.3).

Demand is increasing for year-round supplies of foods produced by traditional extensive agricultural techniques with a strong ethical component (Barrett et al. 1999). This change in demand is clearly consumer-led, and is articulated by powerful retailers, who are increasingly adopting global sourcing policy to satisfy these new demands (Barrett et al. 1999). In the UK, it has been estimated that 65–80% of all fresh horticultural imports from Africa is marketed through the large multiple retailers (Christian Aid 1996).

The policies and practices of the supermarkets are now central to this trade (Barrett et al. 1999). Indeed, in line with growing consumer concerns over food quality and safety, retailers are increasingly obliged to clearly demonstrate that their products meet ever more exacting quality and safety standards. In order to maintain and/or enhance market share and profitability, retailers have increasingly introduced rigorous and exacting quality control policies and procedures and elaborate systems of food traceability (Barrett et al. 1999). Indeed, the use of barcodes enables multiple retailers, at least in theory, to locate the source of food production, to farm-level or even plot-level accuracy (Barrett et al. 1999). In addition to very strict health and hygiene standards, retailers operate random quality spot checks. Quality control is, therefore, meticulous; supermarkets have much to lose in the event of bad publicity, not least their reputation.

Since the 1990s, multiple food retailers dramatically expanded organic food lines in response to consumer demands. Preference for organic production was extended from fresh produce to processed foods. Indeed, some multiple food 'processors have entered the organic market, e.g. by either establishing their own product lines or by acquiring a specialist organic company' (Padel et al. 1999).

By the mid-1980s, multiple retailers recognised that, in order to maintain market share under stiff competition and increased consumer mobility, they had to address the new cultural and social consumer practices associated with the PPT (Marsden 1996). Multiple retailers have thus taken the initiative to accommodate consumer concerns, which go 'beyond biophysical characteristics'. Other considerations, such as food safety and quality, authenticity and convenience, environmental sustainability and 'ethically appropriate methods of production' – especially for animals – have increased in importance (Blandford and Fulponi 1999). As a result, traditional retailers have tended to lose their market share as consumers demonstrate less confidence in their claims of quality (Marsden 1996). Since the mid-1980s, agricultural and horticultural commodity chains have become increasingly buyer-driven. Powerful buyers in the developed world, including multiple retailers and trade organisations, dominate most buyer-driven chains (Barrett et al. 1999).

It is now the major retailers who effectively regulate quality standards (Barrett et al. 1999). According to Morris and Young (2000), 'quality can encompass health, safety, special nutritional ingredients, "naturalness" and environmental effects of crop cultivation'. Retailers have become the key players in translating quality definitions of food during the 1990s (Goodman and Redclift 1991). This private interest re-regulation of the food sector means that states are increasingly dependent on retailers to provide safe and high-quality food. Such power has enabled multiple retailers to develop their own regulatory systems in order to ensure dominance over the supply of food products (Marsden 1997), in what Marsden (1998) labels a new form of private interest regulation (see Fig. 6.1). Supermarkets are thus in a powerful position to influence what is actually grown, how it is grown and by whom, which reflects their needs to keep profit margins as high as possible, as well as ensuring that the needs and demands of their customers are satisfied (Barrett et al. 1999).

Retailer- and processor-imposed changes also offer opportunities for companies that are able to mobilise stakeholders to collaborate in the development of crops with specific quality traits and generate increased added value (Bijman and Joly 2001a). In both the EU and the USA, it is retailers who determine what food processors want from farmers. Retailers are the point of contact between the majority of OECD citizens and the rural economy.

The supermarket sector is most concentrated in the EU, but is also rapidly consolidating in the USA. In the 9 years since Rio, US food retailing chains concentrated dramatically, with the five leading chains moving from 19% control of grocery sales to at least 42% (Vorley 2002). Since 1992, global retail has consolidated enormously and three retailers, Carrefour, Ahold and Wal-Mart, have become truly global in their reach. In 2000, these three companies alone had sales (food and non-food) of US$300 billion and profits of US$8 billion, and employed 1.9 million people. It was predicted that there would be only ten major global retailers by 2010 (Vorley 2002).

The control of agri-food supply chains by clusters of powerful downstream industries has profound impacts on agriculture, especially in weakening the link between farm prices and food prices (Vorley 2002). High levels of concentration in downstream processing and retailing industries mean lower levels of value added going to local communities; 78–85% of value added in the agri-food chain in the USA and UK, for example, is not done by farms (Vorley 2002). During the 1990s, the average annual median return on equity for the US food manufacturing industry was 17.2%, and 18% for food retail. Over the same period, return on equity for US farming averaged 4.5% (Vorley 2002). Estimates for the performance of US agriculture values return on assets in the late 1990s at only 0.4%, compared with nearly 4% for life sciences, 9% for food processing, 10.6% for retail and 16% for food service (Vorley 2002). Indeed, in the USA, advertising is a larger component of the final cost of food than the actual raw commodities prices that farmers receive. Market access for producers to supply chains does not have much to do with classical notions of 'efficiency' (Ikerd 1996). Rather, market access is a feature of the ability to exploit marketing advantage, meeting large processor and supermarket demands for consistency of supply (reliable quality), speed of response, compliance with standards and payment of fees (Vorley 2002). The farmers' slice of the retail cost of a basket of foods sold in grocery stores shrinks further once they have paid for seeds, fertilisers, feed and machinery, finance, labour and land rental costs; again from a very concentrated sector in the case of seeds and chemicals. The size of the food market also shrinks as intense retail competition and concentrated retail buying power is translated into consumer surplus. Farmers have to produce more, but get less (Vorley 2002). Coordination of supply chains by supermarkets raises the requirements for farms and firms to stay in the market. For instance, the example of grape production in Brazil, for which Collins (2000) notes that although production costs are lower and quality higher among small-scale producers, marketing advantage accrues to large-scale producers through their better access to post-harvest cold storage and refrigerated transport services. In Latin America, as well as Europe and North America, there is a growing tendency for supermarkets and processors to use this form of coordination to source sub-regionally or regionally, and occasionally extra-regionally, rather than locally or nationally (Vorley 2002).

Markets are undergoing rapid change, with closed commodity chains rapidly replacing wholesale or spot markets. Highly concentrated food processing, retail and food service industries at the end of these chains are having an increasingly important impact on decisions made on the farm. Downstream processors and retailers are demanding stringent levels of quality, compliance with standards and codes of conduct and post-production service from their suppliers. Whether an apple grower in Kent, or a coffee producer in Peru, the major supermarket chains control access to consumers (Vorley 2002).

The Kenyan bean is a model of globalised, liberalised trade (Lawrence 2003). Homegrown, Kenya's leading exporter of cut flowers and vegetables to the UK supermarkets, supplies Tesco, Safeway and Marks and Spencer. Until recently, Homegrown also supplied Sainsbury and ASDA (Lawrence 2003). Supermarkets e-mail over orders about midday for what they want put on the flight that night, depending on how much their computerised tills tell them they sold the day before. In this way, they can eliminate nearly all financial risk from their end of the chain. Like most other suppliers, Homegrown has no written contract with them specifying quantities. The orders can go up and down dramatically. At Christmas, it can be a 3–4 tonne increase in 1 day. A flexible workforce at the production and packing end is essential. Most of this work force is migrant workers living in half-finished slums. Workers stand at a stainless steel bench for 12 h with a break in the middle. Women, who do most of the work, earn £1.60 a day, for a 6- to 7-day week. Subsequently, after Marks and Spencer highlighted that worker overtime exceeded the 12 h/week specified by their protocol, Homegrown was obliged to hire more staff for shorter hours (Lawrence 2003). Homegrown contracts large intensive and smallholder farmers to produce its beans. It provides the seeds to smallholder farmers and tells them when to plant and spray. According to Homegrown, each crop receives a maximum of four sprays in each crop cycle of 48 days, stipulated by Marks and Spencer, and is checked for pesticide residues. About 35% of beans fail the grade. Due to price competition between supermarkets, there is always pressure to cut costs of production through chemicals and labour. Much of the labour force is casual and has little meaningful health care or benefits (Lawrence 2003). In addition, agents of Marks and Spencer's Ethical Trading Initiative (ETI) found evidence of excessively long hours and damaging exposure to pesticides on some of its supplier farms, and evidence of deception of auditors. While it is encouraging that the supermarkets' umbrella initiative (ETI) is highlighting worker abuses, much of the blame for the abuses are directly related to the production schedules and contractual arrangements that supermarkets have with their suppliers (Lawrence 2003).

6.2.5 Increased control of pesticides and GMOs in the food chain by food processors and retailers

In response to growing mass boycotts and widespread protest against pesticide residues and GM food in the late 1990s, the European food industry has sought to minimise the use of pesticides on high-value agricultural and horticultural produce, particularly fruit and vegetables, and to exclude GM ingredients in their supplies (see Figs. 6.1 and 6.3). According to Levidow and Bijman (2002), 'this cooperative approach has several aims. (1) To maintain consumer confidence in product quality. (2) To establish Europe-wide supply chains which meet common or minimum standards. (3) To make supplies interchangeable. (4) To avoid competition for "non-GM" or "low pesticide" products defined in various ways'. The exclusion of GM ingredients in retail supply chains was given extra zeal after the StarLink affair. StarLink maize, which contains the *Bt* gene, was only cleared for use in animal feed as it can cause allergic reactions in humans. Unfortunately, StarLink maize contaminated crisps and cornflakes in the USA and Homemade Baking brand cakes in Japan (Le Monde diplomatique 2001).

In attempts to maintain customer confidence in product quality, retailers (often in conjunction with processors, input suppliers and farmers) have developed complex food production guidelines. Retailers are increasingly insisting on farmer implementation of organic principles and integrated farming practices, particularly IPM and ICM, as well as comprehensive traceability mechanisms (see Fig. 6.1). Some food companies, such as Unilever, actually prescribe the pesticides that their supplier farmers can use and sponsors research on pesticide reduction methods (Levidow and Bijman 2002). Building on supplier pesticide reduction guidelines, the 'Euro-Retailer Produce Working Group (EUREP) adopted a Good Agricultural Practice (GAP) Protocol in 1997 and revised it in 1999. The detailed guidelines emphasise ICM methods for avoiding pest problems (e.g. through resistant varieties, crop rotation, soil management) and for minimising pesticide usage and its effects (e.g. through biological control, mechanical methods, regular crop monitoring, seed treatments rather than foliar sprays). It also makes distinctions among crop protection methods: seed treatments are preferred to foliar sprays; biological or mechanical methods are preferred to chemicals; and varieties should possess resistance/tolerance to commercially important pests and diseases' (EUREP 1999). For example, Carrefour adopted the GAP Protocol. This ultimately led to products produced using lower amounts of pesticides becoming the company standard (Levidow and Bijman 2002).

In 1998, in an attempt to asway consumer fears, the European food industry adopted tentative measures for voluntarily labelling GM products and mechanisms were established to ensure that food products do not contain GMOs (Levidow and Bijman 2002). Later that year, the EU imposed rules on labelling GMO-based foods (EC 1998).

On the 28th of April 1999, the world's two largest food processors Nestlé and Unilever removed GM ingredients from their supply chains (Genet 1999), at least for food products headed for consumption in the EU (Levidow and Bijman 2002). New EC rules on labelling GMO-based products deterred the use of GM ingredients, especially in their own-brand products (Levidow and Bijman 2002). Concern over the lack of consumer acceptance of GM food ingredients even prompted USA-based Perdue Farms Inc., to consider feeding its chickens and turkeys non-GM maize (The New York Times 1999a). Australian food companies George Weston Food and Sanitarium also reacted to consumer concern by removing GM ingredients from a number of key products (Australian Financia Review 1999). In August 1999, Germany's largest retailer, Edeka, declared total eradication of GM ingredients in its foods (Green Peace 1999). They joined other German retailers such as Tengelmann and Rewe. In order to give the American public choice, the US Consumers Union called for labelling on GM foods (Dow Jones 1999b). Many retailers and processors that have refused to remove GM ingredients from their supply chains have faced a barrage of anti-GMO protests. In Canada, GreenPeace and the Council of Canadians called on consumers to lobby Loblaws and affiliated companies such as Superstore and IGA to label all GM products and eventually remove them from selves because of the unknown threat to humans and the environment (The Halifax Daily News 1999).

According to Levidow and Bijman (2002), 'food companies accommodate public pressures to demonstrate that they minimise environmental pollution - e.g. energy usage, packaging, and agrochemical inputs by their suppliers. They conduct audits across the agro-food-distribution chain, in order to identify means of achieving those aims. Claims for environmental improvements generally promote the entire company's image and product range, rather than promote specific products as "green." This is especially true for retailers with a large portfolio of own-brand products'. In some cases, food processors and major retailers (such as Nestlé and Unilever) have gone further than simply promoting organic, ICM/IPM products, and initiated plant breeding for (non-GM) pest-resistant seeds. According to Unilever, 'We aim to maintain the highest standards at our sites and in the products we sell. Our intention is to produce superior varieties that contain natural resistance to pests and diseases, which reduces the need for agrochemicals' (Unilever 2001). The food industry also sponsors research on pesticide reduction methods. In many cases, products produced with lower levels of less harmful pesticides became the standard (Levidow and Bijman 2002). National government legislation also spurred food retailers into action. In 1997, Albert Heijn (responsible for ¼ of retail food sales in Holland) announced that all its fresh produce would be sourced from suppliers practising ICM or organic production methods. In October 2000, Albert Heijn stated that 'it will seek to sell only residue-free products'. The Dutch Multi-Year Crop Protection Plan launched in 1991 spurred on this affirmative action (Levidow and Bijman 2002). In the UK, the 1990 Food Safety Act legally obliged food retailers and processors 'to do all that is reasonably possible to ensure that their products are safe' (Henson and Northern 1998).

However, while greater concern for the environment and consumer health is highly commendable, expression of this newfound morality in agro-food systems is uneven and often contradictory. Indeed, Unilever publicly announce that it intended to leave the door open for GM-based food ingredients if the public's attitude changed and GMOs were accepted (Unilever 2000). Furthermore, one of the key assumptions underlying the efforts of powerful food system actors to internalise environmental and health externalities into post-productivist food systems is that, ultimately, the consumer is willing to pay a little extra for 'healthier food' and more environmentally friendly production practices.

Several studies support this assumption. In a 1989 Gallop Poll, 50% of those surveyed suggested that they would pay more for organic food: 41% would pay 10% more, 16% would pay 20% more and 10% would pay more than 20% more (Vorley and Keeney 1998). A 1990 survey by OTT found that approximately 20% of the US public had changed their consumption decisions; buying organic food or avoiding treated fruit and vegetables, 65.5% were willing to pay 5% more for certified pesticide-free produce (OTT 1990). Furthermore, Buzby and Skees (1994) found a willingness to pay (WTP) of US$0.15 for a 50% pesticide risk reduction in grape fruit, and a WTP of US$0.69 for a 99% pesticide risk reduction (Buzby and Skees 1994). The uncertainty of risk was found to be a major issue (Vorley and Keeney 1998). Conversely, other studies have demonstrated that consumers often forget their pesticide politics when they enter their local supermarkets (Vorley and Keeney 1998). A 1999 study by the Commission of the European Communities found that only 33% of Europeans claim they would pay a little more for food in order to protect the environment. Roughly the same percentage say they would not pay more on anything to protect the environment. In addition, consumer surveys and retailers' experience have shown a large gap between environmental awareness of consumers and the premiums they actually pay for environmentally safe products. Furthermore, consumers are unlikely to accept a rise in food prices in real terms (Vorley and Keeney 1998).

Another widely held assumption is that consumers with take into consideration the trade-off between the use of pesticides to produce blemish-free fruit and vegetables and reduced pesticide usage and the higher incidence of blemishes and that they will choose to accept less than cosmetically perfect produce in light of perceived risk reduction of lower pesticide residues. While some studies have suggested that consumers will purchase produce that is not perfect if they know that it has less or no pesticide residues (Pimentel et al. 1993), most studies suggest that consumers not only do not want to pay more for pesticide-free produce, they are also generally unwilling to accept blemishes (Weaver et al. 1992). Consumer preference for blemish-free produce is conveyed by retailers and processors down through their supply chains. Large orders for top-quality fruit and vegetable producers have to meet high cosmetic standards as well as a low risk of pesticide residues, in line with consumers' perceived demands (Vorley and Keeney 1998). In US apples, cosmetic quality criteria accounted for approximately 33% of pesticide use (Vorley and Keeney 1998). Between 60% and 80% of pesticides used on Californian oranges are to avoid cosmetic blemishes (Rosenblum 1994) and pesticide expenditure in the export vegetable sectors of Indonesia, Philippines and Thailand may reach over 40% of production costs (Vorley and Keeney 1998). While surface blemishes on fruit and vegetables generally do not affect nutritional content, storage life or even flavour (Pimentel et al. 1993), the use of pesticides on these crops is almost obligatory to meet the exacting standards of processors and retailers. For example, citrus rust mites cause russetting or bronzing on Florida oranges. Unless infestation is high (Pimentel et al. 1993), the internal quality is virtually unaffected (Pimentel et al. 1993). Even though up to 95% of oranges are destined for fruit juice, approximately 80% of the orange area in Florida is sprayed for rust mites, three times a season at US$200/ha, despite the fact that failure to spray results in little, or no, reduction in yield (Pimentel et al. 1993). Tomato fruit worm is another classic example. While 90% of tomatoes are peeled and are used for paste, sauce, ketchup, juice or puree (Pimentel et al. 1993), and only high infestations of fruit worm cause damage to nutritional content (Pimentel et al. 1993), most processors allow no more than 0.5–2% damage to the tomato surface, and some accept only perfect tomatoes (Pimentel et al. 1993). Growers have strong incentive to reduce blemishes by applying pesticides (Pimentel et al. 1993). Not surprisingly, growers apply higher levels of pesticides to ensure that they meet high cosmetic standards and are able to sell their tomatoes (Pimentel et al. 1993). In addition, retailers continually pass on new quality criteria to farmers but retain low prices (Vorley and Keeney 1998). Food standardisation by retailers and high-quality/cosmetic appearance equals higher pesticide use (Vorley and Keeney 1998).

The third widely held assumption is that post-productivist supply chains are more ethical. On the one hand, environmental and consumer pressure groups have secured improved working terms and conditions throughout many commodity networks embedded in developing countries. Ethical, often organic, trade networks have developed that ensure at least relative improvements for workers and growers in developing countries (Whatmore and Thorne 1997). With respect to pesticides, some flower farms have improved occupational health conditions. In both Ecuador and Colombia, several flower companies now take workers' blood samples to check for pesticide exposure, and some have improved medical services and provide masks and gloves for workers (WRI 1998). Conversely, the production of high-quality roses and carnations in Ecuador use an average of six fungicides, four insecticides, three nematocides and several herbicides, often used in intensive greenhouse conditions (WRI 1998). Flowers in Colombia and Costa Rica are treated with methyl parathion, terbufos and aldicarb, which are banned in the USA (WRI 1998). Pressures to maximise output and speed exacerbate health problems, encouraging management to push workers beyond the limits of safety. Inadequate, or below minimum wages, poor living conditions and lack of respect for laws governing maternity leave are also common. Workers' attempts to organise and assert their rights have generally been met with reprimands and dismissals, because replacement workers are easy to find (WRI 1998).

6.3 Key characteristics of pesticide development processes and manufacturers

6.3.1 Re-defining the role of pesticides and emergent biotechnologies in Post-Productivist food systems

The pesticide industry has responded to drivers inherent within the PPT both reactively and proactively. However, its principal response has been reactive, defensive and entrenched. Throughout the late 1970s and 1980s, pesticide corporations generally reacted swiftly and decisively to downplay any governmental or non-governmental rhetoric that could potentially lead to the imposition of food system restrictions that were likely to lead, directly or indirectly, to, present or future, reductions in pesticide corporation profits. The pesticide industry used its enormous political and economic power to influence the regulatory process, at both national and international levels and within both public and private sectors, in order to secure favourable outcomes. Its political power was comprised of two key components. First, the pesticide corporations remained key, and entrenched, actors within national and international agricultural policy communities, as well as having significant influence in other key pesticide-related decision-making arenas. Second, the pesticide industry used its significant economic power to influence the type of economic development in which it stood to prosper.

In many respects, this approach should not be unexpected. The pesticide industry has been, and continues to be, a bastion of western capitalism. Pesticide corporations acted in a manner congruent with the maximisation of profit and/or market share. In addition, the reactive and defensive stance taken by pesticide corporations in the Productivist context within Western Europe, the USA and Japan until the early 1990s, and in the Neo-Productivist context that slowly emerged in the mid-1970s, had served the pesticide industry well. Indeed, it was not until the late 1980s in Western Europe that the pesticide industry was faced with such formidable opposition to the use of pesticides that it has been obliged to consider alternative accumulation strategies that were not so firmly entrenched in the industrial high-input/-output paradigm of prophylactic pesticide use. However, instead of simply reacting to this seemingly insurmountable opposition to pesticides by winding down the production of pesticides and re-investing resources, the pesticide industry sought to redefine the role of pesticides, and emergent biotechnologies, in Post-Productivist food systems.

It is important to note that, whilst the pesticide industry modified its pesticide and biotechnology strategies in response to the evolution of Post-Fordist/Post-Productivist food systems in Western Europe and their globally embedded supply chains, it maintains its generally reactive and defensive position, with regard to pesticides and biotechnologies, in Neo-Fordist/productivist food systems across much of the rest of world. This section discusses attempts by the pesticide industry to redefine the role of pesticides and emergent biotechnologies in Post-Productivist food systems in two key decision-making arenas, namely public (political negotiation and pro-pesticide and biotechnology public relations), and the private sector (negotiations with multiple retailers and food processors) (see Fig. 6.1).

6.3.1.1 Political arenas

Since the mid-1970s, the pesticide industry has been engaged in protracted battles to ensure high levels of profitability and market share for both old and new pesticides and emergent biotechnologies. In the context of Western Europe, consumer and environmental NGOs have attempted to re-regulate, indeed, reconstruct, food systems in such a way as to ensure high consumer and environmental benefits and, conversely, to reduce or eliminate many of the worst consumer and environmental excesses associated with the use of pesticides. Inadvertently, consumer and environmental NGOs have championed the cause of Post-Productivism, and have had considerable success. This success was achieved by the successful infiltration, and in some cases domination, of a range of key food-system-related policy communities and key decision-making processes at the national level.

6.3.1.1.1 National regulatory arenas

Whilst it is imprudent to give the impression that national pesticide regulation throughout the Second Food Regime was homogenous with regard to pesticide regulation, it was, compared to the relative uneven nature of pesticide and biotechnology regulation in Western Europe and the USA since the mid-1980s. The heterogeneous nature of national pesticide and biotechnology regulation can be primarily explained by the relative strengths of the pro- and anti-pesticide lobbies, and the strength or weaknesses of their arguments. Again, these strengths and weaknesses were influenced by both endogenous and exogenous factors. Exogenous factors often include the intensity of national farming systems, past experience (including food scares) and public opinion (greatly influenced by the type and quantity of media coverage). However, as with most political processes, particularly ones that involve vested interests whose decision-making capacity has become institutionalised through years of corporatist arrangements with public sector administrators responsible for either, decision-making or decision implementation, change can be painfully slow, incremental and strewn with seemingly irrational compromises. This resulted in two things: First, unless public opinion, and political expediency, necessitated swift action, significant changes in pesticide regulation, and ultimately pesticide use, did not occur quickly. Indeed, it often took many years of determined lobbying activities to achieve either a full or partial ban on the most hazardous pesticides. Second, the protracted nature of the political process allowed the pesticide industry significant room to manoeuvre. This time was either used to develop strategies to further delay change or develop alternative strategies and more defensible positions.

Whilst it is true that the development of pesticide and biotechnology regulation within Western Europe, the USA and Japan was highly uneven throughout the 1980s and 1990s, over time, there has been a general trend to impose increasingly tighter restrictions. Pesticide use is controlled by national regulation, generally a system of registering each pesticide formulation for a specific use and crop. Approvals are based on evaluations of efficacy, user and consumer safety and environmental impact. Most industrialised countries have also established laws setting the MRLs that are permitted in food, and which apply to food produced domestically and imported.

6.3.1.1.1.1 UK

As the evidence of negative human health and environmental effects of pesticides became overwhelming, public concern was mobilised through the establishment of new environmental pressure groups and a rapid expansion of many existing ones throughout the mid-1970s and 1980s (Winter 1996). However, while the two largest and most important farmer organisations in the UK, the

NFU and CLA, were aware of the growing political importance of environmental issues and the need to develop new dialogue with environmental interest groups, they were obliged to reflect the often resilient and conservative views of their members. Many articles published in the NFU's in-house magazines appealed to NFU members by bashing environmentalists, calling them cranks and extremists, while other articles were more balanced acting as a source of education on practical conservation or, by way of political education, by stressing how much conservation work farmers were actually doing, and had always done (Winter 1996). In 1977, the NFU and CLA went further, issuing a joint statement entitled caring for the countryside. It represented the beginning of a high-profile and sustained publicity campaign to promote farmers and landowners as the natural custodians and stewards of the countryside (Winter 1996).

Furthermore, recognising that times were changing, farmers began collaborating with conservation and environmental organisations, input suppliers, processors and retailers to develop strategies that would enable them to prosper in these changing times. One such collaboration was the FWAG, formed in 1970 in an attempt to reconcile the interests of farmers and conservationists (Winter 1996). This was followed in 1991, by the establishment of the LEAF, which aimed to promote the benefits of ICM to UK farmers and consumers.

The British Government also paid greater attention to pesticides issues after the DDT scare in the USA 1969–72, and subsequent publicity over other pesticides (Goodman and Redclift 1991). Initially, attention was focused on pesticide residues in food, especially by the London Food Commission, UK (Goodman and Redclift 1991). In subsequent years, the British Government banned dieldrin, aldrin and heptachlor, which were achieved by voluntary arrangements on the advice of the government Advisory Committee on Herbicides and Other Toxic Chemicals, representing users, manufacturers and conservationists (Green 1985).

While most Western European governments have increasingly attempted to reduce the use of pesticides and ameliorate their human health and environmental impacts, most governments have been in favour of the development and release of GMOs. Indeed, the British Prime Minister Tony Blair accused the media of 'extraordinary' reporting of the row over GM food. In 1999, the Prime Minister told the Cabinet's weekly meeting that while positive scientific reports were barely reported, the media gave huge space to anything which fed the hysteria (Independent 1999).

In the UK, the Department for Environment, Food & Rural Affairs (DEFRA) coordinates GMO trials, with considerable input from the Supply Chain Initiative on Agricultural Crops (SCIMAC), which was established in 1998 to support the responsible and effective introduction of GM crops in the UK (FOE 2003). SCIMAC is made up of five organisations: British Society of Plant Breeders (BSPB), BAA, National Farmers Union (NFU), UKASTA and British Sugar Beet

Seed Producers Association. Advanta is a member of BSPB (Corporate Watch UK 2002c). Between February 1993 and March 1997, the British Government issued 55 consents for the experimental release of HT GM crops in the UK (see subsection 5.3.6). These releases consisted of small-scale experimental field trials with risk management procedures such as border rows, isolation distances and post-trial monitoring in place for the purposes of confinement. UK trials of HT GM crops caused significant concern within environmental groups.

6.3.1.1.1.2 Sweden

In 1953, Sweden introduced a limited system for pesticide registration in which the producer was required to obtain a registration from the Swedish Plant Protection Institute to market a product containing substances on their official poison list. Registration was usually based on very limited information on the AI and the product, and products that did not contain toxic substances were left outside the system. It was not until 1963 that a registration system covering all pesticides was introduced. The basic premise of this system was that all pesticides were banned unless approved by the appropriate designated government agency (Wahlström 2003). In 1980, formal data requirements were established. Applications that did not fulfil the data requirements could then be rejected because of insufficient data. Registrations, however, had no time limitation and could only be recalled following new information on effects. In 1986, Sweden established a system of time-limited pesticide approvals, a maximum of 5 years. Sweden also followed the substitution principle; old pesticides were substituted when new and less damaging pesticides evolved (Wahlström 2003). Several pesticides were cancelled when less hazardous alternatives were available, such as the 'drins'. Parathion was also replaced by fenitrothion. In 1966, Sweden banned endrin and all uses of aldrin and dieldrin were banned in 1970. Home and garden use of DDT was banned in 1970, followed by chlordane in 1971 (Wahlström 2003). Similar arguments were used for the cancellation of dinoseb and amitrole in 1979 and 1972, respectively (Wahlström 2003). Hexachlorobenzene was voluntarily withdrawn from the Swedish market in 1980 (Wahlström 2003).

In Sweden, the precautionary approach has been systematically applied to pesticide registration since the early 1970s. It is codified in Article 5 of the Act on Chemical Products. 'Anyone handling or importing a chemical product must take such steps and otherwise observe such precautions as are needed to prevent or minimise harm to man or the environment' (Wahlström 2003). When DDT was banned in 1970, the reason was suspected effects in the environment, including eggshell thinning of birds of prey, even though the direct link to DDT was weak and questioned (Wahlström 2003). In 1985, the Swedish Government proposed halving the use of agricultural pesticides. The process was highly successful and reduced pesticide use in agriculture by 65% in 10 years. This was largely due to a change in attitude of the Swedish Farmers Association, which developed a policy of making Swedish agriculture the greenest in Europe (Wahlström 2003). Part of the reduction in pesticide use (kg/ha) was achieved by the substitution of new lower dose-rate pesticides for older higher dose-rate pesticides. For example, TCA and sodium chlorate were replaced by glyphosate and the phenoxy acids were replaced by sulfonylurea derivatives (Pimentel et al. 1993). Dose rates were also reduced by applying pesticides when pest and disease pressures reached economically damaging thresholds and not before.

In addition, national pesticide regulation was tightened to remove some of the most hazardous pesticides from the market. In 5 years, the Chemical Inspectorate of the Swedish Government managed to reduce the number of pesticide products from 600 to 400 (Pimentel et al. 1993). Furthermore, in an attempt to protect public health and the environment, Sweden, Finland, Norway and Denmark introduced pesticide taxes in order to curb the use of pesticides (Pretty 1995; Vorley and Keeney 1998). In 1986, the Danish EPA used pesticide taxes as part of an ambitious programme to reduce pesticide use by half, a target that it achieved in less than 10 years. Taxes worked because they bridged the gap between private marginal cost and social marginal cost. However, the extra cost imposed on society from pesticide use varies with the pesticide, location of use and distance from population centres. These differences imply that the taxes should vary by region and commodity. However, enforcement of such an in-the-field tax requires exact monitoring of pesticide use. Monitoring of this sort is expensive and difficult to implement, which significantly reduces the appeal of taxes as a regulatory tool. Uniform taxes are easier to implement (for instance, as a sales tax), but are less efficient because they do not account for the variation in the social costs of pesticides.

Sweden has also been active in Europe. For example, the Swedish Government urged the EC to invoke the precautionary principle and revoke the registration of paraquat in the EU (PAN-UK 2003b). Paraquat has been banned in Austria, Denmark, Finland, Sweden, Hungary, Slovenia and Switzerland since 1989, and several developing countries including Indonesia, Kuwait and Malaysia (PAN-UK 2003b). The last registration for paraquat in Malaysia expired in 2005. However, the EC concluded that, when used in accordance with manufacturer's guidelines and GAP, paraquat did not pose any significant health risks to operators, and that any threat to wildlife was acceptable (PAN-UK 2003b).

6.3.1.1.1.3 USA

One of the most important battles over pollution in the USA centred on the Great Lakes. The battle over growing pollution, linked to pesticides and other industrial chemicals, began in 1978 and raged on throughout the 1980s and 1990s (Hough 1998). In April 1992, the International Joint Commission for the Great Lakes (IJC) of Canada and the USA called for Canada and the USA to eliminate dioxins, furans, hexachlorobenzene and chlorine from the production of industrial feedstocks (PAN 1994). The IJC insisted that it was 'necessary to shift the burden of responsibility for demonstrating whether substances should be allowed in commerce. The concept of reverse onus, or requiring proof that a substance is not toxic or persistent before use, should be the guiding philosophy of environmental management agencies' in both the USA and Canada. Locally, the Michigan State Medical Society, consisting of 12 000 doctors, endorsed the IJC's 1992 report (PAN 1994). Half of the known persistent toxic chemicals in the Great Lakes ecosystem are organochlorines.

The 1994 IJC report repeats all the previous recommendations for the virtual phase-out of persistent toxic compounds, including the call for a chlorine phase-out, but it placed a stronger emphasis on human health effects. Behavioural problems in children whose mothers had eaten fish from the Great Lakes, breast cancer, endometriosis (a serious menstrual disorder), male reproductive system defects and cancers, and a fall in the sperm count, have been linked in varying degrees with organochlorines. The IJC say that if even one of these links is real, the human cost of persistent toxic compounds will have been too great. They emphasise the urgency of acting now and not waiting for conclusive proof (PAN 1994). Globally, many conventions and conferences have followed or mirrored the IJC's initiative.

Perhaps, equally importantly, it was at this time that the EPA's Science Advisory Board (1990) publicly acknowledged that pesticides caused: (1) Accelerated development of disease-resistant pests. (2) Acute or chronic health effects among workers in the manufacturing process. (3) Acute or chronic effects on third parties due to accidents in manufacturing or transport. (4) Acute or chronic health effects among applicators and farm workers. (5) Contamination of groundwater due to leaching; contamination of surface waters from run-off. (6) Poisoning of wildlife. (7) Acute or chronic health problems to consumers due to residues on food. (8) Contamination of environment due to improper disposal of unused pesticides and their containers (Glicksman et al. 2003).

In response to the EPA's acknowledgement of the inherent dangers of pesticides combined with increasing pressure from the US public to reduce the negative human health and environmental impacts of pesticides, in June 1993, the Clinton Administration made a public commitment to reduce the use of pesticides and promote sustainable agriculture. One of the central tenets of this commitment was the declaration of an explicit target to achieve a 75% implementation of IPM by 2000. In addition, use reduction goals were announced for specific pesticides in key agricultural sectors. Not surprisingly, the response from industry was hostile (Vorley and Keeney 1998). The chemical industry opposed any mention of use reduction as a goal, favouring instead risk reduction based on sound science. Registrants argued that it was inappropriate and possibly illegal for the EPA to embrace use or risk reduction goals since all registered pesticides have been determined by the agency to pose acceptable risks when used in accordance with their labels (Vorley and Keeney 1998). The reduced-use approach was described as a non-starter by the American Crop Protection Association (Vorley and Keeney 1998). Between August 1994 and April 1995, the EPA dropped the reference to 75% IPM and then the need to reduce use or risk (Vorley and Keeney 1998).

In 1996, when President Clinton signed the FQPA into law, he said, 'If a pesticide poses a danger to our children, it won't be in our food'. However, it was not that simple as politics, and not science, still drove pesticide regulation in the USA (Metcalfe et al. 2002). The pesticide industry's standard-bearer on Capitol Hill, Texas Rep. Charles Stenholm, the highest ranking Democrat on the House Agriculture Committee, told the vice president that Democrats from key farming states would pay a price at the polls in November for an aggressive pesticide policy. Stenholm also hinted that Gore's own 2000 presidential hopes in those same states would be hurt (Metcalfe et al. 2002). Later, Stenholm characterised his warning to Gore this way: 'I don't think anybody that has aspirations for national office wants to take credit for stopping technology in agriculture' (Metcalfe et al. 2002).

Some of the first estimations of the impacts of FQPA implementation assume a total ban of both the organophosphates and carbamate pesticides. Two studies funded by the American Farm Bureau Federation found significant impacts on society from FQPA implementation. A study by Taylor and Smith predicted decreases of US$17 billion in countrywide economic output, and the loss of 209 000 jobs. Gray and Hammitt state 'it is difficult to imagine that the benefits of a ban could offset the 10 to 1,000 annual premature fatalities predicted from the income losses that would be caused by elimination of organophosphate and carbamate use' (Metcalfe et al. 2002). Metcalfe et al. (2002) claim to have taken a more moderate approach than previous studies, conducting analysis as though all organophosphate pesticides were cancelled, while carbamates remain a viable alternative. A case study approach analysed the impact of an organophosphate ban on 15 important California commodities (Metcalfe et al. 2002). Results of the economic analysis suggest that the total loss to producers and consumers in California from banning all organophosphate use would be approximately US$203 million. This should only be considered an order of magnitude estimate of the effects, and represents only about 2% of the total revenue generated by the 15 crops in California. The crops most directly affected are broccoli, oranges, almonds and alfalfa. The losses are larger because there are less suitable alternative pest control strategies currently available to producers (Metcalfe et al. 2002). According to Metcalfe et al.

(2002), without the use of these pesticides, valuable bio-resources may need to be converted to farming activities in order to maintain production levels. Pesticides also benefit consumers by reducing the price of certain fruits and vegetables, which makes them affordable to low-income consumers. This is an important observation since reductions in pesticide use could indirectly have a negative health effect by reducing the consumption of fruits and vegetables that contain many valuable micronutrients.

Conversely, the US approach to the introduction of biotechnologies can be described as a relatively straightforward rubber-stamping exercise. Both the US Administration, including public sector biotechnology R&D, and the private sector had extensively collaborated to produce innovative technologies with capacity for market domination and the expected associated financial rewards (Deutsche Bank 1999). A series of industry wide alliances in the USA financed scientific research, organised educational forums, lobbied legislators, regulators and farm organisations, and used their web sites to promote the benefits of GM (The New York Times 1999b). The USA has several agencies responsible for GM: the USDA, FDA and EPA. In 1986, the US Government developed a coordinated Framework for Regulation of Biotechnology, which outlines how the regulatory responsibilities are divided among the three agencies. The USDA has two responsibilities administrated through its Animal and Plant Health Inspection Service. These are the monitoring of GM field trials, and the evaluation of environmental impact of the GM plant, including its potential weediness, potential for genetic pollution and impact on beneficial insects (Geo-PIE 2004). The FDA evaluates food safety aspects of a new GM food crop. GM crops are treated just as any other food commodity and must not exceed the standard allergen, nutrient and toxin content levels (Geo-PIE 2004). This is a process known as substantial equivalence. Because some GM plants manufacture their own pesticides, the EPA is responsible for the safety of pesticide levels in the plants that produce them. Their jurisdiction extends to both human health and environmental impacts of the pesticide (Geo-PIE 2004). The regulatory environment for GMOs in the USA provided conditions for the aforementioned rapid commercialisation and adoption of GM crops. However, because of the relatively seamless introduction of GM crops, in the mid-1990s, many US citizens were unaware of the rapid commercialisation of GM crops and, subsequently, the growing proportion of GM ingredients in their food. According to an IFIC poll in the late 1990s, almost 50% of Americans believed their foods to be GM free, when in reality, almost 60% of processed food contained GM products, including virtually all candy, chocolate bars, ice cream, cookies, salad dressing and so forth (Deutsche Bank 1999).

Many felt that the rush to develop and market biotechnologies has been at the expense of human health and the environment. Little health assessment was undertaken to ascertain potential effects of ingesting GM food crops, and no studies have been performed to measure the impact of chronic ingestion of finished products that contain increasing amounts of transgenic crops (Deutsche Bank 1999). In August 1999, the first national coalition against GM crops in the USA sent a letter to the EPA urging restrictions on the planting of Bt corn until it can be sure that GM pollen will not have adverse effects on the monarch butterfly or other rare species. The coalition included the Union of Concerned Scientists, National Wildlife Federation, Sierra Club, Natural Resources Defense Council, Arizona-Sonora Desert Museum and Defenders of Wildlife (Union of Concerned Scientists 1999). The coalition also called for a stop to field tests and commercialisation of bioengineered crops until they could be thoroughly, and independently, tested for their impact on human health and the environment (Belsie 2001). According to Clapp (2002), even the US EPA admitted that its initial approval of Bt crops was not based on extensive study of risks, and that subsequent approvals have been subjected to more stringent requirements. However, US Commerce Department officials countered these arguments by stressing that biotech products have proven their safety and that after 'thirteen years of US experience with biotech products have produced no evidence of food safety risks beyond those of their natural counterparts' (Reuters 1999e).

6.3.1.1.1.4 Japan and Australia

Whilst Japan, the largest importer of GM ingredients/foods, initially embraced biotechnology as the next step in the continuous evolution of new and improved technologies, sentiments began to change. By the end of the 1990s, a strong coalition called the 'No GMO Campaign' including 100 Japanese organisations ranging from consumers to farmers to religious bodies had managed to persuade one in eight Japanese local governments to eliminate GM foods from school lunches (Reuters 1999f).

In Australia, anti-GMO sentiments have been spearheaded by the Emergency Coalition Against Genetically Engineered Food, formed in 1999 (AAP 1999). Despite their attempts, in December 2002, the Australian Government granted US$16 million over 7 years to establish a new sugar research body, the Cooperative Research Centre (CRC) for Sugar Industry Innovation through Biotechnology to use sugarcane molecular biology to enhance sucrose production and produce valuable end products other than sugarcane, including biodegradable plastics, bio-fuels and pharmaceuticals (Planet Ark 2003c).

6.3.1.1.1.5 India, China and other developing countries

India and China are acknowledged as being the vanguard of biotechnology in the developing world. In 2013, India planted approximately 11 million ha of GM cotton. In the same year, China planted 4.2 million ha of GM crops, including cotton, papaya, poplar, tomato and sweet pepper (James 2015). However, whilst not experiencing significant and influential consumer and environmental backlash against GMOs, China has become increasingly concerned over the trade consequences of a pro-GMO position. China learnt a costly lesson when it was obliged to withdraw GM virus-resistant tobacco from production in 1992, only a few years after its introduction, because of pressure from a major international tobacco importer (Planet Ark 2003a). China has also introduced rules to slow down the GMO approvals procedure, whilst restricting its own trade in GM, such as soybean from the USA (Planet Ark 2003a). It was also rumoured that Beijing suspended commercialisation of GM herbicide-resistant rice, which had already passed the safety evaluation for environmental release.

In Brazil, the Agriculture Ministry approved the release of GMO soybeans in the late 1990s, only to have a Federal Court block it pending an environmental impact study. Because of the study, 18, of the 27, Brazilian states called on the federal government to halt commercial releases of GMOs (GAIA Foundation 1999). In Brazil, public concern about pesticides increased after journalists revealed that tomatoes were being treated with mercury compounds. A large percentage of workers in sugarcane fields in the State of Rio de Janeiro were poisoned by mercury (Knight 1997).

As concern over the impacts of pesticide grew in Malaysia, pesticide corporations increasingly supported the Government's attempts to regulate illegal and adulterated sales of pesticides but were wary of the imposition of a costly registration process (Boardman 1986). In 1971, with the backing of industry, the Malaysian Government instigated a voluntary pesticide regulation process. Subsequently, the Malaysian Government passed the 1974 Pesticide Act. While this was supported in principle by the pesticide industry, they retained several reservations and made numerous stipulations (Boardman 1986). By the late 1970s, pesticides became recognised as environmental and health hazards in Malaysia (Boardman 1986). However, many of the early demands for clean air and water were often dismissed as the ravings of liberal do-gooders (Boardman 1986). By mid-1981, the Ministry had received 1912 pesticide applications, 1087 were registered and approximately 600 were not; 319 products were withdrawn (Boardman 1986). In 1981, one activist stated that the public is equally doubtful of the claims of producers and distributors of agrochemicals and about their efficacy or harmlessness of the products because of vested interests (Boardman 1986). As the environment and pesticides became important issues, newspapers began publishing pesticide calamities including deaths, severe illness, fish poisoning, wildlife disruptions and adulteration, particularly as pesticides overlapped with water pollution and food quality (Boardman 1986). Local Malaysian environmental groups were attached to FOE or Sahabat Alam Malaysia (SAM) and the Environmental Protection Society of Malaysia (EPSM), the most influential, formed by a merger of environmental groups after the Stockholm Conference (Boardman 1986). This group focused on pesticide poisonings, pest outbreaks and resistance problems, and pesticide residues on vegetables (Boardman 1986). Middle-class urban supporters of EPSM used media and international links to fight their cause (Boardman 1986). SAM insisted on tough pesticide regulations and blamed the overuse of pesticides on the lack of public information and lack of good education (Boardman 1986).

6.3.1.1.2 Intergovernmental regulatory arenas

Whilst governments in most developing nations have progressively tightened pesticide and biotechnology regulations, many governments, and their incumbent administrations, have not been exposed to the same degree of concern over the impact of pesticides and biotechnologies on human health or the environment as governments in developed countries. Indeed, until the 1980s and 1990s, many developing countries did not have home-grown NGOs with capacity to effectively lobby their national governments. Likewise, even where organised and politically astute NGOs did exist, many developing country governments lacked the institutional infrastructure and capacity to assess risk and manage chemicals, and the financial resources, or political willingness to impose tighter restrictions on pesticides and biotechnologies. Ultimately, the political and economic power of pesticide corporations in most developing countries is considerable. As a result, the pesticide industry has considerable influence over the regulation and use of pesticides and biotechnologies in developing countries.

In the 1980s and 1990s, organised action to reduce or eliminate the most severe human health and environmental impacts of pesticides and biotechnologies came from international anti-pesticide, human welfare and environmental NGOs, concerned governments from developed countries and key individuals in intergovernmental bodies such as the WHO and FAO. The principal concern of these actors continued to be the elimination or restriction of the world's most hazardous pesticides. To this end, international NGOs have done much to publicise what they believed to be the significant environmental degradation and human suffering caused by pesticides and biotechnologies. Since 1985, and the launch of its 'dirty dozen' campaign (PANNA 1995), PAN International, alongside Green Peace, Friends of the Earth and other key NGOs, have continually lobbied national governments and intergovernmental institutions to the following ends: (1) Ensure that human safety is considered foremost in all pesticide policy decisions. (2) End the use of dirty dozen pesticides whenever their safe use cannot be assured. (3) Generate support for research on, and implementation of, safe pest control methods (PANNA 1995).

Much of the momentum to curtail the worst excesses of pesticide use in developing countries came from the Stockholm Conference of 1972 and the establishment of the IPCS in 1980 (IPCS 2003; WHO 2003b). Other critical momentum came from prolonged anti-pesticide activities (such as PAN's dirty dozen campaign), growing concern over the direct and indirect impacts of pesticides in developing countries and concern over pesticide residues on increasing volumes of imported foods from developing countries (Hough 1998). In 1981, the US Government estimated that 5% of food imports contained pesticide residues above domestic residue thresholds (Peterson and Harris 1981). This prolonged effort culminated in a series of intergovernmental interventions that sought to kerb the worst excesses of hazardous chemicals (including key pesticides). First, EU regulation. Second, the development of the International Code of Conduct on the Distribution and Use of Pesticides in 1985. Third, the Basel Convention on the Control of Transboundary Movements of Hazardous Wastes and their Disposal (1989). Fourth, the Nordic Minister's Council Action Programme on Cleaner Technologies, Waste and Recycling (1990). Fifth, The Barcelona Convention for the Protection of the Mediterranean Sea against pollution (1991).

6.3.1.1.2.1 European Union

Environmental Actions Plans have been one of the principal mechanisms employed by the EU to address environmental problems. Since 1973, there have been seven Environmental Action Programmes: 1973-6, 1977-81, 1982-6, 1987-92, 1993-2000, 2001-10 and 2013-20 (European Union 2017). Pesticide regulation has played a central role in many of these actions plans, specifically with regard to the EU's, previously EC's, initiatives to improve water quality. With the exception of Lindane, by the late 1990s, all organochlorine pesticides had been banned, or severely restricted in most Western European countries. Lindane is a highly toxic pesticide that can pose serious risks to health and the environment, and is a persistent organic pollutant (POP) (Pesticides News 1998c). In common with other organochlorine pesticides, Lindane is fat-soluble and can bioaccumulate through food chains. In recent years, Lindane residues have been found in dairy produce. Residues have been detected in the kidneys, livers and adipose tissue of a wide range of wild animals and birds, and in human body fat. It is highly toxic to aquatic invertebrates and fish (Pesticides News 1998c). It is also an endocrine disrupter, which is capable of imitating certain hormones in humans. There is a significant body of evidence which suggests that where Lindane is used, the incidence of breast cancer is higher (Pesticides News 1998c). Cases of human poisoning by Lindane were also reported in Europe.

Lindane was banned in Sweden, Denmark and France (in agriculture) and was part of PAN's Dirty Dozen pesticides (Pesticides News 1998c). In the UK, PAN first called for a ban on Lindane back in 1985 (Pesticides News 2000). However, despite a safety review in the UK by the Ministry of Agriculture Fishers and Food (MAFF) in 1996, Lindane remained on sale. The decision to continue allowing the use of Lindane provoked fierce responses from UK environmental NGOs. Indeed, FOE suggested that MAFF's review was a disgrace, and criticised the UK Government for not applying the precautionary principle (Pesticides News 1998c).

In the late 1990s, a review of Lindane was undertaken by Austria on behalf of the EU (Pesticides News 1998a). In 1998, the Austrian report listed a number of harmful effects of Lindane including carcinogenicity, hormone disruption, behavioural changes, nervous system damage, immune system damage and birth defects (Pesticides News 1998c). It also concluded that there was a lack of adequate data to support the approval of Lindane, particularly on toxicity to birds, fish, small mammals and aquatic (Pesticides News 1998c). The report also concluded that there was no safe limit of exposure for Lindane and recommended a Europe-wide ban (Pesticides News 1998c; PAN-UK 2000). Subsequently, this gave added momentum to the anti-Lindane campaign across Europe (PAN-UK 2000). In the UK, this was given additional momentum following the death of an 8-year-old Sussex girl, which was linked to the consumption of a Lindane-based ant killer (PAN-UK 2002). The Ban Lindane Campaign called for all Lindane products to be taken off shelves immediately, and for existing stocks to be handed in for safe disposal. PAN-UK called for tighter regulation of pesticides, particularly with regard to children, and the application of the precautionary principle where pesticide products have significant knowledge/information gaps (PAN-UK 2002). Further justification for banning Lindane came from the Codex Commission, which set a new more stringent ADI for the insecticide Lindane of 0.001 mg/kg/body weight. For a 60-kg adult, the maximum daily dose should not exceed 0.06 mg in total (Pesticides News 1998a). Working on the new ADI, the Codex Commission suggested that the ADI for residues of Lindane in the European diet may be exceeded by up to 12 times (Pesticides News 1998a).

The Ban Lindane Campaign found that, although some retailers voluntarily phased out the use of Lindane, residues were still being found in UK-purchased chocolate. Some supermarkets still offered bland reassurances that levels of Lindane found in chocolate were 'safe', even though the Austrian report in 1998 concluded that it is not possible to set a safe exposure level for Lindane. Supermarket responses reveal an alarming lack of control over the chocolate supply chain, which some retailers used as an excuse for not taking action. Safeway said: 'The complex infrastructure makes it impossible to ensure that Lindane is not used'. Sainsbury's quoted a trade association for chocolate suppliers which says 'It is not practicable for purchasers to stipulate that they will buy only beans from cocoa trees not treated with Lindane'. By contrast, the Co-op, Marks and Spencer and Waitrose stated that they are seeking Lindane-free chocolate from their suppliers, and Somerfield stated that its suppliers do not use Lindane 'at the present time'. Co-op found that its Fair Trade chocolate, which can be traced directly back to the cocoa grower in Ghana, is free from Lindane residues. However, it could not give the same assurance for the rest of its own-brand chocolate, which is supplied by a major chocolate manufacturer. Tesco, the largest supermarket, did not reply (PAN-UK 2001).

On the 13th of July 2000, the EU's Standing Committee on Plant Health voted to ban Lindane with support from the UK and other Member States. Statutory withdrawal of Lindane products was agreed from the 20th of June 2002 (PAN-UK 2002). In the UK, the Lindane Campaign Group – Friends of the Earth, the Pesticides Action Network UK, UNISON, the Women's Environmental Network, the Soil Association, Green Network and Breast UK (PAN-UK 2000) – warmly welcomed the Lindane ban. However, the NFU expressed concern that the continual reduction in chemical products could leave farmers and growers unable to protect their crops (Pesticides News 2000).

6.3.1.1.2.2 The International Code of Conduct on the Distribution and Use of Pesticides

In 1979, and then again in 1981, FAO expert consultations, including GIFAP, were convened to discuss environmental criteria for registration of pesticides (Boardman 1986). The 1979 consultation adopted the principle that developing countries 'should be encouraged and supported in the initiation and/or development of pesticide registration schemes which include attention to potential environmental effects' (Boardman 1986). Later, in response to pressures from the UNEP and the WHO, FAO's Director General proposed a Code of Conduct on the Distribution and Use of Pesticides (Murray and Taylor 2000). The Code was one of the first voluntary codes of conduct that sought to protect human health and the environment, whilst remaining firmly focused on enhancing food security. According to Bull (1982), Ciba-Geigy was among the first pesticide corporations to express support for the Code, which was formally adopted by the FAO Conference at its Twenty-Third Session in 1985 (see Fig. 6.3). The Code was deemed necessary due to increasing human health and environmental problems in developing countries, concerns in the developed world over high levels of pesticide residues in the growing volume of food imports from developing countries and the difficulties associated with the use of pesticides in developing countries, particularly with inadequate regulatory infrastructures (Pesticides News 1998c). It was designed to increase international confidence in the availability, regulation, marketing and use of pesticides for the improvement of agriculture, public health and personal comfort (UN 2003a). Its objectives were to establish voluntary standards of conduct for all public and private entities engaged in, or affecting, the distribution and use of pesticides. The code suggested how to distribute the responsibilities between government, industry and others (UN 2003a).

The Code included many important reforms in pesticide marketing and use, such as the improvement of product labelling, namely: adequate information on protective measures, proper dosage and so forth. In particular, there was considerable attention given to technician and user training and the proper use of pesticides (Murray and Taylor 2000). However, whilst the Code was an obvious step in the right direction, for many, it did not go far enough. For example, the Code omitted broader measures to reduce pesticide hazards, including mandatory or enforceable reductions in the use of the most hazardous pesticides (Murray and Taylor 2000). However, the 1985 version of the Code, Article 9, only endorsed the need for pesticide exporters to inform importing countries of the potentially hazardous nature of the pesticides concerned. Upon the Code's release, PAN quickly retaliated by insisting that the wording of Article 9 was inadequate, and that, in fact, it had been substantially adjusted in comparison to previous draft versions. Subsequently, PAN led a determined campaign to amend the Code, insisting that, in order to import pesticides, it should first be necessary for exporting parties to gain 'express approval' from importing country governments, rather than the current wording, which simply required the exporter to notify importing governments of the hazardous nature of the pesticides in question. The pesticide industry fought this proposal suggesting that simply notifying governments of importing countries was enough (Pesticides News 1998c). Ultimately, despite appearing in seven of the eight drafts of the Code, the prior informed consent (PIC) clause was removed from the final version, apparently in face of strong British and American persuasion, and pressure from GIFAP (Murray and Taylor 2000). However, according to PAN, no national delegation had officially requested its deletion, and 30 countries protested at its removal. PAN suggests that 'covert pressure persuaded delegates that the Code as a whole would be at risk if PIC was not removed' (Pesticides News 1998c).

Subsequently, PAN and Oxfam led a campaign to incorporate PIC into the UNEP London Guidelines for the Exchange of Information on Chemicals in International Trade (London Guidelines) by lobbying representatives at UNEP's 1987 London working group meeting. However, the Americans, British and German representatives continued to resist the inclusion of PIC (CSE 2003). Ultimately, with PAN's facilitation, the intransigence of the Americans, British and Germans helped to galvanise opinion among many developing country governments in support of the inclusion of PIC. Indeed, many developing country governments wanted a decision on compliance, including a mechanism for enforcement and liability for contravention, with most, if not all, responsibility shifting to exporters. But, exporting countries did not want to be held responsible for problems related to poor legislation in importing countries (CSE 2003).

In 1989, at the Twenty-Fifth Session of the FAO Conference, the FAO Code was subsequently amended to include provisions for the PIC procedure (Pesticides News 1998c; CSE 2003). Industry's apparent 'U' turn on PIC appeared to be a fear of the alternatives. The lead-up phase prior to the drafting of a bill in the USA during 1991–2 proposing to introduce export controls for pesticides prompted the chief representative of GIFAP to take the extraordinary step of criticising the bill on the grounds that it was contrary to the FAO Code of Conduct. A major concern was the appearance of a draft bill on pesticide export control in the USA, which was very much at variance with PIC in the FAO Code, namely that this draft legislation is export rather than import control oriented. GIFAP had opportunistically interpreted PIC: given the choice between PIC and export restrictions, they saw PIC as the lesser of two evils (Pesticides News 1998c). Northern pesticide manufacturers also accepted a PIC-based binding agreement since it was less stringent than direct bans on pesticide trade (CSE 2003). The Code established voluntary standards of conduct for all public and private entities engaged in, or associated with, the distribution and use of pesticides, and since its adoption has served as the globally accepted standard for pesticide management.

Subsequently, PIC was made legally binding in some European and developing countries (CSE 2003). Unfortunately, the PIC procedure was, and still remains, restricted by financial resources (CSE 2003). For many developing countries, substantial investments in infrastructure were, and in some cases still are, required in order for them to implement pesticide bans or restrictions, and, indeed, to follow the detailed process to ban or restrict the use of certain pesticides (PAN 1998b).

Ultimately, the changing international policy framework and the persistence of certain pesticide management problems urged FAO to initiate the revision and update of the Code. This process started in 1999, with a number of recommendations made by the FAO 'Panel of Experts on Pesticide Specifications, Registration Requirements, Application Standards and PIC'. Government experts, NGOs, the pesticide industry and other UN organisations participated in the revision process. A government consultation subsequently established the basic text for the present revised version of the Code (FAO 2002). The new Code reflected more strongly than ever the responsibility of governments, the chemical and food industry, traders, pesticide users, public interest groups and international organisations in reducing the health and environmental risks associated with pesticide. It promotes IPM and natural pest control systems (FAO 2002). Governments, the pesticide industry and international organisations recognise the voluntary Code, and several governments have incorporated the Code or parts of it, into their national pesticide legislation (FAO 2002). Compliance with the Code is also obligatory for the members of Crop Life International, the international association of pesticide manufacturers. The revised Code calls upon the pesticide industry 'to supply only pesticides of adequate quality, and to pay special attention to the choice of pesticide formulations and to the presentation, packaging and labelling in order to reduce risks to users and minimize adverse effects on the environment' (FAO 2002).

6.3.1.1.2.3 The Basel Convention on the Control of Transboundary Movements of Hazardous Wastes and their Disposal (1989)

The 1989 Basel Convention was adopted in response to concerns about toxic waste from industrialised countries being dumped in developing countries and in economies in transition. Whilst the Basel Convention's principal focus was on the minimisation of waste generation and the environmentally sensitive disposal of existing wastes, it had relevance to the pesticide industry in two key respects. First, the production of many pesticides often involved the generation of toxic by-products, including dioxins. In theory, this meant that the pesticide production process should have been placed under pressure to reduce the production of these toxic by-products. Second, if not used before its use by date, a pesticide then becomes classified as a hazardous waste and subject to the Convention's regulations in respect of its transportation and disposal.

During the 1990s, the Convention's principal focus was the elaboration of controls over the international transportation of hazardous wastes and the development of criteria for the environmentally sensitive management of wastes (UNEP 2002). Indeed, the Secretariat of the Basel Convention (SBC) became increasingly concerned about the trade in, and transportation of obsolete pesticides, particularly DDT and dieldrin between developing countries without adequate compliance with the requirements of the Basel Convention (Pesticides News 1998c). Even though it remains active as an insecticide for many years, DDT is often used, and traded, after its use by date. As part of its activities, the SBC responds to problems identified by countries relating to hazardous waste. The issue of obsolete, often extremely hazardous, pesticides has been raised by numerous developing countries; particularly in Africa. African parties to the Basel Convention request assistance and advice in the case of transboundary movement of obsolete pesticides, as well as when considering disposal of obsolete pesticides in an environmentally sound manner. SBC has a mandate to help solve obsolete pesticide problems in developing countries. Parties to the Basel Convention have been assisted in the export of obsolete pesticides destined for disposal operations, particularly final disposal in OECD countries. The Basel Convention emphasises full implementation of treaty commitments and minimisation of hazardous waste generation (UNEP 2002). Coordination on chemical management in the UN system has also been improved through the establishment of the Inter-Organization Programme for the Sound Management of Chemicals (IOMC).

6.3.1.1.2.4 Nordic Minister's Council Action Programme on Cleaner Technologies, Waste and Recycling

In 1990, the Nordic Minister's Council Action Programme on Cleaner Technologies, Waste and Recycling agreed to give priority to a 'prioritised list of substances and products for which special substitution efforts should be made'. The mandate of the working group was, among others, to look at organochlorines as a priority. In August 1993, the Nordic Council's International Conference for Parliamentarians on Development and Protection of the Arctic region aimed 'to reduce and ultimately eliminate airborne and seaborne pollution such as that from heavy metals, greenhouse gases, and chlorinated hydrocarbons such as PCB and DDT. Participating countries were Denmark, Finland, Greenland, Iceland, Norway and Sweden' (PAN 1994).

6.3.1.1.2.5 The Barcelona Convention for the Protection of the Mediterranean Sea against pollution (1991)

First convened in 1976, the Barcelona Convention for the Protection of the Mediterranean Sea against pollution agreed, in October 1991, to facilitate the progressive elimination of organophosphorus compounds by 2005 (Hough 1998). In all, 17 countries have signed the Mediterranean Action Plan, which aims to cut chemical discharges into the Mediterranean Sea. Participating countries are Albania, Algeria, Cyprus, Egypt, France, Greece, Israel, Italy, Lebanon, Libya, Malta, Monaco, Spain, Syria, Tunisia, Turkey and Yugoslavia (PAN 1994).

6.3.1.1.2.6 United Nations Conference on Environment and Development-Earth Summit (1992)

The need to tighten pesticide regulation was given considerable impetus after the 1992 Earth Summit (see Fig. 6.3). Pesticides were specifically mentioned in several chapters. (1) Chapter 6 'Protecting and Promoting Human Health'. (2) Chapter 17 'Protection of the oceans, all kinds of sea, including enclosed and semi-enclosed seas, and coastal areas and the protection, rational use and development of their living resources'. (3) Chapter 18 'Protection of the quality and supply of freshwater resources: Application of integrated approaches to the development, management and use of water resources'. (4) Chapter 19 'Environmentally sound management of toxic chemicals, including prevention of illegal international traffic in toxic and dangerous products'. Indeed, Chapter 17 required that states consider priority actions, including the elimination of organohalogen (including organochlorine pesticides) compounds that threaten to accumulate to dangerous levels in the marine environment (PAN 1994). In addition, Chapter 19 proposed six key programme areas: (1) Expanding and accelerating international assessment of chemical risks. (2) Harmonisation of classification and labelling of chemicals. (3) Information exchange on toxic chemicals and chemical risks. (4) Establishment of risk reduction programmes. (5) Strengthening of national capabilities and capacities for management of chemicals. (6) Prevention of illegal international traffic in toxic and dangerous products (UN 2003c). In addition, Chapter 19 called for governments to participate in PIC and for an international convention, constituting a legally binding instrument for PIC by 2000 (PAN 1998b).

Governments at the UNCED also requested increased coordination of UN Bodies and international organisations involved in chemicals assessment and management. Just prior to the Earth Summit, a key meeting of intergovernmental institutions recommended the establishment of a specialist intergovernmental body to address the problems associated with the pervasive use of toxic chemicals. Consequently, the Earth Summit's Agenda 21 called on the Executive Heads of the WHO, the UN Environment Programme and the ILO, to convene an international meeting to establish a Forum. The Earth Summit also recommended increased coordination of UN Bodies and international organisations involved in chemicals assessment and management, prescribing that the IPCS should have a central role in coordinating risk assessment and management (Hough 1998).

The Earth Summit acted to stimulate a flurry of important new intergovernmental interventions, and the revision of existing regulatory mechanisms. For example, in 1992, PIC was made mandatory across the whole of the EU (PAN 1998b). In September 1992, the Paris Convention, originally convened in 1974, agreed to a number of measures to prevent and eliminate pollution to protect the North-East Atlantic maritime area against the adverse effects of human activities (Hough 1998). These measures included the following: (1) To reduce, and ultimately eliminate, organohalogen inputs into the maritime area. (2) To adopt measures for the prohibition of the use of organohalogen substances which are unnecessary for the intended use or purpose. (3) To adopt measures for the reduction or elimination of hazardous substances from diffuse sources, including pesticides. Participating countries were Belgium, Denmark, Finland, France, Germany, the UK, Iceland, the Netherlands, Norway, Portugal, Spain, Sweden, Switzerland and Luxembourg (PAN 1994). By October 1993, proposals were also adopted against organohalogen compounds, which included measures to reduce, and to phase out by the year 2005, inputs to the marine environment of toxic, persistent and bioaccumulative substances, in particular, organohalogen compounds having carcinogenic, teratogenic and/or mutagenic characteristics (which included some organochlorine compounds). In 1993, the International Whaling Commission called on all governments to remove existing threats to the marine environment from the emission or discharge of organohalogen compounds that threaten to accumulate to dangerous levels in the marine environment (PAN 1994).

6.3.1.1.2.7 Codex Alimentarius Commission

The pesticide industry has been extremely influential within the Codex framework. Pesticide residue limits set by the Codex Commission have tended to downwardly homogenise global standards and, in so doing, legitimised higher levels of pesticide residues in foods that would have otherwise been the case without their influence.

Having said this, in alliance with sympathetic governments, the pesticide industry's power has not been absolute. Indeed, it can be argued, that aside from the most basic standards required to restrict the room for manoeuvre of smaller commodity pesticide produces and formulators, the Codex Commission has suggested to, and latterly did, impose residue limits on pesticide actives that have the potential to feed back into the production process and restrict application of pesticide products. Even as early as 1977, developing countries raised concerns over the use and misuse of pesticides. Indeed, in the 1977 Codex session, Egypt highlighted a series of pesticide-related concerns and called for 'greater supervision of pesticides destined for export by governments in developed countries' (Boardman 1986). In the early 1980s, Codex was dealing with 122 pesticides and 1689 tolerance proposals (Boardman 1986). By 1981, approximately 70% of governments had accepted Codex MRLs, either full or free (Boardman 1986). By the end of the 1990s, over 2500 MRLs were set for 195 AIs, although standards still did not exist for all crops, or for all pesticides. As previously discussed, exporters from developed and developing countries are required to meet Codex standards under WTO rules, and produce that fails to meet these standards can be refused entry. In effect, this means that producers must adhere to GAP, and that residue limits are within Codex MRLs.

However, while science-based MRLs and downward homogenisation of standards have been dominant for several years in the Codex/WTO regime, the role that non-scientific factors should play in the decision-making process is still being debated (Macfarlane 2002). For instance, not surprisingly, in the post-productivist tradition, the EU continues to insist that Codex adopts labelling, traceability and the precautionary principle based on cultural, social and political considerations. Again, there was no surprise when the US Biotechnology Industry Organization (BIO) responded by accusing the EU of undermining the WTO's SPS and TBT agreements by injecting unscientific and socio-economic factors into Codex's decision-making process (BIO 2004c). BIO suggested that EU proposals 'disrupt international trade, especially US agricultural exports, and provide no additional consumer protection'. Subsequently, draft Codex proposals now contain precautionary labelling and traceability provisions, whilst recognising that GM ingredients are compositionally and nutritionally equivalent to conventional counterparts and raise no safety or nutritional concerns (BIO 2004c). However, BIO and other industry sources insist that costs associated with segregation and labelling are impractical and inequitable. Indeed, BIO believes that 'traceability is unmanageable, costly, and contrary to Codex's scientific principles' (BIO 2004c). Monsanto, who is also in favour of labelling, not surprisingly prefers labelling of GM-free food ingredients (Deutsche Bank 1999). In addition, BIO also stated that 'the precautionary principle is unscientific; allowing government's to apply restrictions to products if there is a perceived uncertainty concerning risk, even if the risk is extremely remote or hypothetical' (BIO 2004c). Further post-productivist measures have included Codex's endorsement of the Ad Hoc Working Group on Problems Relative to Pesticide Residues in Food in Developing Countries call for IPM as a preventive measure, and for consideration of IPM when developing GAP information.

6.3.1.1.2.8 Intergovernmental Forum on Chemical Safety

In 1994, just 2 years after the Earth Summit, the Intergovernmental Forum on Chemical Safety (IFCS) was established (see Fig. 6.3). The aim of the Forum was to convene representatives of national governments responsible for chemicals management, non-governmental (industry, public interest, labour, science) and intergovernmental organisations in order to address the six key Agenda 21 programme areas proposed in Chapter 19. The 'IFCS is a non-institutional arrangement whereby representatives of governments meet together with intergovernmental and non-governmental organizations with the aim to integrate and consolidate national and international efforts to promote chemical safety'. According to PAN, the result has been a 'dynamic programme of work developing higher standards of chemical management, better co-ordination between the chemical management aspects of UN and other key agencies, and a process for building the capacity to manage chemicals in an environmentally sound manner in developing countries'. An important principle established in the IFCS meetings is the commitment to equal access to NGOs (industry, public interest, labour and science) that are free to express their views in meetings, and bring items on to the agenda, though may not vote. The goal is to seek consensus among these groups on strategies to address global chemical safety. Forum activities are often led by one of its partner agencies. For example, the OECD leads the 'Harmonisation of classification and labelling' activity. This activity aimed to achieve a globally harmonised hazard classification and comparable labelling system for chemicals, including material safety data sheets and easily understandable symbols, by 2000. The IPCS and OECD also developed an international science assessment report, a global inventory of ongoing research, and test guidelines, after IFCS participants expressed concern over scientific evidence that certain pesticides may interfere with the endocrine system and cause adverse health effects. In addition, the IFCS Ad Hoc Working Group on POPs, established in 1997, assisted key negotiations towards a legally binding instrument to deal with POPs. Of the 12 identified POPs, nine are pesticides: DDT, aldrin, dieldrin, endrin, hexachlorobenzene, heptachlor, chlordane, mirex and toxaphene (FAO 2001).

The IFCS discussed the issue of obsolete pesticides and chemicals in December 1998. As a result, addressing obsolete chemicals was identified as one of 17 priority areas for discussion at the Forum III meeting in Brazil in October 2000. The document identifying these priorities states that the identification, neutralisation and safe disposal of obsolete stocks of pesticides and other chemicals (especially PCB) must be urgently facilitated particularly in developing countries and countries with economies in transition. With respect to the final disposition of chemicals, the Forum and IOMC organisations should promote the use of less polluting and safer technologies. By 2004, IOMC Participating Organisations should have established relevant action plans for all regions, and at least two countries in each region should have commenced implementation of their National Plan with respect to disposal (New recommendation – Forum II, ISG-3). It therefore falls to IOMC members to support the implementation of these modest targets (FAO 2001).

6.3.1.1.2.9 Inter-Organization Programme for the Sound Management of Chemicals

In 1995, the IOMC was established (see Fig. 6.3). IOMC's role was to serve as a mechanism for coordinating efforts of intergovernmental organisations in the field of chemical safety. The IOMC was designed to be a cooperative undertaking among intergovernmental organisations that, within the framework of their own respective constitutional mandates, work together as partners to promote international work. Scientific and technical work of the IOMC is carried out through the existing structures of the Participating Organisations, either individually or jointly. An Inter-Organization Coordinating Committee, composed of representatives of the Participating Organisations, consults on the planning, programming, implementation and monitoring of activities undertaken jointly or individually by the Participating Organisations. The original six Participating Organisations were UNEP, ILO, FAO, WHO, UNIDO and OECD. UNITAR formally joined the IOMC as a Participating Organisation in January 1998. These organisations have substantive programmes in the area of chemical safety.

6.3.1.2 Education, marketing and food safety initiatives

Whilst the introduction of GMOs in the USA and Canada was relatively problem free, the introduction of GM crops and GM food ingredients in Western Europe met with rapid, well-organised, hostile and effective opposition. GMO hysteria not only reached fevered pitch in the EU but also spread back to the USA, Brazil, Japan and other key markets. In its 1997 annual report, Monsanto, one of the principal companies involved in the introduction of GMOs into Europe, published 'Monsanto's Law', which posited an almost exponential level of genetic discovery and the associated profits available for those brave enough to exploit this potential (Washington Post 1999). Indeed, Monsanto is quoted as stating that, 'the biggest mistake that anyone can make is moving slowly, because the game is going to be over before you start' (Deutsche Bank 1999). This gets back to the contention that the quest for corporate profits has ridden roughshod over questions of public health, freedom of choice and ecological stability (Deutsche Bank 1999). However, 2 years later in 1999 after the disastrous introduction of HT OSR, Monsanto admitted that it had probably irritated and antagonised more people than it had persuaded about the merits of GMOs. What Monsanto described as 'our confidence in this technology and our enthusiasm for it has, I think, been widely seen, and understandably so, as condescension or indeed arrogance' (Washington Post 1999). As a result of the uproar in Europe over GM foodstuffs, Monsanto initiated a Europe-wide advertising campaign targeting key countries, namely Britain and Germany, where opposition to GM foods was perceived as the most acute (see Fig. 6.3). Monsanto spent in excess of US$1.5 million on full-page advertisements in newspapers and magazines that were typically read by members of high socio-economic categories who were perceived as being influential on decision makers. Monsanto therefore selected target countries, customers and media. Advertisements were also posted on the web. The advertisements were text rather than image oriented and several claimed that GM crops required less pesticide and were therefore better for the environment. In parallel to its campaign to persuade European consumers of the merits of biotechnologies, Monsanto actively sought to reduce negative publicity circulated about GMOs which it considered inaccurate. According to a 1998 press release, an issue of the Ecologist Journal was withdrawn from the printers in the face of possible legal action by Monsanto (https://nature.berkeley.edu/srr/Alliance/novartis/monsanto.htm). Monsanto was also alleged to have prevented the screening of a programme on Fox TV, in 1997, on rBGH. The UK Guardian has also been approached by Monsanto to end the criticism of the company. Werner Reisberger, a German who had an anti-Monsanto e-mail server, was also taken to court by Monsanto. Monsanto was also keen to take action against groups, such as GenetiX Snowball, that organised non-violent destruction of Monsanto GM trials. According to Holder, Monsanto's

PR campaign was thought to have failed for a number of reasons. First, the UK Advertising Standards Agency ruled that 'Monsanto's advertisements were misleading and confusing as well as presenting opinion as accepted fact'. This was at a time when public awareness of corporate environmental and human rights records was high. Second, trust in government authorities was low following the BSE crisis. Subsequently, the British public were more easily persuaded by the pro-organic and anti-GMO stance of Prince Charles and the broader anti-GMO lobby than the British Government (Giannakas and Fulton 2002). Third, information to counter Monsanto's claims was exchanged via the web and e-mail extremely quickly. However, after a U-turn in PR strategy, Monsanto made some ground. In 1995, in the Netherlands, Monsanto encouraged the media to attend a workshop six months before GM soya was scheduled to enter the country. The idea was that, by this time, any negative reaction to GM would have subsided. This strategy was extremely effective.

In Switzerland, people voted by 2:1 to accept GM. GM was sold not by Monsanto but by a coalition of pro-biotech groups including farmers, medical workers and patient groups. Scientists and local politicians also spoke out in favour of GM. The role of scientists was particularly important, as they took to the streets to sell biotechnology as their own. The focus shifted away from man and the environment to health and research issues as well as focusing on the economic aspects. Industry is now involved in direct local action by scientists, and PR training programmes are also being organised by the EC. The EC also funded education programmes for adolescents on biotechnology, whilst industry (Monsanto, Novartis and AgrEvo) co-funded pro-biotechnology pullouts for adolescent girl magazines in Germany. The strategy now is how to convince the public that biotechnology is the only way forward, and policymakers that public opinion is the problem, not biotechnology. The industry is already selling biotechnology as green technology and attempting to claim a stake in a sustainable future. In 2002, Syngenta, BASF, Bayer CropScience, Dow AgroSciences, DuPont and Monsanto established the Agricultural Biotechnology Council (ABC). Its aim was to be an 'information and education service on behalf of the UK biotechnology industry, promoting debate surrounding the production of GM crops' (see Fig. 6.3) (Corporate Watch UK 2003a). ABC is the UK arm of the Agricultural Biotechnology in Europe initiative.

6.3.1.3 Voluntary measures to reduce the impacts of pesticide use

6.3.1.3.1 The Global Safe Use Campaign

As global opposition to the negative human health and environmental impacts of pesticides mounted, GIFAP recognised that, in order to delay or avoid further restrictive pesticide regulation legislation, it must act to reduce the negative impacts of pesticides and improve their image. To this end, GIFAP began to work closely with FAO to promote 'new and better' pesticide use practices. In June 1991, the pesticide industry launched one of their most ambitious non-product specific voluntary initiatives to date, the Global Safe Use Pilot Projects. With a commitment of US$1 059 000 from the GIFAP (now renamed CropLife International), the Safe Use campaign was launched with pilot projects in three countries, Guatemala, Kenya and Thailand (Murray and Taylor 2000). The Safe Use programme in Guatemala aimed to 'provide education and training and ensure that pesticides are used rationally and safely in the agricultural production process' (Murray and Taylor 2000). The Guatemalan pilot project focused on a variety of training and education activities, along with efforts to test and distribute more effective safety equipment and establishment of appropriate pesticide waste disposal facilities. Training topics included the protection of humans and their environment, the prevention and treatment of pesticide poisoning, how to dispose of empty containers, and pesticide regulations. Training materials for agricultural technicians, distributors, users, school children, housewives and others were developed and used in a wide ranging effort at public education. The programme also aimed to establish centres for the control of pesticide poisoning (Murray and Taylor 2000). In addition, USAID provided US$4 million to fund a complementary Pesticide Management Activity (PMA), a 3-year project intended to provide additional training, including training in the use of alternative pest control strategies such as IPM (Murray and Taylor 2000).

According to industry sources, by the end of the first phase of the Safe Use programme in mid-1994, the industry had trained 800 government extension agents. These trainees in turn became trainers, and spread the message to other target groups who in turn would have a 'cascade effect' as they 'shared what they have learned with their employees, fellow workers and families' (Murray and Taylor 2000). The industry reported that the 'Safe Use' programme had trained 226 000 farmers and housewives, 2800 schoolteachers and 67 000 schoolchildren, 700 pesticide distributor employees, 330 technical and sales people and 2000 physicians and paramedical personnel (Murray and Taylor 2000).

In a second phase in 1995, the programme narrowed its focus to small vegetable growers. A more personal approach was implemented with field visits, demonstration plots and master trainers who resided in target communities to coordinate pesticide training activities. In this phase, 7169 farmers, 3475 housewives, 105 extensionists, 188 dealers, 152 teachers and 6145 students were trained (Murray and Taylor 2000). The programme reduced its training efforts to concentrate on a target population of some 7000 trainees (Murray and Taylor 2000). Another significant change has been a shift towards greater emphasis on IPM in the industry campaign. While some have complained that this represents an attempt to capture the increasingly popular alternative paradigm and make it pesticide based (Murray and Taylor 2000), it appears that, in some cases, such as Product Stewardship, the industry is promoting strategies that may lead to reduced pesticide dependency (Murray and Taylor 2000) . Having said this, many see this course of action as the lesser of two evils and believe that the pesticide industry should be encouraged to continue and deepen its promotion of IPM, even as critics observe that the industry's version of IPM emphasises pesticide use more than they would like (Murray and Taylor 2000). Figure 6.2 highlights changes in the number of pesticide poisoning cases in Guatemala between 1972 and 1997.

However, whilst it is recognised that the Safe Use campaign clearly represents a serious effort to address the pesticide problem in Guatemala, there have also been several counter claims. Namely, that in 'the pesticide industry's haste to respond to continued criticisms of the negative effects of the technology and their responsibility for the problems it creates, the claims also appear to have overreached supporting data' (Murray and Taylor 2000). However, there are additional, and more fundamental, issues to contend with. The most important issue is the conceptual framework in which the pesticide problem is framed, and the key assumptions inherent within this framework. The pesticide industry begins with the key assumption that pesticides are needed and that, if used properly, they can be safe. According to PAN (2003a), this approach fails to recognise some important features of agricultural reality. First, structural constraints (e.g. poverty and education) prevent poor rural communities (most) from having access to the equipment and information necessary for safe use. Secondly, pesticides, besides being extremely hazardous, are not agronomically or economically justifiable under many prevalent conditions in tropical agriculture (PAN 2003a).

Heavy publicity by the pesticide industry has brought its Safe Use programme under growing scrutiny. For example, as part of its broader interest

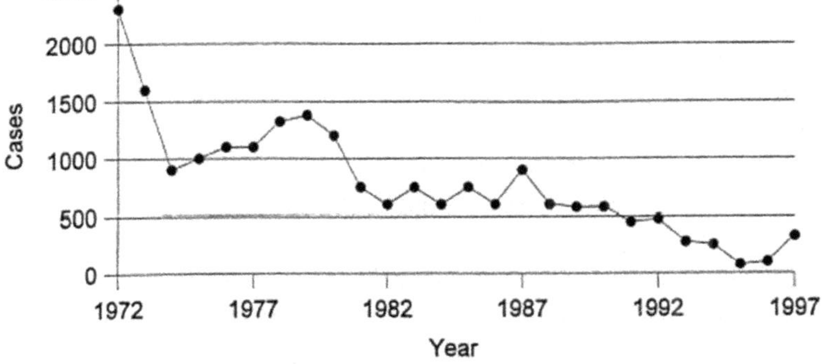

Figure 6.2 Pesticide poisonings in Guatemala, 1972–97. *Source*: Epidemiology Section, Guatemalan Social Security Institute).

in Voluntary Initiatives within the chemical industry, the European International Food Workers Union (IUF) has recently expressed concern about the claims of the Safe Use programmes' successes. Safe Use training, it suggests, in many cases appears to be carried out by pesticide salespeople who have a conflict of interest because they promote the purchase of pesticides as well as their safe use (Murray and Taylor 2000). The IUF also expressed concern that training may have amounted to general talks of 1–2 h to village audiences of farmers and workers, a mode of training less effective than small group, teacher-led training courses (Murray and Taylor 2000). Moreover, the very high turnover among agricultural workers in Central America is likely to undermine the effectiveness of occasional workplace training sessions. Industry reports that the Safe Use project has responded to these kinds of criticism by moving to more teacher-based instruction, installing master trainers in villages. Many have also criticised the Safe Use Campaign on more fundamental epistemological grounds, and remain sceptical about the industry's claim that the use of pesticides is the only rationale approach to the management of pests and diseases. Indeed, there was an overwhelming sense that industry continued to view the pesticide problem as one of public perception rather than as a serious health and environmental hazard problem (Murray and Taylor 2000). For example, a recent GIFAP publication asserts that 'the risks associated with rational use of these products and practices are exaggerated and misinterpreted by regulators, government offices and the public in general' (Murray and Taylor 2000). If the problem is one of public perception, claims of success based on a clear message about types of communication and numbers and types of people reached may be more effective in convincing regulators and consumers in Northern markets that the industry behaves responsibly. If, on the other hand, the problem is defined as a real and serious set of hazardous conditions, the industry's objective should be more focused on altering hazardous behaviour and conditions, rather than consumer and regulator perceptions. In the short term, the dynamics of intensive agricultural production in developing countries is largely dependent on the perceptions of the pesticide industry and key consumer and environmental power brokers in developed countries and intergovernmental institutions.

The Safe Use campaign and related industry voluntary initiatives are part of a larger process of globalisation in which industry actors are struggling to re-regulate the international spaces across which they operate (Murray and Taylor 2000). Neo-liberal economic reform, supported by CropLife International, has lowered tariff and regulatory barriers to TNC's capacity to act within any single nation state's borders (Murray and Taylor 2000). Corporations have moved quickly to devise strategies which allow them to respond to variations in regulation by implementing diverse standards and policies in the different countries in which they operate (Murray and Taylor 2000). At the same time, workers and communities also struggle to re-regulate their worlds, expanding the issues they give attention to, incorporating national and global phenomena into their vision of local problems, and devising strategies which allow them to meet TNCs on more even ground (Murray and Taylor 2000).

6.3.1.3.2 Food systems arenas (retailers, wholesalers, processors, farmers and input suppliers)

The so-called post-productivist transition in predominantly European-based food systems and in commodity systems where consumption is embedded in Western Europe but where production is embedded in developing countries created complex operational contexts in which agribusiness TNCs must operate. Subsequently, the changing and dynamic contexts associated with PPT, often juxtaposed against Neo-Fordist food systems, have necessitated increasingly flexible strategies from pesticide and biotechnology producers. For example, According to Chataway (2001), Novartis (Syngenta) developed a range of pragmatic responses to the PPT in Western Europe by. (1) Attempting to 'anticipate change; influence it where possible. (2) Adapting internally to change and have flexible strategies. (3) Involving regulatory affairs people in the R&D planning process at a much earlier stage than before. (4) Forging new alliances with the food and retail sectors'. Flexible strategies were employed with the capacity to facilitate 'complex mixes of chemical and biotechnology solutions to agricultural problems, which were tailored to social, regulatory, and agronomic requirements of different regions and crops' (Chataway 2001).

A pilot programme with citrus-fruit growers in Florida used the internet to transmit updates on weather, plant growth and insect conditions to Novartis (Syngenta) advisors, who use the data to fine-tune crop protection prescriptions. 'Closed loop' accords are those in which Syngenta would act as a general contractor for retailers or food processors. Syngenta would recruit growers and establish detailed protocols stipulating everything from colour and taste of crops or vegetables and special properties desired for successful transport over long distances, to guarantees on MRLs from pesticides (PAN 1998a). The introduction of proprietary, value-added crops will increasingly involve contractual links between the seed company, the farmer, the grain elevator and the processing company. The ETC Group (1999) suggest that this will usher in an era of 'bio-serfdom', where farmers are systematically eliminated from farm-level management and decision-making as they become renters of proprietary germplasm from the gene giants or their subsidiaries (ETC 1999).

According to Bijman and Joly (2001a), many pesticide corporations did not have a problem with the predictable and incremental evolution of more stringent PPT pesticide policies. Indeed, several pesticide manufacturers suggested that while the costs of developing and registering new and safer pesticides continue to grow, good returns on investment are expected for those companies with the resources to develop them, particularly if the cost acts as a significant barrier to the entry of other pesticide producers (Bijman and Joly 2001a). For example, Rhône-Poulenc estimated that two new blockbuster pesticides Regent (fipronil) and Balance (isoxaflutole) would contribute approximately US$1 billion in sales by 2003 (Assouline and Pierre-Benoit 2001). Rhône-Poulenc had a further five new products based on innovative compounds in the pipeline for 2001–6 (Assouline and Pierre-Benoit 2001). Reformulation of new with older actives was a key strategy for Rhône-Poulenc. This was also the case for many pesticide corporations with a slow turnover of new blockbuster actives (Assouline and Pierre-Benoit 2001). New formulations and packaging also offer potential to reduce the doses of AIs, and meet evolving regulations including those aimed at reducing water contamination and increasing user protection (Assouline and Pierre-Benoit 2001). Rhône-Poulenc also developed High-Throughput Screening in 1998. This allowed the early selection of molecules with positive plant health action at very low concentrations. Rhône-Poulenc also developed a predictive methodology for early assessment of a molecule's toxic and eco-toxic profiles (Assouline and Pierre-Benoit 2001). In addition, most companies have developed wettable granules or soluble sachets (Assouline and Pierre-Benoit 2001) in response to demands for safer and recyclable packaging. Most pesticide manufacturers strongly promote ICM (Wield et al. 2010).

6.3.2 Development of output trait plant biotechnologies, functional foods and nutraceuticals

While input trait biotechnologies have been criticised on the grounds of environmental and customer safety, and the lack of obvious benefits for the processor, retailer and consumer, the second and third waves of biotechnologies may prove considerably more amenable (see Fig. 6.3). Second wave of agricultural biotechnologies refers to those that change the nature of the crop itself, producing desirable and valuable, output traits. The third wave of biotechnologies refers to the production of pharmaceutical and nutraceutical products, known as 'pharming' (UNESCO 2003). Both the second and third waves of output trait biotechnologies are being driven by developments in genomics (UNESCO 2003). Output traits generally focus on enhancing a positive, and potentially valuable, characteristic or suppressing a negative one (Deutsche Bank 1999). Although very few have been released, analysts suggest that output trait biotechnologies have considerable potential (Deutsche Bank 1999), and will inaugurate the real biotech race (New York Times 2003).

One of the first examples of an output trait GM crop was an increased protein soybean, containing a methionine-rich gene from a Brazilian nut. However,

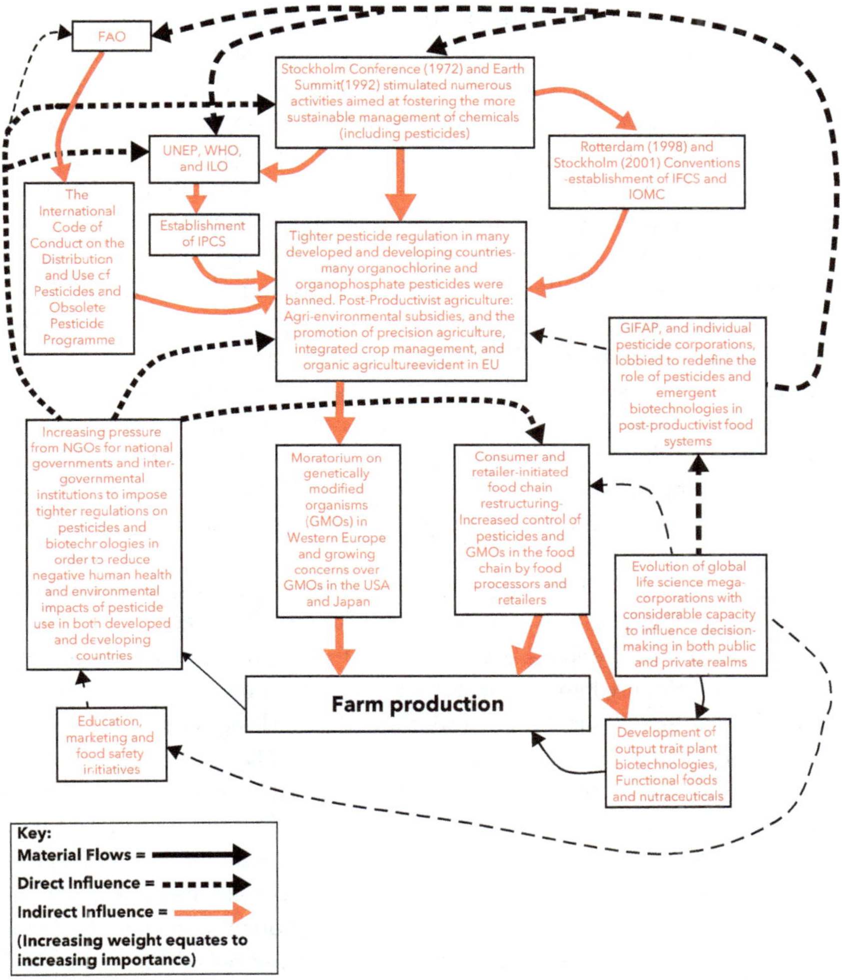

Figure 6.3 Pesticide manufacturers and the Post-Productivist model of agriculture in the Divergent Period.

it was quickly discovered that the soybean contained the allergenic properties of the Brazilian nut, and was not commercialised. In 1994, Calgene introduced the Flavr Savr tomato. This tomato was modified so that it ripened slowly in the field and stayed ripe for longer. These qualities assisted processing and improved the tomato's flavour. Marketed as a tomato paste, and clearly labelled as a GM product, Flavr Savr sold well for a short period. However, when the anti-GM movement grew, and both customer confidence and sales waned, Flavr Savr tomato paste was quickly removed from the market (Deutsche Bank 1999; Wield et al. 2010).

According to UNESCO (2003), plant biotechnologies offer opportunities for production and 'delivery of individualised and holistic services through retail outlets that address diet, drugs and other aspects of treatments'. Indeed, third-generation GM crops could reduce heart disease, cancer and obesity; improve attention-deficit hyperactivity disorder, schizophrenia and dyslexia; and help to improve human quality of life, particularly in old age (Demmig-Adams and Adams 2002). To date, output trait GMs have been targeted either towards enhanced quality animal feed or towards ingredients for healthy foods for human consumption. These include. (1) Omega-3 fatty-acid-enriched maize and canola (New York Times 2003). (2) Lauric acid – enhanced canola called Laurical (Groundup 2003j). (3) High Stearate Soybean and Canola Oils, designed for margarines and shortenings. (4) Medium-Chain Fatty-Acid-Enriched Vegetable Oils. (5) Improved Fibre Cotton. (6) Modified Oil OSR and soybean. (7) Modified Protein Soybean. (8) Modified Quality Potatoes. (9) Modified Quality Wheat. (10) Low-Calorie Oils. (11) Oil-absorbent-reduced potato. (12) Low-Phylate Corn. (13) High Corn Oil for animal feeds (Groundup 2003k).

All the major pesticide and seed corporations are heavily involved in the development of second-generation output trait GM crops. Monsanto is once again one of the lead companies in the race for output trait biotechnologies. Monsanto has already introduced output trait GM crops, including fatty-acid-altered OSR in the USA in 1994 (Groundup 2003k). Using innovative biotechnologies, Monsanto's biotech subsidiaries aim to modify maize and other crops to 'produce proteins used in the manufacture drugs that fight cancer and other diseases' (Groundup 2003j). Monsanto suggests that drug production via plants is much more cost-efficient as compared to fermentation, up to 20 times cheaper. Monsanto is committed to maintaining its current lead in biotechnology, investing 80% of its annual R&D budget to produce biotech traits. At the same time, other pesticide/seed corporations invested only 20% of their R&D budgets in biotech (New York Times 2003). In addition, Monsanto is working on the production of coloured cotton and Golden Mustard, which would produce cooking oil with high beta-carotene (pro-vitamin A) (Groundup 2003j; Genewatch 2003a; CropLife International 2004).

To compete with Monsanto, other pesticide and biotechnology companies also planned to introduce a series of output trait biotechnologies (New York Times 2003). According to Tait (2001), Syngenta favours output trait biotechnologies over input trait biotechnologies because they 'transform commodity crops into specialities that can be segmented in the market place, providing an "engine of value"'. Examples include the development of GM rice with enhanced vitamin A and another that suppresses a protein responsible for allergic reactions. Whilst acknowledging that the suppressed protein rice is targeted towards a niche market, namely kidney dialysis patients in Asia, it sees the move as 'a latch-lifter', stating, 'regulators either have to become less fastidious or deny benefits to patients' (Corporate Watch UK 2003d). In a similar vein, Syngenta was also involved in a collaboration aimed at giving away vitamin-A-enhanced rice to farmers in the global south; something some observers claimed was nothing more than 'a flag of convenience under which biotech companies are trying to win support for GM crops' (Corporate Watch UK 2003d). Other output trait products include biodegradable plastic synthesised in fermentation vats by the bacterium *Alcaligenes eutrophus* (Ball 1996), and sugar beet that produces fructose rather than sucrose (Corporate Watch UK 2002c). Lastly, at the turn of the Century, Advanta was closely involved with NEODIET, a group focused on examining the potential need for improving a selected range of essential nutrients and beneficial factors present in plant foods but where the levels are possibly inadequate for health (Corporate Watch UK 2002c).

DuPont is another pesticide corporation that has realigned R&D to concentrate on food quality traits for food processing and livestock production based on output trait biotechnologies (Groundup 2003g). According to industry analysts, the sale of foods touting healthy properties is expected to soon reach US$29 billion/year, from virtually zero in 1990 (West 1998). In 1996, DuPont introduced high oleic acid (increased stability and reduced polyunsaturated fatty acids) soybean in the USA (Genewatch 2003b). DuPont also developed High Sucrose Soybeans, High Protein Soybeans, High Oleic Soybeans, Low Saturate Soybean Oil, Low Linolenic Soybean Oil and High Oleic Sunflower Oil (Groundup 2003h). In addition, DuPont developed animal feeds with High Available Phosphorous Corn (Groundup 2003h) and Increased-Energy-Availability Corn (BIO 2004b). Two key developments in DuPont's biotechnology pipeline are reduced cholesterol soybean, and proteins capable of fighting osteoporosis and cancer (Dutton 1998). Conversely, Dow has also been actively involved in the production of output trait GM products. One of the most notable has been the development of Natreon Naturally Stable Canola Oil from Nexara Canola Seed, which contains virtually no trans fats, making it very attractive oil for baking, frying, snack foods and so forth (BIO 2004b).

Whilst BASF was late to enter the biotech race, it has endeavoured to quickly make up lost ground (Wield et al. 2010). BASF stated that, whilst it is not explicitly interested in input trait biotechnologies, it is determined to enter the biotechnology race with second- and third-generation products (Strube 2000; Wield 2001), especially those linked to increased nutritional value (Moore 2000). BASF was especially interested in technical-grade starch, seeds with higher levels of vitamin A, E or beta-carotene, plants with long-chain fatty acids for infants' food or with omega-3 fatty acids for the prevention of cardiovascular diseases. BASF sees 'good market opportunities especially for plants with improved ingredients that have an obvious benefit for consumers' (BASF 2000;

Wield et al. 2010). According to Wield (2001), BASF refrained from entering the input trait biotechnology sector because its product range does not include non-selective herbicide, and insecticides are not a core sector of their business.

BASF's vision encompasses the whole agro-food supply system (Wield 2001). Its goal is to leverage its traditionally strong competencies in crop protection, in human and animal health, and basic research to provide a platform to bring second- and third-generation plant biotechnology products to market. According to BASF (2000), 'We intend to become one of the major players in plant biotechnology'.

However, according to Finch (2000), whilst functional foods already have a foot hold in the USA, Western Europe and Japan, widespread customer acceptance is, as yet, unassured. For example, some companies cancelled advertising campaigns and withdrew products due to poor sales (i.e. Flavr Savr). In some cases, sales have not been as strong as expected due to the often high cost of functional foods, that is Benecol is four times the price of other spreads. 'According to a UK survey, 4/5 of people disbelieve health claims made by food manufacturers, while most regard organic food as more healthful' (Finch 2000).

6.3.3 Economies of scale and scope: Perceived synergies between agrochemicals, pharmaceuticals and nutrition

Economies of scale have been a key component of residual-Fordist, Neo-Fordist and Post-Fordist strategies throughout the past 40 years or so. However, whilst acknowledging contrary claims in the literature, economies of scope are best envisaged as a Post-Fordist phenomenon. The supporting rationale for this claim relates to the fact that economies of scale generally embody supply-side changes, such as level of production, whilst economies of scope embody demand-side changes, such as marketing and distribution. Economies of scope are achieved by using an asset to produce different types of outputs; even if the outputs are unrelated. It is claimed that per-unit costs decline when heterogeneous outputs are produced using the same asset. Modular platforms are examples of technology assets that realise economies of scope. To recap, as R&D expenditures and regulatory costs are both sunk costs, there is a significant potential for economies of scale and scope. Because economies of scale or scope mean that larger and more diversified firms have lower average costs, there is clearly an incentive for firms to get large (Fulton 1997; Hayenga 1998).

Economies of scope are said to exist between the food, agriculture and health sectors (ETC 1996). Because of perceived economies of scope, the 1990s witnessed an evolution of the so-called 'life industry' (ETC 1996). Life industries are giant TNCs producing bio-industrial products related to food, agriculture and health. The business of life science ranges from food processing, seed production, plant breeding and agrochemicals to pharmaceuticals and veterinary medicines (ETC 1996). Key examples of life science corporations include all four top pesticide-producing corporations Syngenta (now ChemChina), DowDuPont, Bayer and BASF. The purpose of these companies is to not only conquer the agricultural market (seeds and agrochemicals) which had a value of US$35 billion in 1999 including seeds and plant protection products, but also change food processing through genetically modifying agricultural raw materials so as to meet the special requisites of the food industry. Another objective is to extend their activities to the production of pharmaceuticals, either directly, or through the production of nutraceuticals (Kempf 1999). In 1999, the pharmaceutical industry was worth approximately US$300 billion. According to ETC (2001a), the synergies between biotech, food retailers and pharmaceuticals are likely to grow dramatically in the coming decades as second and third generation of 'consumer friendly' biotech products debut. Emerging technologies such as nanotechnology and neurosciences will also play a role in the corporate quest for 'integrated science'.

Plant protection products and pharmaceuticals are science-based, highly R&D intensive and crucially dependent on patent protection (Bijman and Joly 2001a). Innovation in the pesticide industry operates within a framework determined by the interaction of the regulatory system (for registering new products), the patenting system (for granting temporary monopolies to appropriate the benefits of new products) and market competition (Hartnell 1996). In addition, pesticide manufacturers have to consider the opportunities and risks associated with modern biotechnologies, shifting power in food system from production to consumption, and the need to develop appropriate technologies for IPM and ICM systems (Bijman and Joly 2001a). The greater the regulatory costs the more an industry will concentrate (Fulton and Giannakas 2001). IPR is a non-rival good. Ideas are not scarce and will not face diminishing returns to scale. The protection of IPR is designed to encourage innovation and provide incentives for the development and diffusion of new products and technologies (Romer 1990). IPR may also create economies of scope. This can occur when a gene has been isolated, such as herbicide tolerance, and can be put into a number of crops. The production costs of a number of products together will be less than if the products were produced separately (Fulton and Giannakas 2001).

An escalation strategy is one in which a company spends large amounts on R&D and engages in M&A to achieve a dominant role in the market; that is, it tries to leapfrog its competitors to become the dominant firm (Fulton and Giannakas 2001). Escalation can be a profitable strategy when there is a high degree of substitutability with competitor's products on the demand side, and there are scope economies on the supply side (Sutton 1998). Both of these factors are present in the agricultural biotechnology industry. The combination of demand substitutability and supply-side scope economies present in the seed and chemical industries appears to be linked with escalation strategies. Monsanto has clearly followed an escalation strategy, with Dow, Novartis and others following similar strategies of making acquisitions and spending significant amounts of R&D (Fulton and Giannakas 2001). While 'the agricultural and medical marketplaces are very different, there is growing commonality at the research level. Technologies such as gene sequencing, combinatorial chemistry and high-throughput screening are as relevant to the agricultural as to the human health section' (Financial Times 1998a).

Ultimately, however, it may be that the Food and Beverage industry emerge as the true titans of the life industry. They are likely to become more visible players in the future. In 1999, the total retail value of global food sales was estimated at US$2000 billion, or 6.5× the pharmaceutical industry. The 1997 revenues of the world's largest food and beverage corporation Nestlé (US$45.3 billion) easily surpassed the entire commercial seed industry (US$23 billion), the entire agrochemical industry (US$31 billion) and the animal health industry (US$17 billion). Nestlé's 1997 revenues were more than 3× the revenues of the leading pharmaceutical corporation Aventis (US$13.7 billion). In the near future, with growing emphasis on output traits, crops and animals will be modified for end users: the giant food processing industry (ETC 1999). The top ten food and beverage companies account for sales of over US$233 billion, 16% of global retail food sales (ETC 1999).

Syngenta is the epitome of a Life Science Company and the process by which one is created. This recent history of Syngenta began in the early 1990s with ICI when it spun off its bio-areas (mainly pharmaceuticals and agriculture) to create Zeneca in an attempt to focus on key areas, particularly those with strong synergistic effects. Greater flexibility and effectiveness for both companies were the stated reasons for this step. However, this eliminated the reduced risk sharing which had been a cornerstone of diversified chemical companies. Sandoz, a future component part of the coming Syngenta, did the opposite. In 1995, they divested all lines of business that were not directly connected with pharmaceuticals or nutrition (linked by health) (FAZ 1995a).

In 1996, Novartis, another future component of Syngenta, was formed. According to the Groundup Organisation (2003a), this was as an attempt by Ciba and Sandoz to divest their traditional chemicals focus and enter the new life science race, and concentrate on core competencies such as agrochemicals, pharmaceuticals and nutrition, where high growth and profits were expected. At its inception, Novartis was the world's no. 1 agrochemical corporation, the second largest seed firm, the third largest pharmaceuticals firm and the fourth largest veterinary medicines company (ETC 1996). Shortly after its creation, Novartis acquired Merck & Co., Inc.'s crop protection business, with the aim of achieving 'significant synergies' and revenue enhancements through combining Merck's complementary product range (Syngenta 1997). Novartis's new strategy focused on the food and feed chain and crop solutions, integrating breakthrough technologies, and exploring joint synergies between crop protection and seeds (Chataway 2001). Novartis stated it was keen to develop marker systems and output traits, and emphasised a growing focus on low environmental impact technologies because of difficulty in registering products (Chataway 2001). Novartis stated that 'the common denominator of our business is biology. The research and technology is applied to discover, develop and sell products that have an effect on biological systems, be they human beings, plants or animals' (Financial Times 1998b).

Started in 1998 and completed in 1999, Zeneca merged with Swedish pharmaceutical company Astra to create AstraZeneca with US$10 billion sales. AstraZeneca was ranked second in European sales and third in world sales after Merck and Glaxo Wellcome. The merger was driven by a desire to reduce costs (US$1.1 billion annually, or 6000 jobs), and to obtain better access to the US drugs market (Corporate Watch UK 2002b). Two key goals of the merger were the formation of a leading life science company and to benefit from perceived synergisms (King et al. 2002). Indeed, the AstraZeneca merger witnessed significant synergism in returns to shares (King et al. 2002). In December 1999, AstraZeneca and Novartis merged; Advanta was not part of the deal. Syngenta became the third largest seed business after DuPont (Pioneer) and Pharmacia (Monsanto) (Corporate Watch UK 2002b).

In many respects, Bayer is a relative newcomer to the life science club, attaining the status through strategic M&A. As previously noted, Bayer is comprised of a series of pesticide corporation components, one of which is Hoechst. Hoechst first attempted the transformation to a life science corporation in 1994 when it merged with Schering, forming a joint venture called 'Hoechst Schering AgrEvo Gmbh', otherwise known simply as AgrEvo. At the creation of AgrEvo, it had sale of 3 billion DM and approximately 8000 employees, and was the second largest producer of pesticides in the world, after Ciba-Geigy (Chemische Industrie 1994). The objective of this cooperation was to achieve a critical size in the generally shrinking market for plant protectives, allowing the new venture to meet the challenges of both innovative and price competition. In 1994, Hoechst transformed itself from a centralised group into a strategic management holding company. In its new organisational structure, the Group companies, affiliates and joint ventures enjoy a high degree of independence and decision-making freedom to allow them concentrate on their customers and operating business. The external focus on customers is reflected internally to the full involvement of all employees in the objectives of 'their' company (Hoechst 2003a).

At the same time, Hoechst continued its transition to a life science corporation by divesting profitable cosmetics line Jade and Schwarzkopf because they were not part of their key area of concentration (FAZ 1995a). In 1998, Hoechst spun off Celanese, its big American chemical subsidiary, in order to meet its goal of getting out of the chemical industry by the end of 2000. In December 2000, Hoechst announced its merger with Rhône-Poulenc, creating Aventis, a true life science corporation (New York Times 1998). Substantial synergies were expected as 'existing pharmaceutical products and the research pipelines complement each other well' (Hoechst 1999). Before the merger, Rhône-Poulenc had undertaken significant organisational initiatives in order to enhance its flexibility. The main components of the initiatives were greater decentralisation, increased importance on market knowledge and information gathering for decision-making purposes, and an acknowledgement of the need to approach problem-solving from a holistic food system perspective. However, despite these undertakings, decision-making was still deemed to be top-down and relatively centralised (Assouline and Pierre-Benoit 2001). Rhône-Poulenc had already divested its Rhodia subsidiary in order to focus on life science growth and reduce the corporation's debt, which was thought to restrict the high levels of investments required for a global life science corporation (Assouline and Pierre-Benoit 2001). In addition, Rhône-Poulenc expressed growing concern over the mixed messages received from government and the public regarding GMOs (Assouline and Pierre-Benoit 2001).

However, due to the hostile reception of agricultural GMOs in Europe, Aventis spun off its pesticide and agricultural biotechnology division Aventis CropScience to Bayer in 2001. Some suggest that the bid for Aventis CropScience was part of Bayer's management strategy to resist shareholder pressure to break up its conglomerate structure.

The reorganisation process was aimed primarily at strengthening the focus on core competencies, growth and performance, as well as achieving an entrepreneurial optimisation (Bayer 2003f). Bayer aimed to achieve a sustained enhancement of its crop protection business - a key element of its life science activities - through the acquisition of the FLINT® strobilurin product line from Novartis for Fr. 1.333 billion (€880 million). This promising fungicide from a novel active substance class gave Bayer a further blockbuster product, as the sales potential for the FLINT® line was estimated at about €300 million/year. 'This acquisition is part of our strategy of investing in our core businesses and their growth markets. The move will allow us to enhance the competitive strength of our Crop Protection Business Group and further increase its already significant profit contribution'. The EC approved Bayer as the purchaser, but the transaction, was, in 2001, still subject to the approval of the US FTC, and of several other national competition authorities. Bayer reiterated it was keen to maintain the majority interest and management control in any future health care partnership but that any takeovers in the short term would be limited in scope (Financial Times 1998b). Prior to the acquisition of Aventis CropScience, Bayer had spun off Agfa, in September 1998, and spent US$1.2 billion acquiring the diagnostics division of Chiron, one of the world's largest biotech companies. In the same month, Bayer invested US$465 million in Millennium Pharmaceuticals, the largest, to date, in the field of genomics drugs research (ETC 1999). Concerns about GMOs in Europe have re-centralised GMO marketing decisions in Bayer back to Frankfurt (Bijman 2001a).

Lastly, DuPont sold Conoco, the ninth ranking oil company, to 'rapidly accelerate' investment in the life sciences (Chemical and Engineering News 1998). It used this money to acquire the remaining 50% share of its joint venture with Merck and Co. In 1997, Monsanto spun off its US$3 billion chemicals business as a separate company in order to concentrate on biotechnology (ETC 1999). However, many companies that attempted to combine pharmaceutical and biotechnology components within the life science model have divested their agricultural operations (Chataway 2001; Bijman 2001a,b).

6.3.4 Corporate collaboration: New partnerships and alliances

Since the 1990s, the new life science companies increasingly entered into strategic alliances in the areas of production and R&D. Indeed, strategic alliances for trait licensing, R&D, negotiating regulatory hurdles, patent litigation truces and generic trait agreements were almost as popular as M&A (ETC 2003, 2015). Cooperation between independent life science companies, and between life science companies and academic institutions, became an increasingly common strategy, especially where basic research was concerned and where individual companies felt the need to recognise technological potentials, quicken innovation development and quickly enter new markets (Vorley 2002).

According to Vorley (2002), strategic alliances are generally formed in pursuit of specialised and high-value-added technologies, as well as cost-efficiencies and cost leadership in bulk production. Indeed, global clusters and strategic alliances in agri-food industries are examples of corporate convergence, which is becoming the global norm (see Fig. 6.1). Under conditions of 'cooperative capitalism' transactions become based on industrial relationships rather than on open markets (Vorley 2002). According to ETC (2003), for many life science companies, it became much more profitable to 'cross-license technologies and bury the patent litigation hatchet in order to create global technology cartels that transcend national and transnational regulation; operating below the radar screen of anti-trust regulators'.

However, strategic alliances are only viable when IPRs are well defined, and subsequently, transaction costs are relatively low. IPRs create pressures for either greater vertical integration or strategic alliances and contracting, depending on the nature of the intellectual property and the rights associated with it. As IPRs became increasingly more defined and defensible, strategic alliances became more appealing. Independent companies can efficiently and effectively operate alongside each other, each focusing on their specialty while at the same time having access to the intellectual property of the other firms through contracts, licenses or joint venture agreements. As Table 6.1 shows, the major biotechnology firms have used joint ventures and other business relationships (e.g. licenses, research agreements), in addition to M&A.

Table 6.1 Major strategic alliances in the pesticide industry

Company name	Date of strategic alliance and company/companies involved
BASF	BASF initiated a joint venture with Svalof Weibull called BASF Plant Science in 1999
Bayer	Hoechst (later incorporated by Bayer) initiated a joint venture with Schering called AgrEvo in 1994. In 1997, AgrEvo initiated a 51/49 joint venture with Cotton Seed International (CSI) of Australia, and an alliance with Gene Logic in 1998. Hoechst initiated a joint venture with Rhône-Poulenc called Aventis in 1999. Schering initiated a joint venture with Aventis called Aventis Crop Science in 1999
Dow AgroSciences LLC	Dow Chemical initiated a joint venture with Eli Lilly's Elanco called DowElanco in 1989 and a genomics research alliance with Biosource Technologies in 1998
DuPont	DuPont initiated a joint venture with Westrade Inc., called DuWest Inc., Central America, and DuPont Agricultural Chemicals Ltd., in China in 1990. In 1993, DuPont initiated a joint venture called D.T. DuPont Agricultural Products in Indonesia. In 1997, DuPont initiated a joint venture with Novocheboksarsk (Russia) called AO Khimpron and with Pioneer Hi-Bred called Optimum Quality Grains (before acquiring Pioneer Hi-Bred in 1999). In 1998, DuPont initiated a joint venture called Plaguidas y Fertilizantes Nacionales S.A. de C.V. (Mexico), and one with Agroproductos Corey, S.A. de C.V. (Mexico), and one with CuraGen and Lynx Therapeutics. In 1999, DuPont initiated a joint venture with the John Innes Centre (UK) and initiated a marketing agreement with General Mills Inc., and entered an advanced research and education alliance with MIT in 2000
Monsanto	Monsanto (and subsidiary companies) initiated a joint venture with Cento Integral Agropecuria (Argentina) called CDM Mandiyu and with Seminis (Grupo Pulsar) in 1997. In 1998, Monsanto initiated a joint venture with Anhui Provincial Seed Corporation (China) called Anhui An Dai Cotton Seed Technology Company, one with Mahyco (India) called Mahyco Monsanto Biotech and, a strategic alliance with GeneTrace. In 1999, Monsanto initiated a joint venture with Cargill (USA) called Renessen LLC and a licensing deal with USDA and the University of Massachusetts, and IBM Corp.
Syngenta	Syngenta (and subsidiary companies) initiated a joint venture with VanderHave B.V. to form Advanta in 1996, and initiated a five-year collaborative research agreement with the University of California, Berkeley, and a screening agreement with Alanex in 1998

According to King et al. (2002), Monsanto formed 30 significant biotechnology research alliances in the 1990s (King et al. 2002). Monsanto (and subsidiary companies) initiated a joint venture with Cento Integral Agropecuria (Argentina) called CDM Mandiyu in 1997. The alliance facilitated Monsanto's production of GM seeds in Argentina, which was one of the fastest growing markets for GM crops (Groundup 2003j). In the same year, Monsanto entered into a strategic alliance with Seminis (Grupo Pulsar) and with Millennium Pharmaceuticals to identify patentable crop genes using genomic technologies. The exclusive deal was not limited to a single crop or geographic location; it covered all crops in all countries (ETC 1998).

In 1998, Monsanto initiated a suite of alliances. (1) Anhui Provincial Seed Corporation (China) called Anhui An Dai Cotton Seed Technology Company in 1998. (2) Mahyco (India) called Mahyco Monsanto Biotech. (3) GeneTrace. In 1999, Monsanto initiated a 50/50 joint venture with Cargill (USA) called Renessen LLC, which focused on the production and global distribution of GM (output trait) foodstuffs for processing and animal feed, and a licensing deal with USDA and the University of Massachusetts (Groundup 2003j). See Fig. 6.4 for details.

In 2002, Monsanto formed a strategic alliance with Nissan Chemical Industries, Ltd. (Nissan). The alliance comprised the sale of certain Monsanto herbicide assets to Nissan for use in the Japanese market. The transaction included the transfer of Monsanto's trademarks for Roundup, Lasso and Machete herbicides. In addition, the agreement also included the transfer of related product registrations, labels and certain other assets for use in Japan, as well as a long-term supply arrangement whereby Monsanto would be the supplier to Nissan of these herbicides. The addition of Monsanto products was suggested to be in line with Nissan's strategy to strengthen its business in Japan (PR Newswire 2002). Monsanto, on the other hand, suggested that, the alliance consolidated market access to key geographic areas, while, at the same time, allowing Monsanto in Japan to focus efforts on seed business and biotechnology initiatives (PR Newswire 2002). In the same year, Monsanto and Dow agreed to share biotechnology events in a complex negotiation involving biotechnology licenses. Part of the deal involved Monsanto providing Dow with commercial non-exclusive royalty bearing licences related to its Roundup Ready maize, cotton, soybean and canola. In return, Dow provides Monsanto with commercial non-exclusive royalty bearing licences related to its Herculex I Insect protection maize for use in Monsanto's Latin American maize business (Monsanto 2002). By 2008, Monsanto had licenses with around 200 independent maize and soybeans seed companies in the USA (Howard 2009).

DuPont was equally active, initiating a joint venture with Westrade Inc., called DuWest Inc., in Central America, and a venture called DuPont Agricultural Chemicals Ltd., in China in 1990. In 1993, DuPont initiated a joint venture called

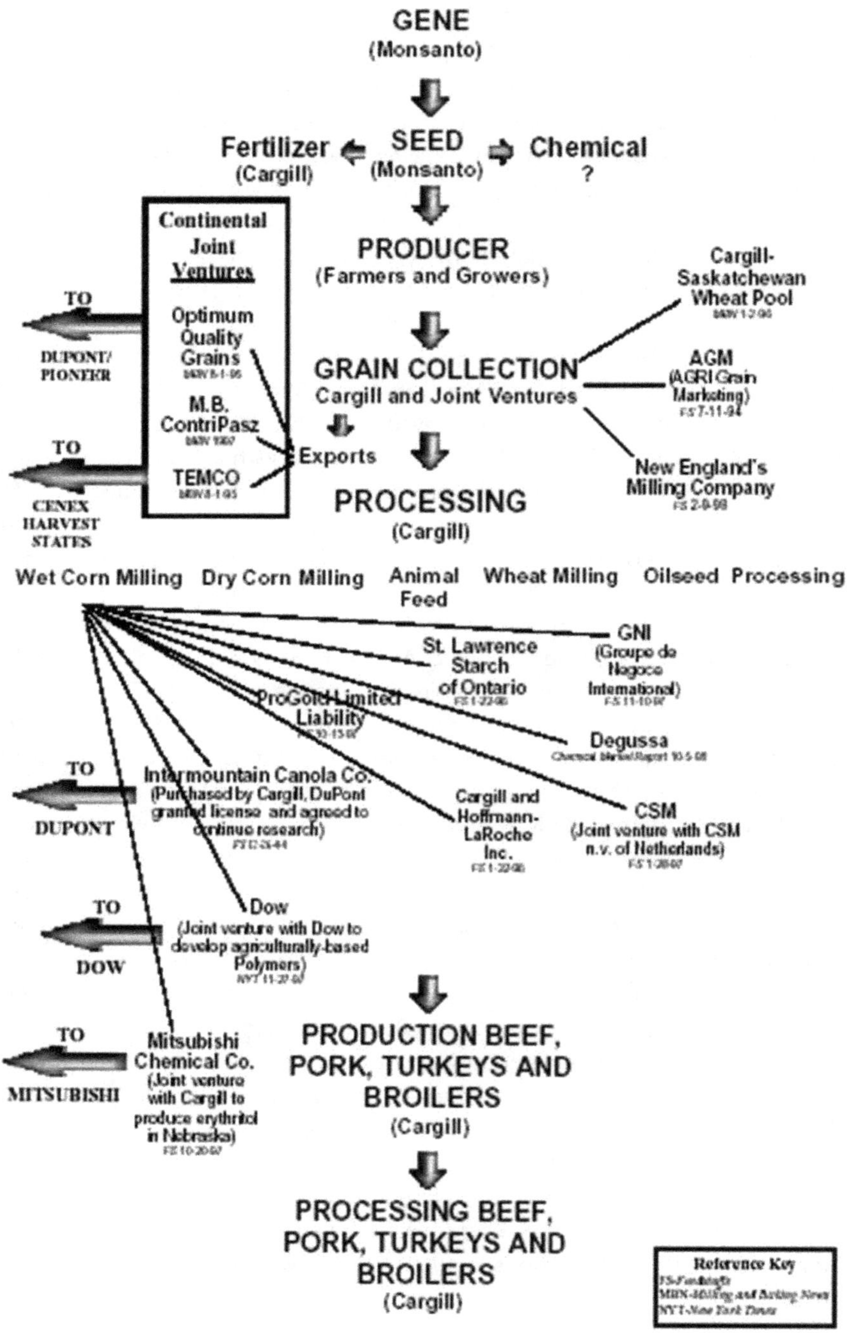

Figure 6.4 Cargill and Monsanto joint ventures and strategic alliances.

D.T. DuPont Agricultural Products in Indonesia in 1993. In 1997, DuPont initiated a joint venture with Novocheboksarsk (Russia) called AO Khimpron and with Pioneer Hi-Bred called Optimum Quality Grains in 1997 (before acquiring Pioneer Hi-Bred in 1999). DuPont acquired marketing agreements for Optimum Quality Grains with Archer Daniel Midlands Co., (ADM) (fourth largest US food company) (Groundup 2003e).

In 1998, DuPont initiated a venture called Plaguidas y Fertilizantes Nacionales S.A. de C.V. (Mexico), Agroproductos Corey, S.A. de C.V. (Mexico) and CuraGen and Lynx Therapeutics, which focused on maize, soybean, wheat and rice (Groundup 2003e). In 1999, DuPont initiated a joint venture with the John Innes Centre (UK) to develop new wheat varieties and wheat products with new agronomic, industrial and food uses (Groundup 2003e). In the same year, DuPont entered into marketing agreements with: (1) General Mills Inc. and ConAgra (50th largest US company and largest food company) for the international marketing of its traditionally bred High Corn Oil (Groundup 2003e). (2) Cargill (largest US grain exporter) to market High Corn Oil domestically. (3) Charoen Pokphand Corp Ltd (Taiwan) to deliver 280 000 tonnes of US High Corn Oil 1997–9 (Groundup 2003e). See Fig. 6.5 for details of the DuPont/ConAgra joint venture.

Meanwhile in Europe, between 1982 and 1992, Hoechst (part of which was ultimately incorporated by Bayer) increased its number of cooperative agreements from 230 to 380 (Chemische Industrie 1992). The most noteworthy collaboration was the joint venture with Schering called AgrEvo in 1994. In the same year, Rhône-Poulenc (part of which was ultimately incorporated by Bayer) initiated a joint venture with Limagrain to create Groupement d'Intérêt Economique, an economic structure for plant biotechnology (Assouline and Pierre-Benoit 2001). This enabled Rhône-Poulenc to build a patent portfolio of more than 40 patents in the fields of herbicide resistance (in oxynils, isoxazoles, asulam and glyphosate), plant disease resistance, quality traits (fatty acid composition) and genetic engineering technologies (Assouline and Pierre-Benoit 2001). Rhône-Poulenc was committed to maintaining a presence in all stages of the creation of a GM crop, from the initial genome study to the delivery to the seed producer, through to the mother plant.

Rhône-Poulenc also formed a 50/50 joint venture with Biogemma, a biotechnology platform backed by French seed companies, to create Rhobio (Assouline and Pierre-Benoit 2001). In 1997, AgrEvo initiated a 51/49 joint venture with Cotton Seed International (CSI) of Australia, and an alliance with Gene Logic in 1998. In 1999, Hoechst initiated a joint venture with Rhône-Poulenc called Aventis. Aventis' GM partners included the Institute of Arable Crops Research (IACR-Rothamsted, UK), Agrinomics LLC (USA) and Stichting Dienst Landbouwkundig Onderzoek. In the same year, Rhône-Poulenc proclaimed that 'its alliance strategy was a key element of its success in the field

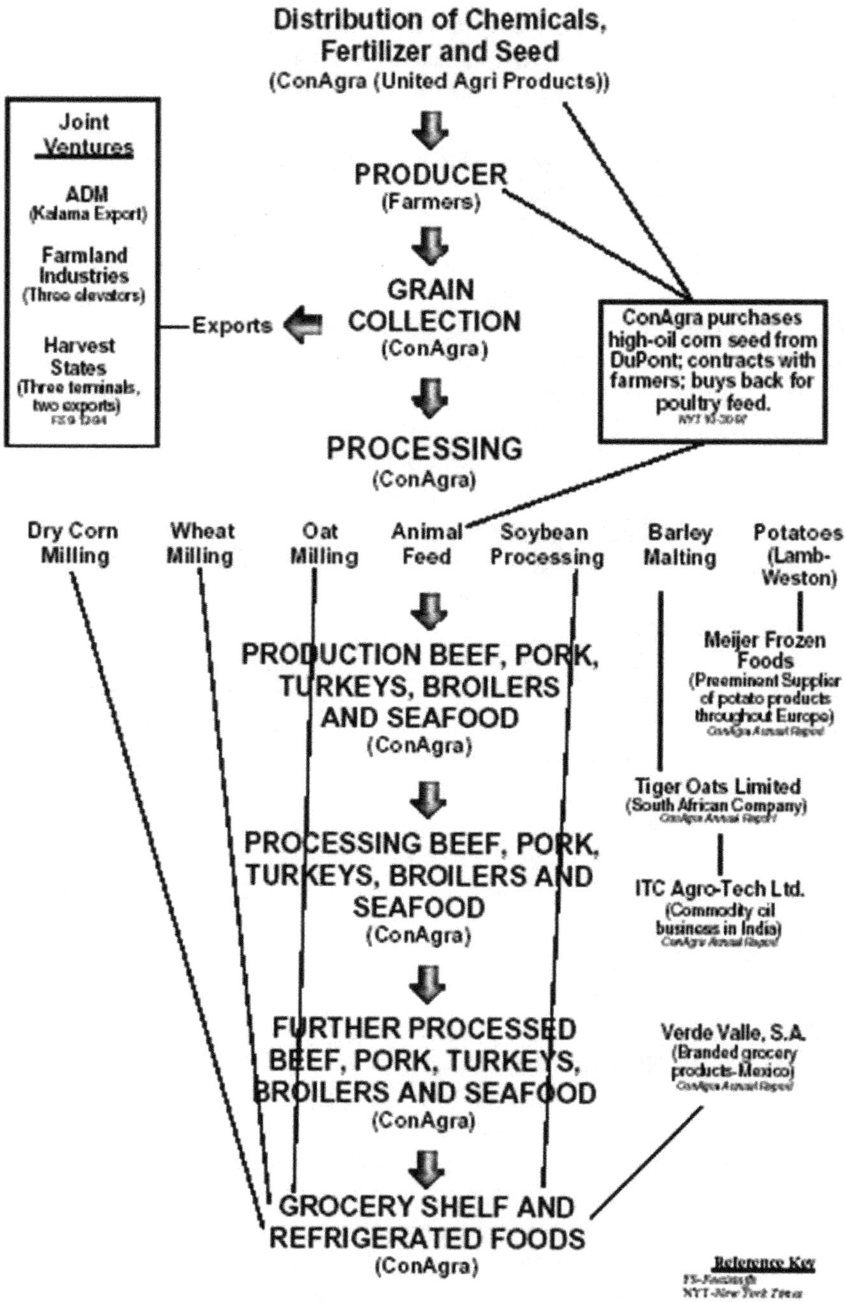

Figure 6.5 ConAgra and DuPont joint ventures and strategic alliances.

of biotechnology' (Assouline and Pierre-Benoit 2001). Furthermore, 1999 witnessed the creation of Aventis Crop Science, a joint venture between Schering and Aventis. Ultimately, Bayer cooperated with US-based Exelixis on the development of selective insecticides, with Paradigm Genetics on genetic codes in weeds, and with German-based LION Bioscience in fungicide research (Søgaard 2001).

Since the 1990s, Zeneca (now ChemChina) favoured collaborations with other companies to create life science opportunities (i.e. in the case of Flavr Savr tomato paste), using IPR and legal contracts to protect its technologies (Tait 2001). Novartis (ChemChina) initiated a joint venture with VanderHave B.V. to form Advanta in 1996, and initiated a five-year collaborative research agreement with the University of California Berkeley, worth up to US$25 million, in 1998 (Corporatewatch 2003a). Whilst Novartis was not able to directly influence the direction of research, it did have first rights to an exclusive license on a fraction of all the research developments in the laboratories, whether or not the research was carried out with Novartis funds (Groundup 2003l). In the same year, Novartis initiated a screening agreement with Alanex and committed over US$100 million to 'new foundation projects that show exemplary entrepreneurial and innovative spirit in future-oriented areas, especially in the fields of life sciences and new technologies' (Groundup 2003m). By 1998, 46 businesses had been launched. Syngenta also funded research and facilities in a number of UK universities including Reading and Cambridge, but withdrew from a major research partnership with the John Innes Centre (Corporatewatch 2003a). Syngenta has the greatest number of representatives on the UK's Biotechnology and Biological Sciences Research Council (BBSRC), a body that exercises considerable control over the direction of research undertaken in UK universities (Corporatewatch 2003a). Latterly, Dow AgroSciences and M S Technologies (a US biotechnology company) granted a non-exclusive licence to Syngenta (owned by ChemChina) for the use of the Enlist E3 (DAS44406) trait in GM soybeans (Agrow 2017). Enlist™ are GM maize and soybeans that are resistant to a new formulation of 2,4-D. Enlist™ is a strategically important herbicide/crop combination as it provides an alternative to the Roundup Ready herbicide/crop combination marketed by Monsanto; especially as Roundup resistant weeds are becoming more commonplace (Pesticides and You 2014). Not surprisingly, anti-pesticide and GM groups have reacted negatively to the development of Enlist™. This negative reaction is partly based on the breakdown of Roundup resistance, and the continuance of the pesticide treadmill and the fact that the new GM maize and soybean rely on 2,4-D, the principle ingredient of Agent Orange and the link to dioxin-based poisoning (Genewatch 2014).

In a similar vein to the Monsanto/Cargill and DuPont/ConAgra joint ventures, Novartis entered into a joint venture with ADM. See Fig. 6.6.

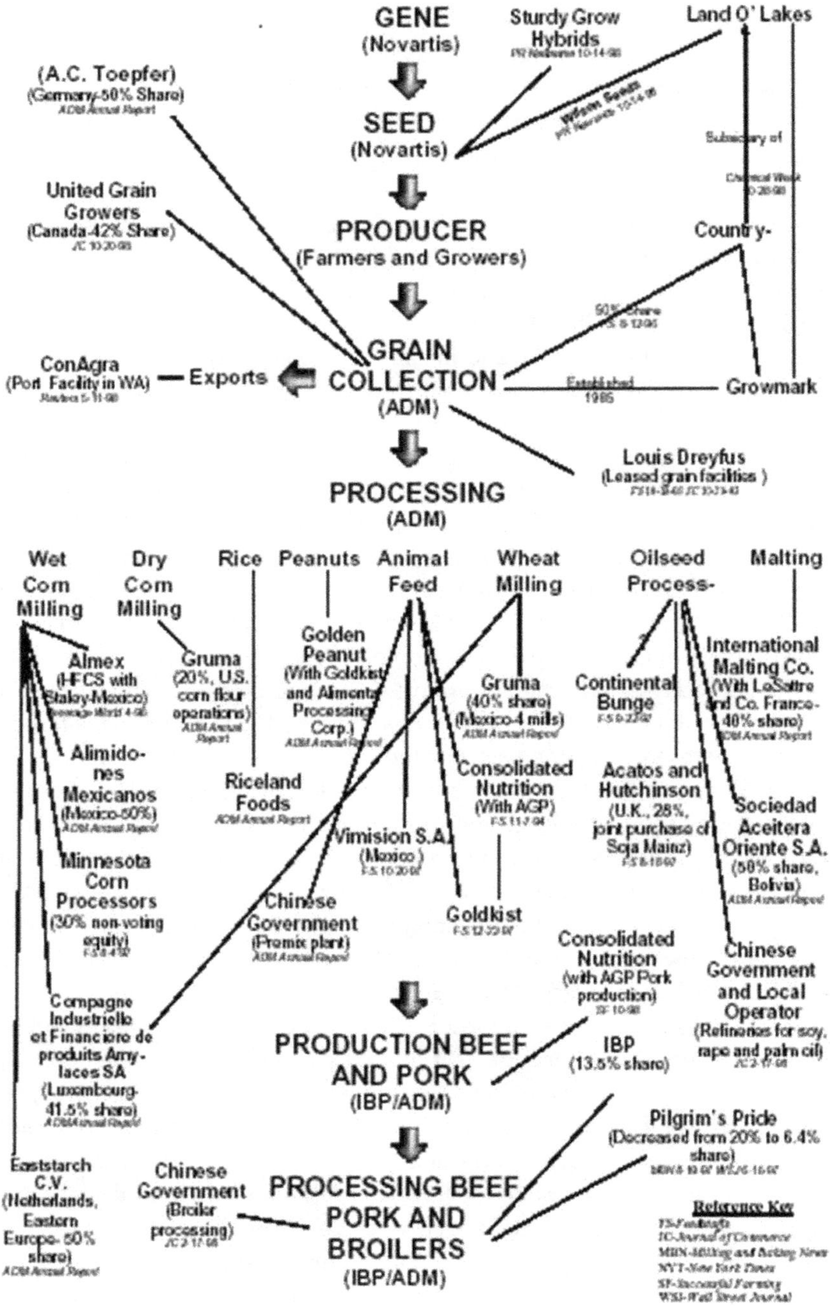

Figure 6.6 Novartis and Archer Daniel Midlands (ADM) joint ventures and strategic alliances.

6.4 Key outcomes of pesticides and biotechnologies

6.4.1 *Increased national and international regulation of pesticides and biotechnologies*

6.4.1.1 *National*

6.4.1.1.1 *UK*

While the PPT is generally associated with recognisable outcomes of post-productivist interventions, it has been institutional changes that have underwritten their success. Once firmly institutionalised, agro-food policy community stakeholders have witnessed a transition in which, because of their relatively conservative positions, much of their influence has been appropriated by environmental, consumer and retailer stakeholders. However, while traditionally powerful stakeholders such as the NFU, CLA and the BAA were initially slow to change, and consequently witnessed an erosion of their power, their recent realignments now reflect more legitimate and defensible positions. Indeed, the changing priorities of the farmers' unions, input sector stakeholders and governmental institutions with at least a modicum of an agricultural and/or environmental remit have been active in forging PPT futures in the UK and the EU. Moreover, whilst still acknowledging the important role of pesticides and biotechnologies, the UK Government has become increasingly committed to a pesticide minimisation policy that seeks to promote a marked reduction in pesticide use and develop at least a modicum of political consensus before the release of biotechnologies into UK farming and food systems (DEFRA 2000).

Pesticides play an important part in the production and storage of food and provide other social and economic benefits (DEFRA 2000). Their use is currently governed by the approval process which, according to DEFRA, balances the risks and benefits of their use on the basis that when correctly used they pose no 'unacceptable risk' to human health and the environment (DEFRA 2000). At present, a small levy is charged on UK sales of approved pesticide products to cover the cost of the authorisation process and subsequent monitoring (DEFRA 2000). The levy is not, however, set with regard to the Polluter Pays Principle (DEFRA 2000). Moreover, the approval process cannot take full account of the environmental effects resulting from pesticide application (DEFRA 2000). Instead, it is largely directed at controlling the availability of pesticides and the application of standard conditions to manage risks (DEFRA 2000). The environmental appraisal is therefore heavily conditioned by judgements about the use of pesticides using standard conditions (but these cannot of course be enforced) (DEFRA 2000). In addition, the approval process provides no real incentive for users to achieve minimum usage or to consider relative environmental risks beyond the statutory requirements (DEFRA 2000). There is thus a case for the introduction of an instrument to provide incentives for more environmentally benign behaviour, based upon. (1) Existing environmental evidence. (2) A precautionary approach (given scientific uncertainty). (3) Application of the Polluter Pays Principle (DEFRA 2000). Given the existence of environmental risks from approved pesticide use, external costs are imposed by such usage.

There is increasing evidence to support the view that reductions in pesticide use can be made without sacrificing adequate crop protection and reducing farm profitability (DEFRA 2000). In addition, falling output prices and the standards being implemented by supermarkets through farm assurance schemes are already changing behaviour and making farmers reconsider their use of pesticides (DEFRA 2000). However, reducing pesticide use represents a substantial challenge to farmers, which requires greater management input and may involve greater risk to the farm business (DEFRA 2000). Nevertheless, some farmers are already reducing pesticide use under the pressures of the market (DEFRA 2000). The Action Plan produced by the UK Pesticides Forum envisages ongoing reductions in pesticide usage as agronomic and ecological knowledge expands, thus representing a dynamic process (DEFRA 2000). According to DEFRA (2000), the new instrument, designed to complement the existing pesticide minimisation policy, would therefore provide additional financial incentives and additional information to pesticide users to reduce pesticide use. The instrument could be implemented as either a tax or a charge, together with funding for the diffusion of information on how to maintain adequate crop protection with lower levels of pesticide use (DEFRA 2000). A tax/charge would seek to influence the users of pesticides through higher prices for pesticides, with diffusion of information helping users to adjust to these higher prices. In an increasingly liberalised market for agricultural commodities characterised by falling output prices, the instrument could also ensure a less costly transition to those farming practices, which will be competitive under future market conditions (DEFRA 2000).

The pesticide minimisation policy formally acknowledges that there remain environmental risks from the use of pesticides, even when used in accordance with specified conditions (DEFRA 2000). There is accumulating evidence of environmental pressures resulting from the current level of pesticide usage, which supports the case for additional action beyond the measures already in place (DEFRA 2000). The following factors support further action, partly based on a precautionary stance. (1) The detection of increasing levels of pesticides (particularly herbicides) in groundwater. (2) The low rate of degradation of pesticides, particularly in aquifers, and delays in the full effects becoming apparent. (3) Practical analytical methods for environmental monitoring of pesticides are only available for about 50% of approved substances so there is considerable uncertainty about the true position. (4) Bioaccumulation and other effects of long-term, low-level exposure are only partially understood but are thought to cause serious adverse effects (e.g., the impact of organochlorines on the otter population). (5) The indirect impacts of pesticides on non-target organisms, though poorly understood at present, may be significant. (6) Some pesticides are identified as endocrine disruptors so that their presence in the environment may affect humans and other species (DEFRA 2000).

6.4.1.2 International

In the sphere of global environmental policymaking over the past 40 years, non-state interests have been key players (Lipschutz 1995). Corporate lobbyists have enjoyed enormous significance with respect to the outcome of global environmental policy, particularly since 1992 and Rio (Clapp 2002). The presence of industry lobby groups at the negotiation of global environmental treaties is routine, and their involvement in the global politics of environmental issues is generally on the rise (Clapp 2002). The dominance of the economic considerations in an industry's stance on global environmental issues can aid understanding of the divergent positions of an increasingly concentrated industry, in different issue areas in which it has a stake. Whilst recognising that corporations are influenced by socio-cultural and political-institutional factors, identifying the factors that influence the profitability of firms is important to understanding firm's economic considerations, and thus their positions on international environmental treaties (Clapp 2002).

6.4.1.2.1 The International Code of Conduct on the Distribution and Use of Pesticides

The FAO has maintained an active programme on pesticides since 1959, which now follows guidelines established by the 1985 International Code of Conduct on the Distribution and Use of Pesticides. The aim is to promote and assist in developing acceptable pesticide management practices within the framework of IPM principles. The FAO runs a substantial technical assistance programme that covers training to improve management; strengthening or establishing pesticide registration and control schemes; and laboratory and related infrastructure. The assistance enables developing countries to evaluate health and environmental hazards and risks, and increase ability to detect, monitor and control illegal import or traffic of banned, restricted or unregistered pesticides. The voluntary code has succeeded in raising awareness of pesticide hazards since its adoption in 1985. However, while most developing countries slowly introduced legislation or regulations related to the distribution and use of pesticides, many did not strictly enforce them (FAO 2002).

A third revision of the International Code of Conduct on the Distribution and Use of Pesticides was adopted by the FAO Council on 4 November 2002

(FAO 2002). According to the FAO, the revised Code sets standards, and outlines responsibilities, for governments, the pesticide, food and equipment industry, pesticide users, traders, environmental and consumer groups, trade unions and international organisations, and strengthens the monitoring and observance of these standards. The Code explicitly calls for the monitoring of pesticide residues in food and the environment, protection of biodiversity, and efforts to minimise the negative effects of pesticides in the water, soil, air and on non-target organisms (PAN 2002). The new Code places greater emphasis on reducing risks and hazards, including a stronger warning against the use of extremely hazardous and highly toxic pesticide, namely, WHO classes Ia and Ib, in developing countries (PAN 2002). Explicit recognition is given to the importance of addressing both human health and environmental risks. According to the FAO, the revised Code promotes practices that minimise potential health and environmental risks associated with pesticides. It addresses the life cycle of pesticides: from their development, regulation, production, management, packaging and labelling, to their distribution, application, use and control and disposal (FAO 2002). The Code calls for product stewardship strategies that address field-level impacts, and the introduction of collection systems for empty pesticide containers (PAN 2002). In addition, the Code calls upon industry 'to supply only pesticides of adequate quality, and to pay special attention to the choice of pesticide formulations and to the presentation, packaging and labelling in order to reduce risks to users and minimize adverse effects on the environment' (FAO 2002). It also requests that commodity and food industry groups use their influence in order to adjust agricultural practices throughout the food system (PAN 2002).

Whilst the revised Code appreciates that pesticide use will continue to be a major factor in agricultural production, it advocates the use of pesticides as part of integrated crop protection strategies, which focus strongly on natural pest control systems (FAO 2002) and which cause minimal ecosystems disruption (PAN 2002). The Code also introduces tougher requirements in relation to the use of protective equipment and application technology, particularly in tropical areas (PAN 2002). Besides saving lives and reducing environmental degradation, applying the Code avoids the buildup of toxic waste. Millions of dollars for cleanup operations could be saved and spent on training, research and application of environmentally friendly pest control measures (FAO 2002). It also recommends that 'manufacturers should recall pesticides that pose an unacceptable risk to people, animals and the environment. Unfortunately, the production and export of highly toxic and cheap insecticides such as organophosphates and carbamates continues' (FAO 2002).

GCPF supports the transition from state-controlled pesticide procurement to a market-driven system, and, almost by default, supports the Code where it can see obvious corporate benefits such as reduced packaging costs, and less tangible benefits such as improved public relations. GCPF also extends training to its distributors to ensure best practice in storage and handling, and appropriate stewardship of products. GCPF companies and regional trade organisations support some container collection schemes that will reduce the number of containers accumulating and needing disposal, and ensures safe handling of those containers collected. However, not all pesticide trade and use in developing countries, even from GCPF companies, can be said to adhere fully to the requirements of the International Code of Conduct on the Distribution and Use of Pesticides.

6.4.1.2.2 Rotterdam Convention on the Prior Informed Consent Procedure for Certain Hazardous Chemicals and Pesticides in International Trade

The principle of PIC was first raised in the early 1980s when FAO of the UN promoted the International Code of Conduct on the Distribution and Use of Pesticides (Pesticides News 1998b). PIC was not included in the 1985 Code, but it was in the amended Code of 1989 (Pesticides News 1998b). In parallel discussions, the UNEP adopted the same wording for PIC in the 1989 London Guidelines on Exchange of Information on Chemicals in International Trade (Pesticides News 1998b). Since then, PIC operated on a voluntary basis through a joint FAO (for pesticides) and UNEP (for chemicals) Secretariat (Pesticides News 1998b).

In 1976, UNEP had already established the International Register of Potentially Toxic Chemicals, a precursor to trade restriction, something that was first discussed as part of the Stockholm Conference in 1972 (Hough 1998). However, UNEP's Register of chemicals was hindered by industry's general unwillingness to disclose proprietary data and to have their products potentially blacklisted by UNEP (Boardman 1986). After a series of resolutions, the UN General Assembly authorised the compilation of a Consolidated List of Products Whose Consumption and/or Sale Have Been Banned, Withdrawn, or Not Approved by Governments in 1982 (Hough 1998). The vote to compile the list was carried by 146 votes to 1, with the USA as the sole opponent (Hough 1998).

In 1992, the Rio Earth Summit (Agenda 21, Chapter 19) called for a legally binding instrument to be in place by 2000. The experience of the voluntary procedure enabled negotiations to be completed 2 years early. International concern about chemical dumping eventually resulted in the signing of the Rotterdam Convention on the PIC Procedure for certain hazardous chemicals and pesticides in international trade (Pesticides News 2000). Adopted and signed by ministers and senior officials on 11 September 1998 in Rotterdam under the auspices of the UNEP and the UN's FAO, the Convention was scheduled to enter into force 90 days after the 50th ratification (FAO 2001). Until this day, governments agreed to apply the PIC procedure on a voluntary basis (FAO 2001). Between the 8th and 12th of October 2001, officials from over 100 governments met at FAO headquarters in Rome to prepare for the entry into force of the Convention and to oversee the implementation of the voluntary interim PIC procedure (FAO 2001). As of 10 February 2004, the Convention had 73 signatories and 59 parties (Rotterdam Convention 2004). Born out of international concern about the dumping of pesticides and other chemicals, the Convention imposes the mandatory PIC procedure to help governments decide whether to accept or refuse the imports of certain hazardous chemicals (Pesticides News 2000).

Countries that export such chemicals will have to respect the decisions of importing parties. It ensures that if a government bans a chemical, it has to notify developing countries the first time the banned product is shipped each year (Pesticides News 2000). The Convention was the first multilateral environmental agreement since 1994 and was adopted 2 years in advance of the goals set in Chapter 19 of Agenda 21 (Pesticides News 1998b). According to FAO (1999b), it offered greater protection to health and the environment, particularly in developing countries and countries with economies in transition, than the previous voluntary procedure, stressing that 'many farmers in developing countries cannot handle highly toxic pesticides safely and this Convention will help to protect them'. In the EU, the process quickly became centralised so that if a member state bans a chemical, the Commission will do the notifying (Pesticides News 2000). The EC (DGXI) operates PIC on behalf of Member States. Member States count as one country for pesticides and chemicals banned or severely restricted in the EU. The EU Council Regulation (EEC) 2455/92 came into force on 29 November 1992, and made Europe the first region to adopt a mandatory PIC scheme (Pesticides News 1998b). The European Chemicals Bureau in Ispra, Italy, operates the European database (EDEXIM) containing information on the status of exports, which require notification. The Development Directorate, DGVIII, supported the voluntary PIC by providing funds for African workshops, for regulators and others (Pesticides News 1998b). The Convention contributes to the reduction of the risks associated with the use of chemicals, particularly in developing countries, and limits the introduction of hazardous chemicals and pesticides into countries that cannot safely manage them (2001). In addition, if a country prohibits imports of a hazardous pesticide, it is also obliged to prohibit national production of the pesticide for national consumption (FAO 2001).

The list of pesticides subject to the voluntary PIC procedure continues to grow. Currently, the PIC procedure includes the following pesticides: (1) 2, 4, 5-T. (2) Alachlor. (3) Aldicarb. (4) Aldrin. (5) Azinphos methyl. (6) Binapacryl. (7) Captafol. (8) Carbofuran. (9) Chlordane. (10) Chlordimeform. (11) Chlorobenzilate. (12) DDT. (13) Dieldrin. (14) Dinitro-ortho-cresol. (15)

Dinoseb. (16) 1, 2-dibromoethane (EDB). (17) Endosulfan. (18) Ethylene dichloride. (19) Ethylene oxide. (20) Fluoroacetamide. (21) HCH. (22) Heptachlor. (23) Hexachlorobenzene. (24) Lindane. (25) Mercury compounds. (26) Methamidophos. (27) Monocrotophos. (28) Parathion. (29) Pentachlorophenol. (30) Toxaphene. (31) Tributyl tin compounds. (32) Trichlorfon. (33) Certain hazardous formulations of Benomyl, Carbofuran, Thiram, Methyl parathion and Phosphamidon (Rotterdam Convention 2017).

Technical assistance was made available for developing countries in order to help develop infrastructure for the safe management of pesticides from a number of industrialised countries, but remains limited (FAO 1998). According to Pesticides News (1998b), developing countries desperately need resources to make PIC work, but the Convention contains no specific provision for technical and financial assistance. Under-resourced regulators may be unable to mount an investigation into incidents, or may not hear of them. Very few countries in Africa even have a poison control centre (Pesticides News 1998b). In addition, the distribution of pesticides is not regulated in most developing countries. This is primarily because the infrastructure is insufficient to control pesticides and there is a shortage of trained farmers and workers.

The Conference also established an Interim Chemical Review Committee (ICRC) with a remit to recommend the inclusion of banned and severely restricted chemicals or hazardous pesticide formulations in the PIC procedure. One of the first tasks of the ICRC was to offer recommendations on bromacil, ethylene dichloride, ethylene oxide and maleic hydrazide, referred to it by the Conference (FAO 2001).

Interestingly, under the PIC procedure, Ethiopia, Malaysia, Sudan, Tanzania and Zimbabwe have prohibited the use of aldrin, dieldrin and DDT for agricultural use, but have reserved the right to import them for assisting in any present or future campaigns to control pest-transmitted tropical diseases (Hough 1998). This is because there is greater priority afforded to public health than pesticide use in food production, and less opposition to the use of pesticides on public health grounds from dissenting parties (Hough 1998). No actor, of any note, has advocated the phasing out of all pesticides for public health purposes, unlike the case of pesticides and optimising crop yields (Hough 1998). While anti-pesticide NGOs do not expect that PIC will solve all the problems of pesticide use in the developing world, most feel that it is an important step towards higher standards (Pesticides News 1998b). The Convention has an essential role in promoting the life cycle management of chemicals and pesticides. It fully complements the efforts of the Basel Convention on hazardous wastes and the Stockholm Convention on Persistent Organic Pollutants as well as the FAO International Code of Conduct on the distribution and Use of Pesticides and programmes on health and occupational impacts, such as those carried out by WHO and ILO (FAO 2001). Some 70 000 different chemicals are available on the market, and 1500 new ones are introduced every year (FAO 2001). This poses a major challenge to many governments who must attempt to monitor and manage these potentially dangerous substances. Many pesticides that have been banned or whose use has been severely restricted in industrialised countries are still marketed and used in developing countries (FAO 2001). Traditionally, when a country or supranation such as the EU banned the sale of an old pesticide within its borders, it was likely that production of the pesticide would continue, and that it would find its way to developing countries, and possibly add to the already considerable stocks of obsolete pesticides in Africa, South America and Asia (Pesticides News 2000). Many pesticides banned or severely restricted in industrialised countries are still marketed and used in developing countries, threatening millions of farmers and the environment (FAO 1998c). It is important to note that, whilst Carson focused on the environmental impacts of pesticides, most of the attention of intergovernmental, governmental and non-governmental organisations is focused on human health impacts of pesticides. An environmental aspect was conspicuously excluded in the draft Rotterdam Convention (Hough 1998).

6.4.1.2.3 Stockholm Convention on Persistent Organic Pollutants

Adopted in 2001, the Stockholm Convention on Persistent Organic Pollutants (POPs) was the culmination of over 20 years of negotiations. The Convention became legally binding on the 17th of May 2004. The 90-day countdown to the treaty's entry into force was triggered on the 17th of February 2004 when France became the 50th state to ratify the agreement (Stockholm Convention 2003). The Convention seeks the elimination or restriction of production and use of all intentionally, and ultimately unintentionally, produced POPs, and stipulates that stockpiles must be managed and disposed of in a safe, efficient and environmentally sound manner. The Convention imposes a global ban on 24 of the world's most hazardous pesticides [Aldrin, Chlordane, Chlordecone, Decabromodiphenyl, Dieldrin, Endrin, Heptachlor, Hexabromobiphenyl, HBCDD, Hexabromodiphenyl ether, HCB, Hexachlorobutadiene, Alpha hexachlorocyclohexane, Beta hexachlorocyclohexane, Lindane, Mirex, Pentachlorobenzene, Pentachlorophenol, PCB, Polychlorinated naphthalenes, SCCPs, Endosulphan, Tetrabromodiphenyl ether, Pentabromodiphenyl ether and Toxaphene (Stockholm Convention 2017)].

Negotiations for a legally binding instrument to deal with POPs began at the turn of the Century. It is also intended that the Convention will generate resources to support these activities. The effects of POPs on health and the environment range from acute toxicity to intergenerational endocrine-disrupting effects. POPs are bioaccumulative and once in the environment cannot be removed. POPs are also transported by climatic and environmental processes over long distances. They tend to move from warmer climate regions, where most obsolete pesticide stockpiles are held, to colder climates, even as far as the poles where they accumulate in the fatty tissues of humans and wildlife at the top of food chains. Extensive monitoring of POPs' body burdens, effects on health and movement in the environment is being carried out in Canada, northern Europe and the USA. Little or no work has been carried out in the tropics where POP insecticides have been extensively used and now form a significant proportion of obsolete pesticide stockpiles.

The POP insecticide DDT also continues to be stored and used in several tropical countries for the control of malaria vectors. In the meantime, there is concern that sources of new DDT are limited and there is often no information on whether the new DDT meets WHO standards (FAO 2001). In the high- and moderate-risk regions of Colombia and Peru, for example, the risk of malaria doubled when spraying ceased in the 1990s (Hicks 1992). The spread of the disease means it has reappeared even in the USA and Europe (Attaran et al. 2000). By the year 2000, the number of countries using DDT declined to 23. It is produced in only three countries and is becoming difficult to obtain. Besides killing a child every thirty seconds, malaria is a recurring disease for many. Children who survive malaria past infancy suffer an average of six bouts each year, making it the most common reason to miss school; adult sufferers miss an average of ten working days a year (UNICEF 1999). The infection rate had fallen significantly over the decades, primarily because of DDT sprayed inside homes and on mosquito breeding sites. But as a UNICEF (1999) report describes it, 'DDT was widely discredited in the 1960s because of its harmful effects on the environment'. So the disease is nearly back to where it was 50 years ago. The tragedy is not being ignored. Roll Back Malaria was launched in October 1998 by UNICEF, the WHO and the World Bank to 'prevent and control this centuries-old scourge' (UNICEF 1999). The WHO also wishes to prevent the transfer of DDT from obsolete pesticide stockpiles to active use in the health sector, without the most stringent controls, and to prevent long-term reliance on DDT in the health sector. The DDT working group has been advised to liaise with the pesticide industry and its representative organisations with regard to ongoing, high-quality DDT supplies while they continue to be needed.

The POP negotiations were relatively smooth and swift by comparison to the Biosafety Protocol. The talks began in mid-1998 and the Convention was open for signing in May 2001, with most of the text ready by December 2000 (Clapp 2002). DDT still has exemptions for malaria control. While developing countries, the main users, were unhappy about the additional costs imposed by more expensive alternatives to POP pesticides, they did not hold up the process. The pesticide industry has been supportive of a treaty to phase out POPs since the mid-1990s. Groups present at the negotiations included: the American Chemistry Council, formerly the Chemical Manufacturers

Association, the European Chemical Industry Council (CEFIC), Global Crop Protection Federation, International Council of Chemical Associations and the World Chlorine Council (Clapp 2002). Pesticide industry actors accepted the international community's resolve to implement strong measures that eliminated the production, use and trade of eight agricultural pesticides that had once been extremely important to the agrochemical industry (Clapp 2002). However, because most of the POP pesticides no longer enjoy patent protection in industrialised countries, they are relatively inexpensive relative to new and patent-protected alternatives, and, as a result, their production and use in developing countries has been growing in recent years (Clapp 2002). The phase-out of production of these chemicals will thus create a market for newer alternative chemicals that still enjoy patent protection, and deal a blow to generic pesticide producers (Clapp 2002). Indeed, many of the large pesticide manufacturers did not bother to send delegates to observe negotiations over the elimination of their own POP pesticides because they either did not produce them anymore or their production was not profitable (Clapp 2002). DuPont and Syngenta have acquired many of the corporations producing, or that once produced, POP pesticides (Clapp 2002).

However, once signed, the EU has pushed for a strong legal commitment to the total elimination of POP chemicals, and promoted the use of the precautionary principle to guide the addition of new chemicals to the POPs list. The USA and the pesticide industry continue to argue against the total elimination of all POPs, suggesting that the idea was impractical and that careful management of remaining POPs was preferred compared to their elimination. In addition, the pesticide industry and the USA continue to battle against inclusion of the precautionary principle in the text of any agreement, favouring sound science instead (Pruzin 2000; Clapp 2002). The addition of new pesticides to the POPs list, particularly those still under patent protection, is of grave concern to the pesticide industry.

In India, nine of the ten 'manufactured' POPs on the initial list have already been banned (e.g. Aldrin, dieldrin, chlordane) or were never manufactured in the country (e.g. mirex and toxaphene). The sole exception is DDT, which is used and exported for malaria control, but manufactured only by the government and in decreasing quantities. It is important to understand why ICMA would so aggressively oppose a treaty that does not affect its current markets (PAN 2003b). The restrictions and ultimate bans laid out in the Stockholm Convention represent a new reality for Indian agrochemical companies, which have faced minimal restrictions over the years as government policy linked pesticides and fertilisers to increased food production (PAN 2003b). The provision in the POPs treaty to add new chemicals beyond the 12 currently on the list may be the major source of industry opposition. The Indian chemical industry has a particular investment in two pesticides, endosulfan and Lindane, which are widely considered likely candidates for addition under the treaty. Endosulfan alone accounts for over 10% of total insecticide consumption in India, and has recently come under severe pressure owing to the health effects it has caused in communities living around cashew plantations in the Kasaragod district in South India (PAN 2003b). To counter this public pressure, a new Association of Endosulfan Manufactures in India has been formed and is campaigning aggressively to protect endosulfan production and use (PAN 2003b). The Indian chemical industry's efforts to block adoption of the Stockholm Convention flies in the face of an emerging international consensus supporting global elimination of POPs (PAN 2003b). New internationally linked networks of civil society such as IPEN (the International POPs Elimination Network) bring credible global information and experience, through its worldwide membership, to the doorstep of affected villages in India (PAN 2003b). Civil society is now participating in international policy fora such as the IFCS and the UN Environment Programme (PAN 2003b). Regional and national public interest can engage with and counter industry arguments effectively (PAN 2003b). By calling on the Indian Government to avoid ratification of the Stockholm Convention, ICMA is attempting to block inevitable progress. Indian chemical companies are refusing to accept real progress in protecting human health and the environment (PAN 2003b).

In addition, the Montreal Protocol includes the ozone-depleting pesticide methyl bromide, widely used for soil and storage fumigation in developing countries. Developing countries are required to phase out methyl bromide by 2015. Methyl bromide was one of the first chemicals over which a global consensus was reached (Hough 1998). A global agreement was made in 1992 to freeze production and consumption of methyl bromide at 1991 levels beginning in 1995. In September 1997, the Ninth Meeting of the Parties to the Montreal Protocol committed 160 governments to a complete phase-out of methyl bromide production and use. Subsequently, methyl bromide has been joined by a growing list of POPs. Table 6.2 provides a snapshot of some of the UNEP POPs.

6.4.1.2.4 Obsolete Pesticides Programme

Obsolete pesticides are a growing problem across much of the world, particularly in tropical areas of developing countries and economies in transition. According to FAO (2003a), obsolete pesticides are pesticides that can no longer be used for their intended purpose or any other purpose and therefore require disposal. Problems with dumping banned pesticides in developing countries became an issue in the early 1980s (Boardman 1986). While UNIDO has no activities specifically supporting prevention and disposal of obsolete pesticides, it provides technical support for activities related to chemical production and use, and hazardous waste management in developing countries.

Table 6.2 Production data for the UNEP POPs pesticides

Organochlorine pesticides	Initial producer	Year of introduction	Global use (tonnes)
Aldrin and Dieldrin	Shell	1948-50	>130 000
Endrin	Velsicol and Shell	1951	>2 300
Chlordane	Velsicol	1947	>150 000 (1950-92)
DDT	Ciba-Geigy	1946	3 000 000
HCB		1945	>100 000 As By-Product 600-6000
Heptachlor	Velsicol	1952	n/a
Mirex	Allied Chemical Co.	1946	n/a
	Hooker Chemical	1958-9	>1500
Toxaphene	Hercules Powder Co.	1946	1 333 000

Source: Adapted from Vancouver (1995) and ATSDR (1997).

In 1994, the FAO launched its Obsolete Pesticides Programme (see Fig. 6.3). The project was primarily funded by the Netherlands, with additional support from UNEP Chemicals for Latin American country inventories and from Japan for Asian country inventories. The Programme aimed to assist in the disposal of obsolete pesticides and prevent the occurrence of future pesticide stockpiles. According to FAO (2003a), the main reasons for this buildup of poisonous pesticide waste include: (1) The banning of pesticides while they are in storage. (2) Prolonged storage of products with a short shelf life. (3) Difficulties in forecasting outbreaks of pests such as locusts. (4) Excessive donations (inappropriate, untimely and uncoordinated). (5) Poor assessment of pesticide requirements. (6) Inadequate storage facilities and poor stock management. (7) Wrong or ineffective pesticide formulations. (8) Lack of staff trained in storage management. (9) Aggressive sales practices. Between 20% and 30% of quantified obsolete pesticide stockpiles are POPs, and therefore, according to the POP Convention, the responsibility of dealing with POPs stockpiles and sources, and of finding replacements for existing POPs uses, rests at the country level. In December 1999, Basel COP5 agreed a Declaration on Environmentally Sound Management to assist developing countries manage their own wastes through technology transfer, capacity building and so forth (DEFRA 2000).

The FAO took the lead in coordinating action on obsolete and unwanted pesticides, and, by the late 1990s, had undertaken a preliminary inventory of stocks in Africa, the Near East and Latin America (Pesticides News 1998a). However, the quantification of obsolete pesticide stocks is difficult because of the wide distribution of pesticides and the remote location of many of the storage points. According to FAO (2002), at the turn of the Century, there were 500 000 tonnes of obsolete pesticides, 100 000 tonnes in Africa and the Near East, 200 000 tonnes in Asia, and 200 000 tonnes in Eastern Europe and the former Soviet Union. These obsolete pesticides pose a serious threat to the health of millions of people and the environment in nearly all developing countries and economies in transition. Most stores are in the centres of urban areas or close to public dwellings. Some stocks are over 30 years old and are kept in poor conditions with few or no safety precautions. Banned organochlorines such as dieldrin, aldrin, chlordane and heptachlor and highly toxic organophosphates such as parathion, ethoprophos, dichlorvos and monocrotophos make up a high proportion of the stocks (Pesticides News 1998a). According to GreenPeace (2002), approximately 33% of obsolete pesticides were manufactured by Bayer and Shell; the remainder were made by Union Carbide (Dow), Sumitomo, Sandoz, Rhône-Poulenc (now Bayer), Du Pont and Monsanto, among other companies.

The situation is particularly dangerous in Poland with 65 000 tonnes and the Ukraine with over 23 000 tonnes. Among the highly toxic and persistent pesticides identified were Aldrin, DDT, Dieldrin, Endrin, HCH, Lindane, Malathion, Parathion and many others, many of which are leaking into the soil and groundwater from corroding concrete storage bunkers. According to Pesticides News (1998b), at the turn of the Century, the Netherlands had between 200 000 and 400 000 contaminated sites, which cost an estimated US$50 billion to clean up. In fact, it is estimated that there is one polluted site for every 30–50 inhabitants in the industrialised countries (Pesticides News 1998b). By 2004, IOMC Participating Organisations were expected to have established relevant action plans for all regions, and at least two countries in each region should have commenced implementation of their National Plan with respect to disposal (Pesticides News 1998b). FAO (2002) underlined that the financial assistance of the chemical industry should be sought. However, the disposal process for obsolete pesticides is hazardous, difficult, administratively complex and expensive, and requires expert knowledge and equipment. At the turn of the Century, the cost of disposal was between US$3000 and 4000/tonne of waste. The cost of disposing obsolete pesticide stocks in Africa alone was estimated to exceed US$100 million (FAO 2002).

According to the UN, donor countries, aid agencies, agrochemical companies and recipient governments are all responsible for the steady accumulation of obsolete pesticides in developing countries. Those responsible for obsolete pesticides are: (1) National agencies. These include government departments (primarily agriculture and health); crop production boards (e.g. cotton, coffee, etc.). (2) Nationalised agricultural input supply agencies – local authorities, state farms and so forth. (3) Multilateral agencies and intergovernmental organisations. These include the World Bank, the EC, regional development banks, UN organisations (in particular FAO and WHO) and specific projects such as the FAO Emergency Prevention System for Transboundary Animal and Plant Pests and Diseases (EMPRES). (4) Bilateral agencies. These primarily include development assistance agencies from industrialised countries supporting southern countries. Examples include the USAID, the United Kingdom Department for International Development, the Netherlands Directorate-General for International Cooperation, the Danish International Development Agency, the Swedish International Development Cooperation Agency and the Canadian International Development Agency. (5) Private sector. The distribution of pesticides in developing countries is increasingly undertaken by private companies, which import, distribute and sell pesticides locally. In many developing countries, certain key sectors, such as major export crops and health, remain under state control so that the private sector supplies pesticides only to smallholder farmers and domestic users.

Environmental NGOs are insistent that pesticide corporations are responsible for the removal of obsolete pesticides. GreenPeace called for the chemical industry to 'move beyond the "responsible care" rhetoric and take genuine responsibility for its products from cradle to grave' (GreenPeace 2002). Accordingly, pesticide companies should be obliged to take full responsibility for the removal and safe destruction of the obsolete pesticides in their country of origin (GreenPeace 2002). For example, GreenPeace insisted that Shell Chemicals, which was virtually the sole producer of endrin, dieldrin and aldrin between 1952 and 1990, should be accountable for their removal and safe disposal (GreenPeace 2002). Most uses of dieldrin had been banned in most developed countries years before Shell ceased its production in 1990 (GreenPeace 2002). While Shell International was acknowledged for contributing approximately US$300 000 towards the disposal of dieldrin in Mauritania, it has not been as active in other parts of the world (GreenPeace 2002).

Prominent NGOs are also deeply involved with the problem of obsolete pesticides. PAN has centres in Africa, Asia, Europe, Latin America and North America and affiliated organisations around the globe. The International POPs Elimination Network (IPEN) also links over 200 organisations on issues related to POPs and other chemicals. The close association between POPs and obsolete pesticides means that IPEN is concerned with the safe elimination of stockpiles while ensuring that they do not recur.

According to GCPF (2000), which currently represents 12 major pesticide manufacturing industries and several regional industry organisations, it is committed to a process of product stewardship that effectively means implementation of the International Code of Conduct on the Distribution and Use of Pesticides. Dealing with obsolete pesticides is seen by industry as 'part of its commitments to the principles of product stewardship' (GCPF 2000). Prevention of obsolete stocks is one of GCPF's key concerns. GCPF has a project team that coordinates industry activities on obsolete pesticides, which states that, 'if a company owns a product, the responsibility for the product clearly remains with that company' (GCPF 2000). In contrast, responsibility of purchased stocks lies with the present owner. GCPF member companies are supposedly willing to help find appropriate solutions for products they originally manufactured or supplied. While GCPF offers training in the safe management of pesticide, not all pesticide trade and use in developing countries, even from GCPF companies can be said to adhere fully to the requirements of the International Code of Conduct on the Distribution and Use of Pesticides. Indeed, the pesticide industry's compliance with Articles 8 and 10; Distribution and Trade; and Labelling, Packaging, Storage and Disposal, respectively, remained very weak. GCPF companies and regional trade organisations have also supported a number of container collection schemes (product stewardship schemes) that will reduce the number of containers accumulating and needing disposal. However, these schemes rarely operate in areas with poor market infrastructure, where pesticide trade is least profitable and there is reluctance to invest in safe use programmes. Unfortunately, these areas are often the most vulnerable to accumulation of obsolete pesticides. Similarly, the technologies that can make pesticide use safest and reduce obsolescence to a minimum (for example, plug-in packs for tractor-mounted sprayers) are expensive and, therefore, difficult to introduce into developing countries. GCPF has, however, made a commitment to provide support for the disposal of obsolete pesticides that were manufactured or supplied by its member companies. Backed by the obsolete stocks project team, the GCPF coordinator supposedly facilitates multi-stakeholder approaches, including representatives from the pesticide industry, government/stocks owner and others. The level of assistance is decided on a case-by-case basis, and is ultimately dependent on the particular pesticide corporation involved. Generally, this commitment translates to payment of the destruction costs which amount to approximately 25–30% of the total disposal cost. In some cases, GCPF companies have paid all the removal and destruction costs for some chemicals, but in others no contribution has been made. No pesticide manufacturers, other than GCPF companies, contribute to disposal.

However, according to Greenpeace (2002), most pesticide corporations have been slow to either acknowledge partial responsibility for obsolete pesticides or assist in their disposal. According to GreenPeace (2002), Bayer refused requests for help with 75 tonnes of obsolete pesticide stocks from the Nepalese government in 1999. Subsequently, as a result of a Greenpeace campaign, Bayer offered 'assistance' via 'CropLife' (previously GCPF) but up to 2002 had not committed itself to taking any concrete steps (GreenPeace 2002). Indeed, in 2002, Greenpeace called on the pesticide industry to compile a full inventory of global stockpiles and to take full logistical, technical and financial responsibility for their retrieval and safe disposal, in line with the regulations of the Stockholm Convention (GreenPeace 2002). To date, the agro-chemical industry has made limited contributions to the disposal of obsolete pesticides.

While initial concern had been focused on the continued production of banned pesticides in developed countries and their subsequent sale in developing countries, recent concern has been increasingly focused on new formulators and producers of 'generic' (off patent) pesticides in China and India, as well as other Asian countries and several Latin American countries. The products distributed from these generic manufacturers tend to be based on older technology (e.g. organochlorine and organophosphate insecticides) and the products are often reported to be of low quality.

In addition to the work of the FAO, several national governments have also actively supported the environmentally sound disposal of obsolete pesticides. For example, in addition to supporting FAO's Obsolete Pesticide Programme, the Netherlands has also funded numerous national disposal programmes including those in Ethiopia, Seychelles, the United Republic of Tanzania and Zanzibar, Yemen, and Zambia. The Netherlands have also funded a detailed inventory of stocks in the United Republic of Tanzania and Pakistan. Denmark has supported the use of a cement kiln for the destruction of obsolete pesticides in Mozambique and the Finnish development agency (FINIDA) contributed $824 000 towards the removal and destruction of pesticides from Nicaragua in 1998. GTZ also ran a pilot programme on obsolete pesticides from 1991 to 1999. USAID has contributed to the disposal of 60 tonnes of dieldrin from the Niger, to the reformulation of 86 000 L of carbaryl in Senegal, and to a disposal project in Ethiopia. Examples of World Bank activities in the field of POPs and obsolete pesticides include. (1) Global Environment Facility (GEF) funding for a UNEP-implemented regionally based assessment of persistent toxic substances. (2) Establishment of a POPs trust fund to encourage the involvement of developing countries in the elimination or phasing out of POPs. (3) Efforts to help phase out DDT use for malaria control. The World Bank also financially contributing towards cleaning up ground contamination in Yemen resulting from the burial of pesticides on a World Bank/IFAD-sponsored farming project during the 1980s.

At the turn of the Century, it was estimated that if the removal of obsolete pesticides continues at the same pace as in the past, it would take more than 30 years to clean up obsolete stocks in Africa and the Near East. This would only include the removal of metal drums and other containers, but not the more difficult disposal of contaminated soil. By the year 2000, around US$24.4 million had been spent on the removal of pesticides in Africa and the Near East alone. In some cases, countries have allocated their own resources to dispose of the obsolete pesticides within their national borders, namely, South Africa, Jamaica and Qatar. South Africa disposed of 603 tonnes of its own waste and waste from Namibia and Swaziland. However, despite attempts by the FAO, international donor governments, NGOs, philanthropic institutions, and national governments directly affected by the problem of obsolete pesticides, by 2003, less than 5% of estimated obsolete pesticide stocks had been disposed of (WB 2003). In response to the slow pace of obsolete pesticide stock removal, the WWF and PAN-UK launched the Africa Stockpiles Programme. If appropriately resourced by donor governments, this programme is expected to cost US$250 million over a period of 15 years (GEF 2002).

6.4.1.2.5 Moratorium on genetically modified organisms in Western Europe and growing concerns over GMOs in the USA and Japan

Initially, GMO were accepted, without debate, into agricultural exporting economies for both industrial (e.g. cotton, rape seed) and food crops (e.g. tomatoes, potatoes, maize and soya beans). In Western Europe, by contrast, public interest in food quality, food health and the traceability of food products (i.e. post-productivist concerns) was raised throughout the 1990s by a number of 'food scares', such as listeria in cheese, growth hormones in beef and poultry, and BSE ('mad cow disease') in beef (World Bank 2003). More specifically, a British researcher claimed on TV that eating GM potatoes could stunt rats' growth. Few scientists found his data compelling; calling his claims 'misleading', and his own institute suspended him. Nevertheless, the episode riveted public attention on GM crops' potential risks (Stipp 2000). In addition, a Cornell study showed that lab-fed monarch butterfly caterpillars died after eating pollen from Bt maize (Stipp 2000). While new and contrasting data has undermined the validity of the Cornell study, anti-GM activists have continued to use the study to their advantage (Stipp 2000). Consequently, opposition to the uncritical acceptance of GMOs in food emerged, with some diffusion of that concern back to those countries where GMO crops were first introduced.

Genetically modified food (GMF), therefore, has become a contested technology for the PPT. Different interest groups developed opposing discourses, with the shaping of public opinion, and the political response, at stake. One survey, conducted in 1995, found that 49% of Swedes, 70% of Germans and 78% of Austrians would not buy a bioengineered food product. In an EC survey 56% of the respondents felt that GMF was dangerous. In contrast, an average of 70% of Japanese, USA and Canadian consumers would buy such products.

In 1996, Monsanto imported RR Soybean foodstuff into the EU. This caused uproar from the public and campaigners. In February 1997, Austria and Luxembourg banned the sale, and France prohibited commercial growing, of Novartis' GM European corn borer resistant maize, followed in March with a ban by Italy and Spain (Deutsche Bank 1999). On 2 April 1997, the EC voted 407:2 to ban the importation of unlabelled GM maize from the USA (Deutsche Bank 1999). Indeed, the undisclosed transgenic origin of much of the soy lethicin in chocolate led the Swiss government to confiscate and destroy 500 tonnes of locally produced chocolate around Easter 1997 (Deutsche Bank 1999). In the same month, the European Parliament called on the EC to suspend market consent for Novartis' modified maize (Deutsche Bank 1999). However, in September 1997, Italy withdrew its ban on cultivation of Novartis' GM maize, while in November 1997; France announced a moratorium on commercial cultivation of all GM crops with the exception of Novartis' maize (Deutsche Bank 1999). In addition, in December 1997, France announced a moratorium on commercial use of all GM crops containing antibiotic resistance marker genes with the exception of Novartis' maize; followed by a moratorium on growing of GM crops (beet and rape) that have wild relatives in Europe in July 1998 (Deutsche Bank 1999). In September 1998, France's highest administrative court suspended authorisation to grow Novartis' modified maize, and Austria's main supermarket chains took GM products off their shelves (Deutsche Bank 1999). In October 1998, Greece banned importation of GM rapeseed, and the UK announced a de facto three-year moratorium on GM insect-resistant plants (Deutsche Bank 1999). In November 1998, France banned import and sale of two varieties of GM rapeseed, and in February 1999, the European Parliament called for labelling of GM crops and foods. It also calls for a ban on antibiotic resistance marker genes in modified crops (Deutsche Bank 1999).

In March 1999, UK retailer Sainsbury's formed a consortium of major European retailers, including Marks & Spencer (UK), Carrefour (France) and Superquinn (Ireland), to jointly buy non-GM food products. Greece also announced that all applications for experimental plantings of GM crops presently pending would be rejected (Deutsche Bank 1999). In April 1999, Nestlé UK and Unilever UK, national subsidiaries of the two largest food producers in the world, announced the phase-out of GM ingredients in their products. Tesco, the UK's largest food retailer, announced that its own brands would be GM free (Deutsche Bank 1999). According to Levidow and Bijman (2002), 'the rise of own-brand labels has made retailers more vulnerable and responsive to consumer concerns . . . because their own brand product lines are designed to symbolize product quality'.

Ultimately, in May 1999, the EC suspended approval processes for all new GM crops (see Fig. 6.3) (Deutsche Bank 1999). Subsequently, food processors began to abolish GM commodities from their supply chains. For example, in December 1999, potato processor McCain announced that it would not accept GM potatoes (Farmers Weekly 1999). The decision to suspend the approval process for GM crops resulted in significant costs for biotechnology corporations. For example, according to the Washington Post (1999), Monsanto stock lost more than a third of its value in 1998/1999. In addition, many of the larger shareholders became increasingly uneasy about the impact of the EU decision, many doubting the resilience of their life science company model (Washington

Post 1999). According to The Financial Times, European opposition to GM cost Syngenta tens of millions of dollars (Corporate Watch UK 2002a). However, realising the sensitivity of the GM issue, and the importance of securing a future market for GM products, Syngenta voluntarily limited the sale of GM maize in Spain, which was authorised before the moratorium, as they wished to introduce the technology progressively (Planet ark 2003b). Concern about GM food quickly spread from Europe, where it bordered on panic for more than a year, to North America and Asia.

The USA continued to protect its biotech industry through lobbying the EU and other against labelling GM products (Pechlaner and Otero 2010). However, concern about GMOs in North America have materialised in two distinct forms, consumer opposition and food industry pragmatism. In Canada, a telephone survey in Montreal during October 1999 revealed that 650 people would not eat GM foods, and 530 thought that they should be banned (The Gazette 1999). Several anti-pesticide, consumer and environmental groups quickly developed anti-GM propaganda. These include. (1) The Rural Advancement Foundation International. (2) The Sierra Club. (3) GreenPeace. (4) The Council of Canadians (Schweiger 1999b). Concern also materialised in certain farming groups. For example, the Christian Farmers' Federation of Ontario, Canada, called for mandatory labelling of GM. They suggested that, 'if GM food is going to be a success, we have to market it, not hide it' (The Evening News 1999). This sentiment was also echoed in the USA where the US National Nutritional Foods Association (NNFA) called for labelling of GM foods. They stated that the public has a right to know exactly what it is eating (NNFA 1999). Furthermore, in June 1999, a petition, signed by nearly 500 000 Americans, urged Congress to require labels on GM foods (Reuters 1999b).

The food industry also took a pragmatic view of the GM issue. While many involved in the processing sector advocated a slow and considered response to GM concerns, particularly because of the costs associated with the segregation of GM and non-GM commodities, many quickly developed mechanisms to segregate GM produce and capture the growing premiums for non-GM food ingredients and foods. For example, processors such as Archer Daniels Midland responded by attempting to segregate GM-from Non-GM commodities (Stipp 2000). In 1999, ADM advised its farmer supplier to segregate GM for non-GM commodities (Schweiger 1999a). Gerber removed GM ingredients from its baby foods (Stipp 2000). Whilst in favour of labelling, not surprisingly, Monsanto insisted that it was non-GM foods that should be labelled; retailers could have non-GM and organic sections (Dow Jones 1999a). Conversely, the retail end of the food system increasingly reacted to growing concerns about GM foods by removing them from their supply chains; Whole Foods and Wild Oats were the first big US stores to ban GM ingredients (LA Times 1999). The retail sector has benefited immensely from concerns about GM foods. It has been able to enhance it public image by reacting quickly to public concern by sourcing non-GM foods, even if those foods are only a small percentage of their turnover; while, in the process, increasing their profits. However, perhaps more importantly, retailers and, to a lesser extent, processors have been able to strengthen their position within the food system relative to the farm inputs and production sectors.

While public concern has significantly impacted on the procurement activities of food system processors and retailers, the resolve of the US Administration remains firm. In 1999, the Under Secretary of State stated that 'the Europeans have an absolute fear, unfounded by any scientific basis, in accepting these (GMO) products'. USDA Secretary Dan Glickman equated the European reluctance to accept agricultural biotechnology to the 'trade disputes with beef and, more so, bananas' (Deutsche Bank 1999). Subsequently, on the 13th of May 2003, the USA announced that it would sue the EU through the WTO unless Brussels ended a de facto moratorium on GMOs (Planet ark 2003b). In the meantime, many large corporations, both diversified and purely lifescience, have divested their agricultural biotechnology components because of growing uncertainty about the acceptance of GMOs. These include Pharmacia (which divested Monsanto), Hoechst, Schering and Shell (Vorley and Keeney 1998).

Concern over the use of GM crops in Western Europe also spread Japan during 1999. Despite, warnings from the USA that labelling GM foods could mislead consumers into thinking that GMs pose a health risk, were there was none, the Japanese Government introduced GM labelling in April 2001 (Reuters 1999c). While concerns about GMOs in Japan spread quickly to other parts of Asia (Zanatta 1999), including China and India (World Bank 2003), most Asian countries have not imposed GM labelling regulations (Reuters 1999g); with the exception of countries such as South Korea, Australia and New Zealand (Zanatta 1999). In South Korea, a consumer survey indicated that 94.7% of people thought that labels on GM ingredients were necessary (Genet 1999). While the Indonesian Government refrained from passing labelling laws, it established a Food Safety Commission to supervise GM food products (Agence France Presse 1999). In 1999, the Thai Government announced that, in a bid to protect valuable exports, it intended to establish GMO-free agricultural zones. Whilst the import and export of GM crops is banned in Thailand, GM is allegedly still grown unofficially, threatening exports with majors trading partners such as the EU (JEN 1999).

As in the case of Western Europe and North America, food processors and retailers responded rapidly to consumer concern over GM food ingredients. On the 1st of September 1999, Japan's largest maker of soybean protein food products, Fuji Oil Co. Ltd., said that it would stop using GM soybeans in April 2000 due to consumer concerns over GM crops (Genet 1999). Kirin, the

Japanese brewer also stopped using GM products (Stipp 2000). In December 1999, Nisshin Flour Milling Co. Ltd., Japan's largest flour miller, announced that it would stop using GM maize and soybeans in its flour products (Reuters 1999h). In addition, Kibun Food Chemifa Co. Ltd, Japan's largest soybean milk maker announced that it would stop using GM ingredients in its soybean milk products (Reuters 1999h).

6.5 Summary

Chapter Six focuses on the impact of the global Post-Fordist developmental context on the strategies employed by pesticide corporations from the mid-1970s to 2017. It suggests that, in countries such as the USA, Western Europe and Japan, consumer demand for food products has become highly differentiated, and that, alongside Fordist and Neo-Fordist-styled industrial food networks producing cheap, mass-produced, standardised and generic food products, new networks of food provision have arisen based on high-value added quality products. Within the context of the Post-Productivist Paradigm, pesticide companies responded to several key external drivers, namely. (1) Increasing pressure from NGOs for national governments and intergovernmental institutions to impose tighter regulations on pesticides and biotechnologies in order to reduce negative human health and environmental impacts of pesticide use in both developed and developing countries. (2) Weakened legitimacy of pesticide use in developed countries. (3) Promotion of Post-Productivist agriculture [agri-environmental regulations, precision agriculture (PA), ICM, Integrated Farm Management (IFM), organic agriculture] by Western European nations. (4) Evolution of rapidly changing and highly differentiated globalised food markets. (5) Increased control of pesticides and GMOs in the food chain by food processors and retailers. This chapter primarily focuses on the pesticide industry's attempts to redefine the role of pesticides and emergent biotechnologies in Post-Productivist food systems. This battle has been fought on three fronts. (1) Negotiations with private sector food system actors. (2) Negotiations with national, regional and international public sector policy-makers. (3) Marketing and educational initiatives. First, the pesticide industry attempted to redefine the role of pesticides and emergent biotechnologies in light of growing demand-side pressures from food processors and food retailers. For example, the development of IPM, ICM and IFM strategies have been financially supported by the pesticide industry in an attempt to legitimise the use of pesticides in the sustainable production of quality and high value-added food products in both developed and developing countries. Second, the pesticide industry has been instrumental in maintaining support for the use of all but the most hazardous pesticides, and almost unilateral support for the development and application of emerging biotechnologies, in national, regional and intergovernmental institutions and policy arenas. Specifically, support has been forthcoming for the development of productivity enhancing input trait biotechnologies, particularly those that facilitate reductions in the use of pesticides and other synthetic inputs and those with the potential to enhance the livelihoods and food security of marginal agro-ecosystems. In addition, solid support has also been expressed for second and third wave of output trait plant biotechnologies, the so-called functional foods and nutraceuticals, which focus on enhanced consumer value, such as; improved flavour, enhanced nutritional content and plant products containing human medicines. To this end, most pesticide corporations have focuses on economies of scope, particularly the perceived synergies between agrochemicals, pharmaceuticals and nutrition. In an attempt to maximise profitability and minimise risk in increasingly diverse, volatile, and knowledge/information intensive consumer markets, new ventures were increasingly coordinated through corporate collaboration, usually in the form of joint ventures and alliances. Third, pesticide corporations have attempted to improve their image, and that of pesticides, through educational, marketing, food safety initiatives, and other voluntary measures to reduce the impacts of pesticide use, including the Global Safe Use Campaign.

While it is true that the involvement of pesticide corporations in Post-Productivist styled agro-food circuits has led to less intensive and more sustainable systems of production, two major concerns remain. First, much doubt remains over the true sustainability of integrated systems of agricultural production, many believing that they are nothing more than a clever marketing ploy for the continued use of pesticides and new plant biotechnologies, and evidence of the immense persuasive power of TNC corporations in public sector policymaking arenas. For example, in the case of pesticides, while, at first glance, important gains have been made in pesticide regulation over the past 30 years, key players in the pesticide industry have primarily only sacrificed older, off-patent pesticides; many of which were becoming increasingly unprofitable because of stiff competition from generic pesticide producers and from the buildup of biological resistance. With the exception of short-term public concern over emergent biotechnologies, the pesticide/life science industry has increasingly secured regulatory approval for the release of both input and output trait biotechnologies. Second, there is growing concern over increased corporate integration and control of food systems, particularly the distribution of value added.

Chapter 7

The evolution of the pesticide industry under the 'Sustainability Paradigm' (1974–2017)

7.1 Key characteristics of urban-centred industrial agro-food systems and broader economic, social and political contexts

Chapter 7 begins by introducing the concept of sustainability and its evolution. It then proceeds to address how the pesticide industry responded to constraints and opportunities associated with the integration of so-called sustainability goals into agro-food system regulation.

Published in 1972, by the so-called Club of Rome, 'Limits to Growth' was a landmark publication, which focused the world's attention on what they perceived as unsustainable levels of population growth and per capita consumption. Based on computer-generated future world system scenarios, Limits to Growth concluded that humans were soon likely to overshoot the earth's resource carrying capacity (Meadows et al. 1972). According to the Club of Rome, avoiding this scenario necessitated immediate action to achieve global equilibrium between human demands and environmental resources, focusing on the stabilisation of population and per capita consumption levels.

The term 'limits to growth' was derived from the writings of Thomas Robert Malthus, John Stuart Mill and Adam Smith (Johnston et al. 2000). These authors believed that there were basically five major factors that would determine, and therefore ultimately limit, growth on this planet. These factors were: (1)

population; (2) agricultural production; (3) natural resources; (4) industrial production and (5) pollution (Meadows et al. 1972).

Whilst many of the claims of the Club of Rome have since been contested, these sentiments struck a chord with a growing number of people during the 1960s, 1970s and 1980s. Many of the new environmentalists favoured a more ecocentric model of economic development. This model rested on the supposition of a natural order in which all things moved according to natural law, in which the most delicate and perfect balance was maintained up to the point at which man entered with all his ignorance and presumption (O'Riordan 1981). Thereafter, the web of life was broken by a degenerative succession of disturbed harmonies, leading to the ultimate destruction of man himself (O'Riordan 1981).

Ecocentrism preaches the virtues of reverence, humility, responsibility and care; it argues for low impact technology; it decries bigness and impersonality in all forms; and it demands a code of behaviour that seeks permanence and stability based on ecological principles of diversity and homeostasis (O'Riordan 1981). The ecocentric root of modern environmentalism is nourished by the philosophies of the romantic transcendentalists of the mid-nineteenth century. The transcendentalist philosophies initiated two lines of thought, which have influenced modern environmentalists. One is the theme of self-reliance within the context of a small, recognisably interdependent community, a theme that has stimulated the imagination of utopianists ever since (O'Riordan 1981). The other is the case for a bioethic advocated by nature moralists such as Muir (O'Riordan 1981).

7.2 Key external drivers of pesticide evolution

7.2.1 Growing concerns over the loss of biodiversity and natural habitats

Since the mid-1800s, the loss of biodiversity and natural habitats has been of growing concern, particularly in North America and Western Europe (see Fig. 7.1). According to the Convention on Biological Diversity, biodiversity or biological diversity means 'the variability among living organisms from all sources including, inter alia, terrestrial, marine and other aquatic ecosystems and the ecological complexes of which they are part; this includes diversity within species, between species and of ecosystems' (Engels 1995).

The bioethic principle forwarded by Muir and others emerged in environmental policymaking (O'Riordan 1981). It has been applied in order to protect wilderness areas, national parks and wildlife habitats from the pressures of exploitation (O'Riordan 1981). Indeed, the bioethic principle was extended into the discipline of economics, where the theory of discounting attempts to reflect the uniqueness and irreplaceability of natural areas (O'Riordan 1981).

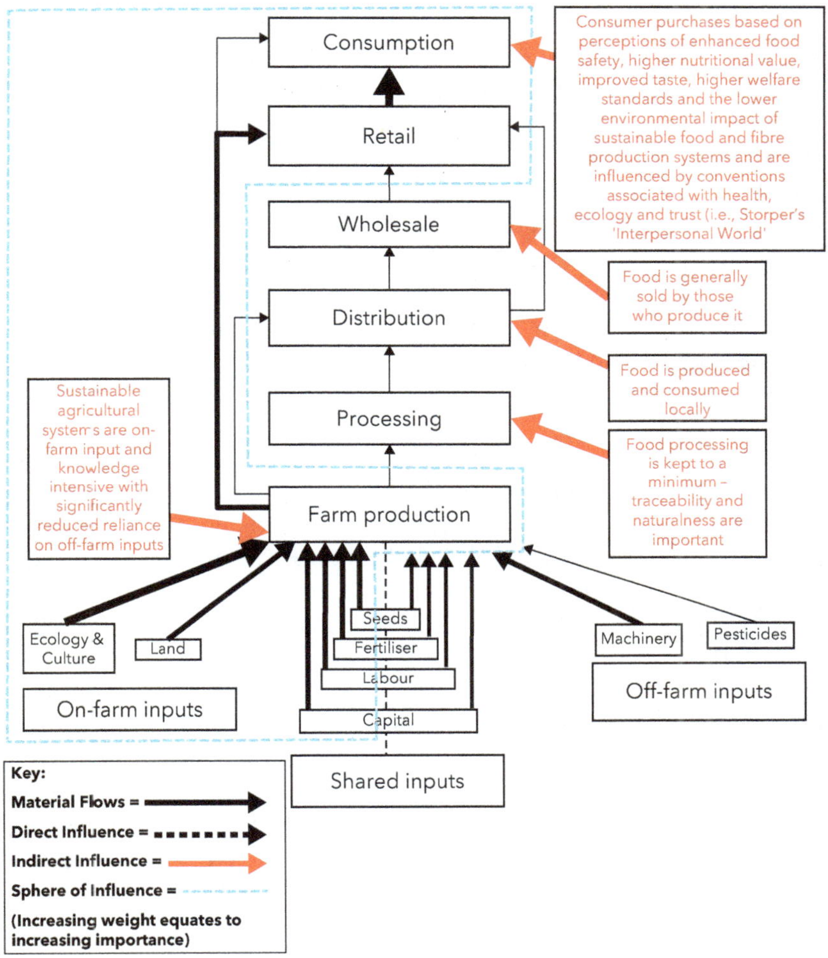

Figure 7.1 Sustainability Paradigm food system dynamics during the Divergent Period (1974-2017) of pesticide and pesticide industry development.

Biodiversity is attributed with a diverse range of values, including: (1) Aesthetic (for natural beauty, variation and inspiration). (2) Recreational. (3) Existence. (4) Bequest. (5) Scientific. (6) Genetic variation in wild species (specifically for agricultural and industrial uses). (7) Bioremediation (neutralising pollution and buffering other environmental changes). (8) Evolution. (9) Medicines. (10) Ecosystem services, including fisheries, water filtration, CO_2 sinks and indicators of ecological change or stress (Selman 2000). Costanza et al. (1997) estimated the value of 17 key ecosystems services to be in the range of US$16-54 trillion/year (Costanza et al. 1997).

However, whilst there has been recognition of the value of biodiversity, the contemporary phase is marked by the highest rate of species extinction in recent history (Pimm et al. 1995). Species extinction rates are suggested to range from 5000 to 150 000/year (Goodland 1991). Key habitats are also being lost. Between 1980 and 1995, over 200 million ha of land was deforested (FAO 1997). Most deforestation occurs in poor developing countries, principally in the highly species diverse tropical zones of Africa and Asia. It is estimated that, global deforestation between 1990 and 2020 could equate to the extinction of between 15 000 and 50 000 species (UNEP 1992). As the area of natural habitats decline and the area of agricultural lands slowly expands, concerns are increasingly turning to declining agrobiodiversity (Woomer and Swift 1992; Greenland and Szabolcs 1994; Altieri 1995; Marsden et al. 1993; Morris and Bate 1999).

Concern about nature and the environment first manifest itself in Great Britain during the nineteenth century, when naturalists came to believe that cruelty and over-collecting were becoming the greatest threats to wildlife. Bodies such as the Royal Society for the Prevention of Cruelty to Animals and the RSPB were founded to fight these practices. The RSPB grew out of the Fur and Feather Group, set up in 1889, which was founded on the issue of the feather trade, which killed birds cruelly. In 1904, a royal charter outlined aims to encourage conservation and discourage wanton destruction of birds. In 1895, the National Trust for England and Wales was registered, concerned not only with the preservation of historic buildings but also with the conservation of nature and landscape. The National Trust Act 1907 became the foundation stone upon which the Trust's future was built. Its main point was that land belonging to the Trust could not be sold, mortgaged or given away but it could be leased with the permission of the Charity Commission. In 1912, the Society for the Promotion of Nature Reserves (SPNR) was inaugurated, created by Charles Rothschild, who saw the need for an integrated string of reserves across the country. The SPNR aimed to: (1) Collect and collate information regarding areas of land in the UK, which retained 'primitive conditions' and contained rare and local species liable to extinction. (2) Prepare schemes showing which areas should be secured as nature reserves. (3) To obtain such areas. (4) To preserve for posterity as a national possession in some parts of the UK, its floral, fauna and geographical features.

Following on an international conference in Fontainebleau, France, the International Union for the Protection of Nature (IUPN), founded on 5 October 1948, was the first global conservation organisation (IUCN et al. 1991). In 1956, the IUPN changed its name to the International Union for Conservation of Nature and Natural Resources (IUCN), changing it once again in 1990 to the World Conservation Union (IUCN) (IUCN et al. 1991).

7.2.2 Concerns over the productivity, stability and resilience of industrial agroecosystems

In addition to concerns over the loss of natural habitat and species, there is growing concern over the impact of rapidly declining biodiversity on the productivity, stability and resilience of industrialised agroecosystems (Woomer and Swift 1992; Greenland and Szabolcs 1994; Altieri 1995; Marsden et al. 1993; Morris and Bate 1999). It is generally accepted that there is a negative relationship between agricultural intensity and biodiversity (i.e. as agricultural intensity increases, biodiversity decreases) (Andrew 1991; Giller et al. 1997). Whilst agricultural industrialisation has spared large areas of natural and semi-natural habitats in the developing world from degradation (WCED 1987; IUCN; UNEP; and WWF 1991; Goklany 1999), declining agrobiodiversity continues to undermine the stability and resilience of agroecosystems (Woomer and Swift 1992; Greenland and Szabolcs 1994; Altieri 1995). Associated concerns include: (1) Nitrate and pesticide pollution of watercourses. (2) Long-term environmental effects on soil structure (caused by the excessive use of nitrogen fertilisers, pesticides and intensive soil management techniques). (3) Concerns over animal welfare, food quality and safety. (4) Degradation of valued landscapes and wildlife habitats, including the potential ecosystems impacts from the use of GMOs (Goodman and Watts 1997; Gasson and Potter 1988; Wrathall 1988; Ward 1993; Pimentel et al. 1995; Watson 1998). Indeed, several blame Green Revolution failures on the inappropriate imposition capitalist or free market ideologies. Pingali et al. (1997) provide an illuminating insight into the environmental impacts of the Green Revolution in Southeast Asia, which caused a loss of local autonomy and indigenous knowledge that far outweighed any benefit from increased agricultural production. There are also concerns that gains from agricultural intensification are beginning to plateau and that, in the long term, the agricultural productivity and environmental sustainability of intensive farming systems may rapidly deteriorate (Marsden et al. 1993; Morris and Bate 1999). For example, Pimentel and Pimentel (1986) suggested that the substantial amounts of energy used in the manufacture, transport and application of pesticides severely compromised the energy efficiency and sustainability of farming (Pimentel 1986).

7.2.3 Emerging concept of sustainable development

A number of key events highlighted the need for action to address environmental degradation on an international basis: (1) The Paris Biosphere Conference and the Washington D.C. Conference on Ecological Aspects of International Development, held during the 1960s. (2) The Torrey Canyon oil tanker disaster (1967). (3) Paul Ehrlich's book 'The Population Bomb' (1968). (4) The first Earth

Day in 1970. However, it was not until 1972, and the United Nations Conference on Human Development (Stockholm Conference), that the sustainable development agenda was catapulted onto the international political arena (Johnston et al. 2000). This milestone conference, comprising delegates from 113 countries, was the first time that environmental issues had been discussed at an international level. Unfortunately, only two heads of government attended, Prime Ministers Indira Gandhi of India and Olof Palme of Sweden. There were two key outcomes of the Stockholm Conference. First, and foremost, was the Stockholm Declaration, complete with an Action Plan containing 109 recommendations for national and international actions in areas of conservation of natural resources, education, human settlements and pollution. Second, the Stockholm Conference resulted in the creation of the UNEP in 1972, created with the purpose of providing a programme focus for a Global Environment Monitoring System. The Stockholm Conference coincided with a spate of influential publications. These publications included the apocalyptic 'Blueprint for Survival', 'Small is Beautiful' by Schumacher (Green 1985) and the aforementioned 'Limits to Growth' (Meadows et al. 1972), and signalled the emergence of deep ecology.

With its origins in the 1968 Biosphere Conference in Paris, one major topic of the 1970s Man and Biosphere Programme of UNESCO was an ecological assessment of pest management and fertiliser use on terrestrial and aquatic ecosystems. Its aim was to enhance productivity with a minimum impact on non-target species and the environment (Boardman 1986).

In 1980, the International Union for the Conservation of Nature, the UNEP and the World Wildlife Fund published their World Conservation Strategy: Living Resource Conservation for Sustainable Development. Their strategy promoted the concept of environmental protection in the self-interest of the human species. Developed over a period of 3 years, and with the involvement of more than 450 government agencies, international bodies, including FAO and UNESCO (Hough 1998), and NGOs from over 100 countries, the World Conservation Strategy (WCS) aimed to advance the achievement of sustainable development through the conservation of living resources. It was intended to stimulate a more focused approach to living resource conservation and to provide policy guidance on how this could be achieved. It provided both an intellectual framework and practical guidance for the conservation actions necessary; and called for global coordinated efforts, for concerted action at national and international levels, and for global solidarity to implement its programmes. The three objectives of the strategy include the maintenance of essential ecological processes and life support systems, the preservation of genetic diversity, and the sustainable utilisation of species and ecosystems. The strategy argued that excessive use of pesticides promoted resistance, destroyed natural enemies, turned formerly harmless species into pests, killed non-target species and contaminated food and feed (IUCN, UNEP, WWF 1980). In 1992, an independent WWF report called on governments to reduce pesticide use on environmental grounds, and to follow the Dutch, Danish and Swedish example (Hough 1998).

Unfortunately, the environmental problems that had given birth to the Stockholm Conference, and, subsequently the WCS, continued, and thus prompted the UN to establish the World Commission on Environment and Development (WCED) in 1983. WCED's function was to examine critical environmental and development issues and to propose mechanisms by which the global community could address them (WSU 2003b). Madam Gro Harlem Brundtland, prime minister of Norway, was appointed to head the commission. In 1987, the WCED, often called the Brundtland Commission, presented the UN a report entitled 'Our Common Future' comprised of over 75 expert studies on a full range of environment and economic development issues. Its definition of sustainable development is often used: 'Sustainable development is development that meets the needs of the present without compromising the ability of future generations to meet their own needs' (WCED 1987).

The report concluded that the integration of environment and development, which was called sustainable development, was the only sound and viable means of ensuring both our environment and development future. This concept implies that there are limits to environmental resources and biosphere's ability to absorb human activities. These limits have roots in technological inadequacies and inequitable social organisation. Thus, sustainable development must entail: a process of change in which the exploitation of resources, the direction of investments, the orientation of technological development and institutional change are made consistent with future as well as present needs (WSU 2003b).

The Commission's document is a conceptual/moral report on the state of global affairs in terms of environmental degradation, international economic inequality and poverty, and the inability of current national and international institutions to deal effectively with the challenges of securing equity for future generations. It provides a useful discussion of the interconnectedness of economic and environmental changes, of the interdependence of nations in both economic and ecological terms, and of the interconnectedness of regional ecosystems. The Commission concludes with recommendations for achieving global sustainable development (WSU 2003b). The Commission envisaged the 'possibility for a new era of economic growth, one that must be based on policies that sustain and expand the environmental resource base (WSU 2003b). Issues of concern include: (a) population and human resources; (b) food production, distribution and terms of trade; (c) species and ecosystem preservation; (d) sustainable energy paths and consumption patterns; (e) industrial production; and (f) rapid urbanisation (WSU 2003b).

Acting in response to WCED recommendations, and mounting public concern over the world's continuing environmental problems, the UN decided to convene a world conference on Environment and Development on the twentieth anniversary of the Stockholm meeting. The location for the Earth Summit would be Rio de Janeiro, Brazil (see Chapter 6 for an overview of the Earth Summit).

7.2.4 Emergence of new national and intergovernmental institutions, facilities and conventions with an overt environmental agenda

Since the Stockholm Conference of 1972, a spate of specialised, national and international institutions were established in an attempt to institutionalise concerns over the environment and unsustainable types of development, and coordinated environmental activities. In the international arena, the UNEP; GEF; United Nations Commission on Sustainable Development (CSD); Convention on Biological Diversity (CBD) and the Global IPM Facility are some of the most important institutions to evolve.

7.2.4.1 United Nations Environment Programme

Established in 1972, UNEP was the first international public sector institution with a purely environmental remit (see Fig. 7.2). The mission of the UNEP is to provide leadership and encourage partnership in caring for the environment by inspiring, informing and enabling nations and peoples to improve their quality of life without compromising that of future generations (UNEP 2003a).

UNEP acts as a catalyst, advocate, educator and facilitator to promote the wise use and sustainable development of the global environment. To accomplish this, UNEP works with a wide range of partners, including UN entities, international organisations, national governments, NGOs, the private sector and civil society (UNEP 2003a). UNEP's work encompasses five key objectives. (1) Assessment of global, regional and national environmental conditions and trends. (2) Development of international and national environmental instruments. (3) Strengthening of institutions for the wise management of the environment. (4) Facilitation of the transfer of knowledge and technology for sustainable development. (5) Encouragement of new partnerships and mindsets within civil society and the private sector (UNEP 2003a).

UNEP independently operates the International Register of Potentially Toxic Chemicals, which provides an international source of information on the potential hazards of chemicals used by other intergovernmental organisations and states (Hough 1998). UNEP also hosts several environmental convention secretariats. These include the Ozone Secretariat and the Montreal Protocol's Multilateral Fund, CITES (the Convention on International Trade in Endangered Species of Wild Fauna and Flora), the CBD, the Convention on Migratory

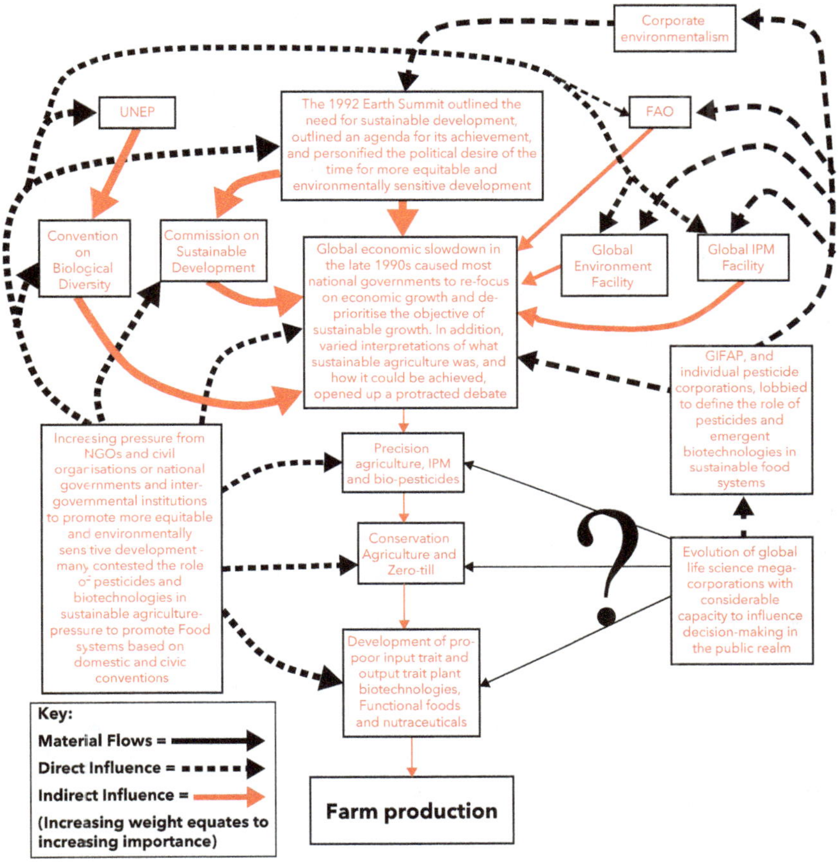

Figure 7.2 Pesticide manufacturers and the Sustainability Paradigm model of agriculture in the Divergent Period.

Species and a growing family of chemical-related agreements. Key chemical-related agreements include the Basel Convention on the Transboundary Movement of Hazardous Wastes and the Stockholm Convention on Persistent Organic Pollutants (POPs) (UNEP 2003a).

In 1985, with more than a decade of experience in collecting, analysing and managing environmental data, UNEP established the Global Resource Information Database (GRID). This acts as a focal point for storage, management and distribution of data. Moreover, GRID is an element of Earthwatch - the UN system-wide mechanism - which in collaboration with the international scientific community, contributes to the gathering, integration and analysis of data and information to produce comprehensive assessments of environmental states and trends to enable early warnings of environmental threats and to identify policy responses and management options (UNEP 2003b).

In 1995, UNEP's Governing Council adopted a decision to gather information and develop a strategy for reducing or eliminating the production and release of POPs. This included consideration of a legally binding treaty to regulate POPs. This work was in parallel to the UN Economic Commission for Europe, which was developing a protocol on such chemicals, and was completed in 1998 (Clapp 2002). In 1997, UNEP decided to push for a legally binding agreement on POPs. Negotiations on this treaty began in Montreal in June 1998 by the International Negotiating Committee on POPs (Clapp 2002).

7.2.4.2 Global Environment Facility

In 1989, at the Annual Meetings of the Board of Governors of the World Bank and the IMF, the French Prime Minister suggested establishing a fund of voluntary grants devoted to addressing global environmental problems. He also committed France to a contribution of 900 million French francs over a 3-year phase. At the same session, the Federal Republic of Germany also pledged its support for this initiative. These proposals gave political momentum to the creation of the GEF. It was the World Bank's intention to establish GEF as a 3-year pilot project. This was a milestone in itself as it was the World Bank's first acknowledgement of environmental problems and the need to respond to them. Furthermore, the Bank broke with its usual solitary nature by entering a cooperative agreement with two UN agencies, UNDP and UNEP. However, as far as the World Bank was concerned, there remained a clear political willingness to avoid the creation of a new full-fledged international organisation. It was decided that existing institutions (UNDP, UNEP and the World Bank) would manage the fund as Implementing Agencies (GEF 2003a).

The importance of the fact that the GEF was first created as a pilot needs to be stressed. In the early stages, participating governments did not have the political will to reach an agreement on a formal structure, strategy and future for such an entity. There was, yet, no consensus on the Funds future. Some European countries saw the pilot as a learning phase and precursor of a future organisation. They saw the GEF as a mechanism 'intended to grow into a more mature instrument for addressing global problems' (GEF 2003a). Other countries, especially the USA, viewed the GEF as a temporary institution, which would become obsolete once the World Bank integrated the global environment in its portfolio (GEF 2003a). Indeed, the cooperation between the three agencies was meant to bring practical and political advantages. Some countries favoured a strong role for the UN agencies, while others supported the leading role of the World Bank. The creation of the GEF was a means of satisfying both groups. As a matter of fact, the idea of bringing together the UN and the World Bank Group broadened the support for the new fund. Moreover, the

GEF served to bring together the UN and Bretton Woods's institutions, which was a first for the UN system (GEF 2003a).

In October 1991, agreement was reached between UNDP, UNEP and the World Bank to formalise their arrangements for operational cooperation. This agreement detailed the responsibilities of each of the three Implementing Agencies. They were expected to collaborate each in accordance with their respective comparative advantages (GEF 2003a). GEF was eventually established in 1991 (see Fig. 7.2). Its mandate was to work in partnership with the private sector, NGOs and international institutions, to provide financial assistance in developing countries for projects and programmes that protect the global environment. GEF became the designated financial mechanism for international agreements on biodiversity, climate change and POPs. It also supports projects that combat land degradation and that protect international waters and the ozone layer (GEF 2003b).

In 1992, the Earth Summit, in conjunction with negotiations of the Framework Convention on Climate Change and the Convention on Biodiversity, acted to catalyse two parallel discussions on the future of GEF. First, one discussion focused on the restructuring of GEF into a permanent and autonomous entity. Governments of developing countries, and the provisions of Agenda 21, provided explicit guidelines for a restructured GEF. There was a specific desire for the creation of an executive agency that would ensure universal participation in its work and transparency and democracy in its decision-making and operations (GEF 2003a). And, one that would assist in the process of formulating national development objectives, priorities and plans in developing countries and act as the financing entity for the climate change and the biodiversity conventions (GEF 2003a). In addition, the restructured GEF should respond to the findings of the Independent Evaluation of the Pilot Phase of the GEF report conducted in 1993. The report concluded that instead of resulting in inter-agency synergy and enhanced leadership, GEF had contributed to a worsening relationship between the executing agencies, which, in turn, led to the differential interpretation and implementation of GEF's remit. The second line of discussion focused on the replenishment of GEF's funds. During the GEF's 1991–4 pilot phase, recipients received US$35.66 million of GEF funds approved for biodiversity protection, US$29 million to forestall climate change and US$30 million for its micro-credit facility. However, this was in stark comparison to the US$56.9 million that World Bank's borrowers used to purchase agrochemicals from US and Western European pesticide corporations.

In 1994, the Instrument for the Establishment of the Restructured Global Environmental Facility was adopted, complete with pledges of a further US$2 billion, spread over 4 years (GEF 2003a). GEF was endowed with the remit to finance the incremental costs related to global environmental protection in the four focal areas as identified in the GEF Pilot phase. Other matters relating to the global environment and to Agenda 21, such as soil degradation, were included within the scope of application of the GEF upon the condition that they relate to one of the four focal areas and produce global environmental benefits (GEF 2003a).

GEF also developed an interest in the health impacts of pesticides, particularly those which accumulate in the food chain and may act as endocrine disrupters, suppressing immune systems and affecting reproductive and developmental systems. GEF has funded a number of projects addressing sources of POPs, mercury and pesticides (DDT). These efforts include the Regionally Based Global Assessment of Persistent Toxic Substances project to provide guidance and set regional priorities and Reducing Pesticide Runoff in the Caribbean. Others projects concern the search for alternatives to DDT for vector control in Mexico and Central America, and country case studies in persistent toxic substances, and POPs as they relate to food scarcity and indigenous groups in Arctic Russia (GEF 2003a).

GEF's Small Grants Programme, administered by UNDP, has directed more than US$42 million in over 1200 grants to community groups and NGOs that help rural communities develop locally appropriate and innovative solutions to protecting the natural resources on which they depend (GEF 2003a). One example is a Kenyan project for household raising and selling of swallowtail butterfly pupae for export. The programme also promotes the conservation of biodiversity, such as small organic produce farms in Poland. GEF-funded activities in 133 nations are steadily building capacities to prepare national inventories, strategies and action plans in cooperation with the United Nations Framework Convention on Climate Change and the CBD.

7.2.4.3 The United Nations Commission on Sustainable Development (CSD)

The CSD was established in 1992 as a functional commission of the UN Economic and Social Council (ECOSOC) (see Fig. 7.2) (UN 2003b). It has five explicit aims: (1) Monitor and report progress on the implementation of Earth Summit agreements at the local, national, regional and international levels. (2) Consider information provided by governments regarding the activities they undertake to implement Agenda 21. (3) Review progress in the implementation of the commitments set out in Agenda 21 and the Programme for Further Implementation, including those related to the provision of financial resources and transfer of technology. (4) Receive and analyse relevant input from competent NGOs, including the scientific and the private sector. (5) Enhance the dialogue, within the framework of the UN, with NGOs and other major groups, including the scientific community and educators, as well as with other entities outside the UN system (Green Year Book 2003).

7.2.4.4 The Convention on Biological Diversity

Like several international mechanisms, the CBD also arose directly from the Earth Summit of 1992 (see Fig. 7.2). However, whilst the CBD has a clear mandate to ensure conservation and sustainable use of biological diversity and the promotion of a fair and equitable sharing of the benefits arising from the utilisation of genetic resources, the inception of the CBD was underpinned by two conflicting ideologies. First, the homogenising tendency of market-based economic and biological reductionism and the commitment to promoting sustainable management of cultural and biological diversities of heterogeneous places. Indeed, its conceptualisation was influenced by the market-based paradigm of environmental management, particularly its endorsement of biotechnology-related IPRs, a provision insisted upon by US delegates to the original treaty negotiations (McAfee 2003). This market orientation has continued to influence interpretation and implementation of the treaty. For example, the 'benefits of biodiversity', which parties to the treaty have pledged to share, are constructed in CBD documents as those benefits, primarily profits, resulting from the commercialisation of genetic resources (McAfee 2003). 'Nevertheless, for US trade negotiators and biotechnology industry lobbyists, the CBD has opened a Pandora's Box of particularisms that threaten the universal, global criterion of economic efficiency that frames the WTO regime'(McAfee 2003). For many, the importance of the Convention is its unintentional role as an alternative political arena to the WTO for the discussion of genetic modification and IPRs (see Fig. 7.2) (McAfee 2003). For many, the WTO, dominated by the USA and private sector interests, has become the principal mechanism by which social and environmental concerns can be subordinated to the objectives of trade liberalisation and economic growth. In principle, at least, the CBD allows for heterogeneity in the resources, ecosystems and priorities of independent nations (McAfee 2003). Second, CBD provisions also call for core sustainability principles of 'fairness' and 'equity' in genetic-resource exchange and for the *in situ* conservation of biodiversity (McAfee 2003).

7.2.4.5 Global IPM Facility

In 1967, the first FAO/UNEP Panel of Experts on integrated pest control (IPC) met. Partly as a response to the growing imperative to develop alternatives to pesticides, such as IPM (Boardman 1986), in 1974, the IPC Panel of Experts established a global programme with biennial sessions (Hough 1998). The Panel has described its role as enhancing awareness and adoption of IPM practices, particularly in the developing world (Hough 1998). The FAO programme for community IPM in rice-based cropping systems in Asia began in 1980. Between 1980 and 1987, Phase I of the Programme focused on the verification of IPM technology and the development of pilot extension activities with farmers plus 'strategic extension campaigns' to promote IPM understanding and application (FAO 2003b). In parallel to the work of FAO and UNEP, after growing pressure to improve its pest management activities and reduce pesticide use, since 1985, the World Bank has published a series of policies on pest management. In 1988, the Bank took the initiative one-step further by convening a panel of experts to advise the Bank on pesticide issues. In the meantime, in 1987, FAO organised a workshop in Thailand, during which it called for government backing for IPM schemes and the ending of state pesticide subsidies (Environmental Liaison Centre 1987). In the same year, FOA initiated the second phase of its IPM Programme, which was scheduled to run from 1987 to 1993. This phase of the Programme focused on capacity building through the introduction of season-long training of trainers and the 'IPM Farmer Field School' approach, which was able to reach hundreds of thousands of farmers.

Phase 3 of the Programme began in 1993 and continued until 2000, and focused on the development of national IPM programmes in countries such as Indonesia, India, Vietnam and Cambodia, based on season-long training and farmer field schools. This phase also witnessed the evolution of 'farmer led' activities including: farmer to farmer training; farmer-led research; action research; health impacts research and other programmes (FAO 2003b). According to Vorley and Keeney (1998), season-long training in Indonesia dramatically improved the success of IPM. In addition, Indonesia established IPM as its national policy. The implementation of IPM led to the removal of 57 of Indonesia's most hazardous pesticides and removal of all pesticide subsidies (Hough 1998). Removal of direct and indirect pesticide subsidies can reduce the overuse and dependency on pesticides and enhance adoption of IPM or other sustainable production systems.

By 1995, much of the IPM work undertaken by intergovernmental organisations was brought together under the umbrella of the Global IPM Facility, which is based in Rome, and hosted by FAO (see Fig. 7.2). The Global IPM Facility evolved in response to two major undertakings. First, the establishment of the Facility was the logical next step after the highly successful inter-country programme on IPM rice in Asia. Organised by FAO, this facility was a showcase for FAO's successful IPM approach. Subsequently, this facility stimulated substantial interest from agricultural policymakers, and initiated numerous requests for assistance in establishing farmer-centred IPM pilot activities in a number of countries, particularly in Africa. Consequently, FAO, UNEP, UNDP and World Bank assembled a task force to outline the prospective Facility's remit and secure funding.

Second, the new Facility responded to the Earth Summit's recommendation that IPM practices be incorporated into agricultural policies and programmes (FAO 2003c). The establishment of IPM programmes would enable governments meet two key objectives of Chapter 15 of Agenda 21 on sustainable agriculture. (1) To improve and implement programmes to put IPM practices in the hands of farmers, through farmer networks, extension services and research institutions. (2) Establish operational and interactive networks among farmers, researchers and extension services to promote and develop IPM (FAO 2003c).

Ultimately, the Global IPM Facility's goal was to reduce excessive and costly pesticide use, which posed a threat to human health and the environment. It hoped to achieve its goal by pursuing four key objectives: (1) Creating awareness and an enabling environment for IPM. (2) Helping in the development of pilot field programmes. (3) Accessing other programmes and information. (4) Establishing technical and policy IPM linkages. The FAO supports agricultural production that reduces reliance on pesticides as far as possible through the implementation of IPM (FAO 2003c). This is promoted through FAO's own programmes as well as through the Global IPM Facility in which FAO is a partner with the World Bank, UNEP and UNDP, with financial support from the governments and people of the Netherlands, Norway, Switzerland, the UK and the USA.

The Global IPM Facility has undertaken activities and initiated partnerships in Asia, Africa, Latin America, Middle East, Central Asia and Eastern Europe (FAO 2003c; Hamburger 2001). Through its Community IPM Programme, FAO initiated farmer field schools in 12 countries in South and East Asia, which has succeeded in training over 2 million farmers since 1990 (Hamburger 2001). Indeed, because of the approaches' success, for example, IPM plots in Ghana increased net returns by 32%, many countries broadened IPM to Integrated Production and Pest Management programmes (Hamburger 2001). In addition, the World Bank now funds an expert from the Global IPM Facility to serve as Pesticide and Pest Management Specialist at World Bank headquarters, providing guidance to World Bank staff on how to comply with the Bank's pesticide management policy (Hamburger 2001). The Global IPM Facility draws upon local, national and international expertise, knowledge and resources to facilitate programme identification, design and implementation. This approach provides rapid and effective responses to the needs of farming communities and national IPM programmes and leads to efficient use of government or donor resources (FAO 2003c).

According to FAO (2004), 'IPM programmes have the potential to generate substantial benefits for farmers, rural communities, consumers and national economies'. Investment in farmer-driven programmes that assist farmers in finding solutions to their pest problems has proven more successful than past programmes that merely 'focused on scientific research' (FAO 2003c). While the pesticide industry has been active in a variety of organisations designed to develop more effective problem-solving strategies, its role has not been one of unequivocally advancing the process.

7.2.5 Emergence of an environmental agenda in agricultural policymaking

According to Winter (1996), there have been five phases of environmental concern. These phases were: (1) the natural history and humanitarian phase (1830–1890); (2) the preservationist phase (1870–1940); (3) the scientific phase (1910–70); (4) the popular phase/political phase (1960–83) and (5) the regulatory phase (1980–present).

Throughout each phase, environmental concerns have been catapulted onto the political stage and subsequently incorporated into the policy process, whether or not the concerns were appropriately redressed. However, it was not until the regulatory phase that an overt environmental agenda was institutionalised within the agricultural policymaking processes of developed nations and intergovernmental organisations (Buttel 1992; Amin and Thrift 1994; Winter 1996; Potter 1998). At this time, agri-environmentalists were able to cite an expanding body of authoritative research highlighting the impact of intensive agricultural practices on soil erosion, and air, soil and water pollution, declining biodiversity and degraded landscapes (Potter 1998). In Germany, Waibel and Fleischer (1998) calculated the cost of pesticide contamination in groundwaters and rivers to be at least 130 million DM per year.

7.2.6 Promotion of sustainable agriculture

Growing concerns over the sustainability of agriculture, particularly the environmental and human health dimensions, catalysed interest in the promotion of more sustainable forms of food production (see Fig. 7.1). However, definitions of sustainable agriculture adopted by national and international institutions differ considerably, and, subsequently, so do the policies and prescriptions with which they aim to achieve it. However, sustainable agriculture is generally defined by contrasting it to conventional agriculture, with discussion often been in black and white terms. For instance, Vorley and Keeney (1998) differentiate sustainable agriculture from conventional agriculture by suggesting that the former is information intensive and the latter is input intensive (Vorley and Keeney 1998).

Conventional agriculture is all that is perceived to be unsustainable in the modern-agrifood system (reductionism, technological intensity, domination of nature, exploitation, dependence, competition, etc.). Sustainable agriculture has thus gained an unfortunate reputation of pointing the finger at the faults of mainstream farming and its suppliers (see Fig. 7.2). Understandably, the pesticide industry reacted defensively, and farmers have often aligned themselves with the industry, who are familiar and friendly to them and who produce something which they can use (Vorley and Keeney 1998). Farmers have, after all, received scant advice on how to reduce their dependence on pesticides, aside from the adoption of GMOs (Vorley and Keeney 1998). According to Vorley and Keeney (1998), industry definitions of sustainable agriculture have become reactive endorsements of the status quo; conventional agriculture is sustainable agriculture (Vorley and Keeney 1998). For example, DowElanco stated that 'sustainable farmers optimise inputs (financial resources) to maximise profits' and BASF defined sustainable agriculture as 'that level of productivity that allows agricultural enterprises to be economically competitive' (Vorley and Keeney 1998). As Turner et al. (1994) suggest, perceptions of sustainability are often based on fundamentally different moral theories, from biocentrism to anthropocentric productivism. Such a huge vista allows the industry to get away with viewing sustainable development through the prism of its own worldview (Buttel 1992).

Within Western Europe, there has been a strong commitment to developing agricultural policy that is more sustainable and environmentally friendly (see Fig. 7.2) (Cobb et al. 1999). With specific reference to pesticides, in 1991, the EU passed the Plant Protection Products Directive 91/414, which created a harmonised union-wide framework for pesticide regulation. Two years later in 1993, the EC compiled a list of 854 AIs used in the EU (Rosenthal 2002). By July 2000, the EC announced that 320 pesticide AIs would be de-registered in the EU starting in 2003, and that it planned to review a further 170 AIs in the near future. According to PANNA (2003), this could result in the removal of over 60% of the AIs used in the EU in 1993 by 2008 (PANNA 2003). In addition, sustainable agriculture is a central pillar to the Fifth, Sixth and Seventh Environmental Action Plans of the EC (Cobb et al. 1999).

However, the promotion of sustainable agriculture is often contested as nothing more than a smoke screen for the continued financial support of rural areas, and particularly rural voters in Western Europe. While the EC has made significant progress, many criticise European Environmental Action Plans for not attaining their goals. For example, more incentives should have been forthcoming for organic agriculture and conservation grade farming practices (Cobb et al. 1999). The promotion of less intensive agriculture has also been criticised as a path of least resistance when over-production of key agricultural commodities and their exportation have consistently caused the EC considerable embarrassment during WTO negotiations.

Indeed, not only has the promotion of sustainable agriculture allowed the continued financial support of rural communities and food systems in Western Europe, but the principles underlying post-productivist agriculture have allowed the EC to restrict agricultural and food imports on the basis of environmental, food safety and labour and animal welfare (Cobb et al. 1999). Interestingly, the CAP Review Group suggested that reforms should be designed to increase farm competitiveness, while sustainability objectives came second (Cobb et al. 1999).

7.3 Key characteristics of pesticide development processes and manufacturers

7.3.1 Creating space for pesticides in sustainable agricultural systems

It is the vagueness of the sustainable development concept, and more specifically that of sustainable agricultural systems, which allows for considerable diversity in its interpretation (see Fig. 7.2). On the one hand, an extreme ecocentric position in the debate insists that technological development has been the cause of many of the world's environmental problems. This position insists that the move towards sustainable agriculture should depend less on modern technologies, particularly those with a highly negative environmental impact, actual or perceived. This position suggests that pesticides and biotechnologies are incompatible with the concept of sustainable agriculture (Hansen 1996). In addition, there are many who believe that MNCs and TNCs are themselves incompatible with sustainable development (Viederman 1997). They contend that attainment of sustainable development depends on 'commitment to community and place future generations, democracy and participation, equity, and the alleviation of poverty' (Viederman 1997). This vision is juxtaposed against corporate ethics that promote capital mobility that embeds and dis-embeds economic activity based on absolute advantage, destabilising communities in the process (Viederman 1997). Conversely, whilst acknowledging the negative environmental impacts of certain technologies, an extreme technocentric position insists that innovative new technologies are essential if the global agro-food system is to both feed a growing world population and conserve valuable natural resources (Harvard University 2000). The diversity of both conceptual and operational interpretations of sustainable agriculture and sustainable food systems bestows agro-food system actors with substantial room for manoeuvre. In turn, orthodoxy, whether ecocentric or technocentric, breeds defensive, linear thinking, and intellectual inbreeding, which in turn causes actors to push harder on more or less familiar solutions (Vorley and Keeney 1998). This situation is aggravated by the institutionalisation of worldviews of pesticide corporations and environmental organisations. The selective funding of research, acquisition of information and employment of staff all tends to self-reinforce their dominant paradigms (Vorley and Keeney 1998).

The concept of sustainable development and its interpretation and application by international and national government apparatus is of considerable importance to the pesticide industry. It is, therefore, within these institutions that the pesticide/life science industry has attempted to exert its influence. Industry has fought, and continues to fight, in order to carve a role for both pesticides and biotechnologies in sustainable agricultural systems. In the case of pesticides, industry sought to find a role for both old and new pesticides.

7.3.1.1 Corporate environmentalism

In the late 1980s and early 1990s, especially in the run-up to the Earth Summit, many CEOs of large MNCs became much more proactive and outspoken visionaries on the subject of sustainable development (see Fig. 7.2). Indeed, by 1992, Ciba-Geigy, DuPont and Dow leaders had become key members of the Business Council for Sustainable Development and were proposing fundamental change in the way companies did business (Vorley and Keeney 1998). Many acclaimed that progress towards sustainable development was important as it could generate both new opportunities and competitive advantage (Vorley and Keeney 1998). Indeed, as Vorley and Keeney (1998) suggest, several pesticide industry CEOs' endorsements of sustainable development verged on the evangelical. Virtually every pesticide corporation developed sustainability rhetoric. For example, Ciba-Geigy suggested that corporate sustainability consisted of three equally ranked components, economic success, social responsibility and environmental responsibility (Vorley and Keeney 1998). In the case of Novartis, emphasis was placed on flexibility and getting the most appropriate combinations of possibilities to deliver sustainable development and at the same time maximising returns to the company (Chataway 2001). Novartis tried to achieve low-impact, high-output agriculture. The strategy was particularly focused on the adjustment of the screening process to detect novel modes of action, which could be characteristic of new and more environmentally sustainable products (Chataway 2001). A need was also recognised for new marketing strategies, as farmers would need to be convinced of the efficacy and benefit for the environment and human health, particularly when faced with stiff competition from cheap generic compounds (Chataway 2001). Environment, biodiversity and sustainability discourses were all seen as central to the development of the pesticide industry (Chataway 2001).

According to Bloom (1995), these visions responded to a growing divergence between the perceived goals of industry and employees' personal values, particularly to social concerns about the worsening state of the environment and its exploitation by industry. In addition, Ciba-Geigy established the Ciba-Geigy Foundation for Cooperation with developing countries, which ultimately evolved into the Novartis Foundation for Sustainable Development (Groundup 2003m). American pesticide and life science corporations, such as Dow and DuPont, have also developed noble sustainability rhetoric.

Indeed, according to Dow (2000c), Dow was named as the leading sustainability-driven company in the chemical industry by the Dow Jones Sustainability Group Index, allegedly placing Dow among the top performers in the area of sustainable development (Dow 2003c). DuPont preached the rhetoric of sustainability in order to improve the image of its activities, particularly those associated with the pesticide and wider chemical industries. It focused on public relations campaigns as mechanisms by which to establish the company as an environmentally responsible manufacturer. DuPont even adopted the language of sustainability, such as the company's pledge to build 'a growing partnership with nature as it moves deeper into world food systems' (Groundup 2003b). As part of its environmental strategy, DuPont built working links with the non-profit environmental sector. It is, for example, one of many corporate sponsors of the WRI (Groundup 2003b). DuPont is also a member of the Wildlife Habitat Council, a programme working on the development of sustainable habitats. This programme was founded in 1988 and worked in 600 plant sites worldwide, 36 outside the USA. In Texas, DuPont has twice won a Wildlife Habitat Council, Habitat of the Year award. The site also operates an environmental education centre (Groundup 2003b). In response to growing human health and environmental concerns, in 1998, DuPont funded research into hormone disruption and eight other topics including air pollution and cancer (Groundup 2003i). Indeed, CropLife promotes the plant science industry as one that 'invents, develops, manufactures and sells products and services designed to improve the global production of food, feed, fibre and other useful products in a sustainable way' (CropLife International 2002). Population growth, urbanisation and rising expectations still underpin the pesticide industry's justification of pesticide use in the twenty-first century (Vorley and Keeney 1998).

In addition, the pesticide and life science industry also faced internal pressures for corporate responsibility (PANUPS 2003). According to PANUPS (2003), during annual meetings, 'shareholders strongly signalled the three largest agrochemical companies to acknowledge the environmental and health risks of their products'. Socially responsible investment firms have led most of the pro-environmental and pro-consumer activity. Indeed, shareholders insisted that both Dow and Monsanto disclose detailed information about the handling of dangerous pesticides and by-products.

Indeed, according to PANUPS (2003), Bayer shareholders reprimanded the board for inappropriate and insufficient response to the tragic poisoning of 42 school children in Peru, of which 24 died. This tragedy occurred when Bayer's pesticide Folidol (methyl parathion) was allegedly mistaken for milk powder. It was alleged that the bags contained a white powder with no discernable odour, and were labelled in Spanish and were without appropriate pictograms. Most of the local population were illiterate (PANUPS 2003).

In April 2003, shareholders at Monsanto's annual meeting expressed concern about the company's handling of unregistered, carcinogenic and obsolete pesticide stocks. Harrington Investments Inc. called on Monsanto to disclose its policies and procedures for exporting probable or likely carcinogens and pesticides not registered in the USA to developing countries. The training and educational information that Monsanto provides to farmers and farm workers was also requested (PANUPS 2003). Bad publicity, particularly in developed countries, relating to human health tragedies and environmental degradation has prompted pesticide industry PR initiatives to limit adverse effects, particularly in the guise of tough pesticide legislation, which increases costs and reduces profits. Sustainability initiatives have included the industry's 'Life-cycle approach to product stewardship', focused on container collection and disposal in developed countries: Responsible Care®; and Safe Use (disposal) in developing countries (CropLife International 2002).

Some CEOs also began to question the wisdom of industrial agriculture and the embarrassing surpluses it created. They called for a social consensus on the application of new technologies, and for application of the Polluter Pays Principle (PPP). They called for environmental protection standards so high that enlightened firms with a talent for innovation and a willingness to accept risks would be motivated to surpass them if only for the competitive advantages they stood to gain in the process (Krauer 1992). The PPP is one of a small number of environmental principles that has received wide international acceptance (Baldock and Bennett 1991). Although industry has been opposed to every piece of polluter pays legislation, some industry leaders are attracted to the concept as a cornerstone of free market environmentalism, and a more efficient alternative to command and control regulation. Ciba expressed its support for the PPP in 1989 (Vorley and Keeney 1998), and in 1993, Dow Chemical suggested that full cost accounting may be the most important step towards sustainable development (Vorley and Keeney 1998). However, because of the likely impacts on blockbuster products, the pesticide industry made little progress in the search for, or development of, mechanisms that would consider the full human health and environmental costs of their products (Vorley and Keeney 1998). Efforts have been largely directed at discrediting the work of Pimentel from the USA and Pingali from the Philippines rather than generating their own estimates (Vorley and Keeney 1998). Indeed, according to Vorley and Keeney (1998), in 1995, a discussion with corporate leaders culminated in the conclusion that 'full-cost accounting is already in place because national pesticide regulations take products with unfavourable cost/benefit profiles off the market'.

However, while the sustainability rhetoric continues to sound admirable, and high-profile PR exercises help to promote at least a hint of corporate responsibility, many interpret the promotion of corporate responsibility and corporate environmentalism as a half convincing PR veneer, which facilitates business as usual (Groundup 2003i) and avoids expensive and damaging lawsuits (PANUPS 2003). As Edwards (1995) points out 'what looks like progress is very often simply the immune system of the status quo defending itself'.

7.3.1.2 Integrated Pest Management, Precision agriculture and Sustainability

7.3.1.2.1 Integrated Pest Management

Whether or not pesticide and life science corporations are committed to corporate responsibility and corporate environmentalism, for much of the time, sustainability has remained predominantly at the stage of rhetoric. Initiatives undertaken have, on the whole, attempted to paper over the cracks and have shied away from questioning the fundamental role of pesticides in the contemporary period. However, IPM is one important exception to this trend. While the pesticide industry has been able to secure the continued production and prophylactic use of often highly hazardous pesticides in much of the developing world, it came to recognise that consumers and politicians in many developed countries were no longer willing to accept the perceived and actual dangers of pesticides in the food system (see Fig. 7.2). However, whilst the merits of organic agriculture had been touted for decades, governments in developed and developing countries still doubt its capacity to maintain high levels of agricultural production. In addition, strong agribusiness actors have generally resisted swift and radical changes to their industry. IPM was the obvious solution to the problem (Dent 2000). A growing number of intergovernmental institutions, national governments and NGOs support IPM. The FAO supports IPM as it reduces reliance on pesticides. As mentioned earlier, the FOA promotes IPM through its own programmes as well as through the Global IPM Facility. Indeed, the Global IPM Facility launched a project called Integrated Pest and Plant Management-2015. The central focus of the project was the promotion of IPM on a global scale by the year 2015. Integral to the strategy is a multi-stage policy reform process that would begin with a phase-out of Category 1 pesticides by a specific date. Once IPM measures have been adopted as substitutes (consistent with step ii in Level 1 above), the project proposes to phase out Category 2 pesticides by a subsequent target date, and then proceed down to the least toxic pesticides still deemed necessary.

Most aid agencies have now adopted strategies that avoid the procurement of bulk quantities of pesticides and instead support IPM. The USDA also supports IPM. Indeed, statutory support is required by Section 303 of the FQPA (1996). Indeed, the US National IPM Initiative had the goal of implementing IPM on 75% of the nation's cropland (Metcalfe et al. 2002). However, while most OECD countries have promoted IPM, they have focused on low-level efficiency approaches (Vorley and Keeney 1998).

IPM has been supported by most pesticide corporations, and has been showcased by the pesticide industry as an approach that satisfies both post-productivist and sustainability goals, whilst maintaining high use levels of pesticides (Pan-UK 2003a). Bayer has been particularly responsive to reinventing pesticides as key inputs into post-productivist and sustainable crop production systems. Indeed, the company has laid down a special three-point programme. First, the application rate of crop protection products should be reduced. Second, new AIs must be safer than their predecessors. Lastly, products must be suitable for IPM (Bayer 2003g). Bayer has been proactive in promoting the need for a wide range of technologies for ICM and the sustainable intensification of agriculture. Newer product groups have managed to combine the advantages of lower application rates with improved efficacy. Examples are pyrethroids (insecticides), azoles (fungicides) and sulfonylureas (herbicides). Compared with older products, they require only one-tenth of the application rate - and still show the same, if not an improved effect (Bayer 2003g). In attempts to reduce the human health and environmental impacts of its products, Bayer invested significant resources into the discovery or acquisition of new classes of pesticides with new modes of action (Heissler 1995; Søgaard 2001). It is also pursuing the development of new seed treatments, often in conjunction with biotechnology, that are suggested to have the benefit of lower farmer costs, zero pesticide residues and reduced environmental impact due to the tiny amounts of AI and the placement of the pesticides (Søgaard 2001). Wrapped in its enthusiastic marketing zeal, Bayer has gone as far as suggesting that farmers in developed countries have long stopped operating according to the motto 'the more the better'. Today, the principles of integrated crop production are well established in farming practice. Crop protection products are used selectively and according to the 'damage threshold' principle. In Germany alone, the application rate per hectare averaged over all crop protection products has decreased by practically 50% in the last 5 years (Bayer 2003g). Bayer and GTZ were involved in a project in Guatemala that aimed to teach ICM to men, women and children. According to CropLife, the activity worked in peanut production, where yields and earnings tripled. Biodiversity was enhanced both through the application of ICM and IPM and the reduced need for shifting cultivation into virgin forest areas (CropLife International 2002).

Syngenta also worked hard to develop a new generation of less objectionable pesticides. For instance, Syngenta released abamectin, an acaricide/insecticide, with a unique mode of action. This unique mode of action confers abamectin with the ability to control pest insects and spiders resistant to other AIs and therefore integrates well with resistance and IPM programmes (Syngenta 1997). Syngenta stated that 'abamectin supplements our insecticide portfolio in an almost ideal way, adding an acaricide to our product range. It is a naturally derived product from fermentation. This acquisition illustrates our commitment to sustainable agriculture, and will be an important element in our ongoing IPM programmes and anti-resistance strategies' (Syngenta 1997). According to Syngenta (1997), combining biotechnologies with chemicals provides growers with 'more options for developing IPM strategies and dealing with potential resistance challenges'. Amistar has been one of the most important new fungicide releases in Northern Europe. Amistar has a superior environmental and toxicological profile compared to competing products, and is applied at lower rates and has less effect on the water table and non-target species (Tait 2001). Conversely, it is not always the newest and safest pesticides that are repackaged as IPM and sustainable agriculture compatible. Indeed, older products are often re-launched in this manner. For example, Hoechst's endosulfan experienced growing sales in developing countries (Pan-UK 2003a). Organophosphates were key components of many IPM programmes in developing countries (Metcalfe et al. 2002).

Whereas adoption of IPM has been encouraging in many parts of the developed and developing world, there are many obstacles to its successful adoption. Major deterrents to successful implementation of IPM include the following: (1) Lack of trained extension personnel. (2) Lack of support for extension services. (3) Lack of training for farmers, extension agents and others. (4) It is also slow and expensive (Rola and Pingali 1993). (5) Often, a common view about objectives and operational aspects of IPM is absent. This prevented effective implementation of IPM in Indonesia (Teng 1990). (6) There is often only a vague demarcation of responsibilities between research, extension and technical support. (7) Often, there is inter-institutional rivalry. (8) Scientists may not know what knowledge extension workers or farmers require.

Ultimately, the productivist model of chemical dependent agriculture is institutionalised and much simpler than IPM. Outcomes of IPM are more uncertain compared to a packaged approach. IPM has a high knowledge requirement. As Rola and Pingali (1993) suggest, while the development of pesticide resistance and increased health risks may have initiated the search for sustainable pest management technologies in developing countries, in order for IPM technologies to be accepted and adopted, they must first be compatible with the management capabilities of farmers. For example, Rola and Pingali (1993) found that 31% of IPM farmer cooperators maintained the belief that all insects were enemies, and that 80% sprayed when they saw insects as they believed that their crops would be damaged. IPM is also labour intensive (Rola and Pingali 1993). In developing countries, labour-intensive agriculture is less attractive in high-waged areas (Rola and Pingali 1993). In addition, the pesticide industry has argued that, where the economic sustainability of farming systems is threatened, many cheap, off-patent, pesticides are ideally suited for low cost production of staple crops. Compared to the cost of newer, less hazardous, pesticides, the price of generic pesticides is low enough to ensure their cost-effectiveness in many situations in developing countries. According to Metcalfe et al. (2002), organophosphate pesticides remain important components of IPM programmes as they: provide the rapid response necessary for a quick knockdown of pest populations; promote beneficial insect populations relative to carbamate and pyrethroid alternatives; and help to maintain low pest resistance (Metcalfe et al. 2002).

Social pressures may also limit the adoption of IPM. Landlords often insist that their tenants maintain their land free of weeds, potentially requiring tenants to use even greater quantities of herbicides (Vorley and Keeney 1998). In Filipino and Thai rice production, farmers will use pesticides even when they are not warranted because it is perceived as being up to date, or because their neighbours were seen spraying (Escalada and Heong 1993). In China and Vietnam, farmers who do not spray are often classified as uncooperative and may even be subjected to certain sanctions by villagers (Escalada and Heong 1993).

While cheap generic pesticides can be important components of successful IPM approaches, they can also undermine IPM because farmers can apply cheap generic pesticides prophylactically as insurance sprays. This is particularly important where pesticide subsidies have recently been removed (Rosenthal 2002; Vorley and Keeney 1998; Tait 2001). In order for the approach to be most effective, IPM should be practised community wide. Indeed, Thompson goes as far as suggesting that 'farmers should practice IPM out of a moral duty to their neighbours. All it takes is one or two farmers to over-use an insecticide and accelerate resistance to the insecticide that will affect all his neighbours' (Thompson 1971).

Interestingly, several major pesticide corporations are becoming increasingly interested in the provision of IPM, ICM and IFM advice. However, this is likely to face problems of economic pay-off, as farmers are generally not used to paying for services, and once knowledge has been transferred, farmers can continue using it without having to return to the original source. Other elements that reinforce path dependency in agricultural pest control are uncertainty at the farm level and coordination problems in moving from one system to another (Cowan and Gunby 1996). In many respects, it may be partnerships with food and commodity producers, processors and retailers with an interest in reducing pesticide residues that are more likely to lead to sustainable production (Hamburger 2001).

To date, while IPM, ICM and IFM studies have generated impressive results, such as pesticide use reductions up to 40% and increased profits, pesticide corporations still continue to push the intensive use of their pesticide products (Bull 1982; FAO 2002; FAO 1998). While the pesticide industry has been active in a variety of IPM and sustainable agriculture forums, its principal impact has been to slow, rather than accelerate, change. For example, discussions between GCPF the World Bank and other key food system stakeholders over the development of a collaborative public/private sector approach to the promotion of IPM and sustainable agriculture failed due to public sector concerns about pesticide corporations' conflict of interest between reducing pesticide dependency and meeting sales targets. Whilst the World Bank has signalled support for sustainable development through its role in the GEF, its interpretation of sustainable food systems may more closely match that of the pesticide industry than that posited by environmental NGOs. In addition, the Bank has also stressed the importance of pesticides within the framework of IPM, and, according to PAN, has sought to establish a role for the pesticide industry in the Global IPM Facility. In response, industry continues to argue that IPM is used by some as a cover for a progressive abolition of pesticides (Boardman 1986).

Concerns have also been raised regarding the perceived gap between pesticide industry rhetoric and actual behaviour. Indeed, PANNA criticised the World Bank's 1996 pest management policy because it was weaker than other international commitments and failed to incorporate farmer-led and ecologically oriented approaches to pest management and a real commitment to reducing reliance on chemical pesticides (Hamburger 2001). Whether or not traditional pesticides can be legitimately described as key components of sustainable food systems, their use continues to be promoted by influential national governments and intergovernmental institutions. National and international institutions have talked much about the benefits of IPM but have been frugal when allocating financial resources for IPM programmes. Indeed, the US budget in 1996 for IPM research was only US$192 million (Vorley and Keeney 1998). The World Bank is also guilty of supporting the adoption of IPM on the one hand, and under-funding IPM programmes on the other. For example, between January 1988 and January 1995, the World Bank financed US$250.75 million worth of global pesticide purchases, whilst only investing around US$81 million in IPM, with US$52 million of that total going to only two projects. Six companies (Rhône-Poulenc – France, BASF – Germany, Zeneca – UK, Sumitomo – Japan, FMC Corp. – USA and Helm – Germany) received US$3 million or more in Bank-approved agrochemical sales over the 3-year phase between 1993 and 1995. Another five received US$1–3 million (Bayer – Germany, Roussel Uclaf – France, Cyanamid – USA, Air Lloyd – Germany and Hoechst – Germany). The company at the top of this list, Rhône-Poulenc – France, was the big winner in terms of sales. In 1993–5, it received US$18.6 million, or 33% of the value of all Bank-approved contracts benefiting the G-7 agrochemical industry; BASF received over US$12 million.

7.3.1.2.2 Precision agriculture

PA is a farming management concept based on site-specific crop management, which uses inter- and intra-field variability to target the application of crop inputs, namely fertiliser and pesticides where they are needed. PA is hi-tech and knowledge intensive and aims to optimise returns on inputs while conserving resources

(McBratney et al. 2005). The concept of Precision Agriculture has been around and practised since the 1980s. However, with recent technological advances and greater affordability of unmanned aerial vehicles, Global Positioning Systems (GPSs) and machinery-mounted sensors, PA has become even more popular, especially with large-scale farmers. Indeed, in 2001, John Deere was factory fitting a GPS link in its farm machinery; until then, the fitting of machinery-mounted GPS monitoring systems were a post factory add-on (ETC 2016). As with biotechnology, this PA revolution did not go unnoticed by the life science corporations. In many ways, this was a match made in heaven. The life science companies had the seed, pesticides and technical knowledge related to crop production and the machinery manufacturers had hardware (machinery) and software to deliver inputs down to the exact centimetre (ETC 2016).

Indeed, as early as 2007, Syngenta and John Deere had created a fully integrated Force CS insecticide delivery system. Later, in 2013, John Deere began collaborating with DuPont Pioneer and Dow AgroSciences. In the case of DuPont, the collaboration with John Deere led to the linking of DuPont's Pioneer Field 360, a suite of precision agronomy software with 'John Deere Wireless Data Transfer, JD Link and MyJohn Deere' (ETC 2016). Similarly, the collaboration between Dow AgroSciences and John Deere aimed to assist farmers to link information collected through John Deere's hardware and software with expert advice on the precision use of Dow AgroSciences pesticide products. John Deere also began working with BASF to develop 'sustainable yield enhancement solutions' (ETC 2016). In 2014, Bayer CropScience began collaborating with John Deere to develop digital tools to move from 'precision to decision' (ETC 2016).

However, John Deere was not the only machinery manufacturer collaborator on the block. By 2015, AGCO (including famous tractor names such as Massey Furguson, Deutz-Fahr, Fendt and Valtra) and CNH (including famous tractor names such as Case and New Holand) were also collaborating with life science companies. For example, AGCO began working with DuPont on a global wireless data transfer system, with Monsanto to include Monsanto's Precision Planting system in AGCO's planters, with BASF on 'Farm Management Information Systems', and with Bayer on 'drive the Future Farm in Zambia' project (ETC 2016).

Lastly, whilst collaborations between life science companies and agricultural machinery manufacturers have primarily revolved around PA, other types of collaboration have recently arisen. For example, in 2015, John Deere Insurance Company and BASF began collaborating to offer 'Risk Advantage' Insurance (ETC 2016). Indeed, whilst still a relatively risky proposition, given global climate change, the farm insurance industry is big business and rapidly expanding. In 2016, the global worth of the farm insurance industry was USD$30 billion, four times the value of 2005 (ETC 2016).

7.3.1.3 Conservation agriculture and zero-till

Pesticides have also been heralded as the cornerstone of sustainable agricultural practices such as zero-till and conservation agriculture (see Fig. 7.2). Zero-till refers to planting crops in previously unprepared soil by opening a hole, narrow slot, trench or band of the smallest width and depth needed to obtain proper coverage of the seed; at least 30% of the soil surface remains covered with residues of commercial or cover crops. The modern development of zero-till began after the discovery of paraquat in 1955 by ICI, and its commercial release in 1961 (Ekboir 2002). ICI funded research to find out if cultivation was still necessary and hired agronomists and mechanical engineers to develop zero-till (Ekboir 2002). ICI also assessed the technical and economic potential of the market for paraquat in several countries (Ekboir 2002). In the late 1960s, ICI established a research team in Australia and another in Brazil in 1972 (Ekboir 2002). ICI's involvement in the USA was limited because it had licensed paraquat to Chevron Chemicals and could not develop the market for itself (Ekboir 2002). In the late 1970s, Monsanto released glyphosate. Whilst even better suited to zero-till agriculture than paraquat, glyphosate was too expensive (Ekboir 2002). However, when Monsanto reduced the average price of glyphosate from US$40 in the 1970s to US$10 in the 1990s, zero-till with glyphosate was economically viable and zero-till expanded rapidly (Ekboir 2002). By the year 2000, the global area of zero-till was already around 59 million ha (Ekboir 2002), climbing to 111 million ha by 2009 (CropLife 2014).

Whilst reliant on glyphosate, the expansion of no-till production is widely acknowledged as being more cost-effective and environmentally benign than conventional approaches to crop production (CropLife International 2002; Planet ark 2003a). However, whilst the spread of zero-till has generally been welcomed, concerns continue to be raised regarding the dependence of zero-till on glyphosate. Indeed, according to the Manitoba Co-operator (1999), in Australia, several annual ryegrass populations are now resistant to five different herbicide groups, including glyphosate, and therefore pose a threat conservation tillage across the continent(Manitoba Co-operator 1999).

7.3.1.4 Biopesticides

Biopesticides include a range of naturally occurring substances that control pests. They can be bacterial, viral or fungal, and include behaviour-modifying chemicals, and natural enemies. Biopesticides are not new, for example, bacillus popilliae was used in the late 1930s and early 1940s to control Japanese beetle (Pimentel and Pimentel 1979).

To date, commercialisation of biopesticides has generally been the domain of small pesticide companies in developing countries, which produce microbial pesticides or facilitate the multiplication of beneficial insects for distribution to farmers. Several Asian countries, including India and Central Asian Republics, are world leaders in biopesticide development. Biopesticide technologies include the use of neem products in India, and the use of beneficial insects such as Trichogramma, *Bracon* for insect control on cotton.

However, large pesticide corporations in developed countries have generally been slow to commercialise biopesticides. Some of the principal problems associated with the development and commercialisation of biopesticides are: (1) High production cost. (2) Variable quality. (3) Short shelf lives. (4) High costs associated with ecologically desirable target specificity. (5) Poor reliability in the field. (6) Generally lower levels of customer return due to residual activity of the biopesticide (Vorley and Keeney 1998). In addition, because of technical problems, biopesticides are also extremely difficult to market alongside synthetic organic pesticides (OTA 1995). However, because the registration of biopesticides is much less stringent, speedier and cheaper than synthetic pesticides, they are much easier to market as sustainable inputs to post-productivist or sustainable farming systems, and, recently, pesticide corporations have begun to increase investment into R&D and commercialisation (OTA 1995).

According to OTA (1995), the average development costs for biopesticides is US$5 million compared to US$100 million for a synthetic organic pesticide. In addition, it only takes an average of 3 years before a biopesticide can be marketed compared to 12 years to market for a synthetic organic pesticide. In 1992, the market for biopesticides was worth USD$68 million. By 1997, this had increased to USD$380 (Gaugler 1997) and a staggering USD$3.36 billion by 2016 (MarketandMarkets 2016). According to MarketsandMarkets (2016), global sales of biopesticides are projected to reach USD$8.82 Billion by 2022.

In a similar vein, the pesticide industry argues that many of its latest pesticides are derived from natural sources, with varying degrees of manufacture. Through the use of biopesticides, pesticide corporations can claim to offer the least controversial and potentially one of the most sustainable methods of pest and disease control. Until its collapse, California-based 'AgraQuest Inc.' led the development and manufacture of natural pest management products. It even entered a strategic partnership with specialty chemical company Rohm and Haas in order to facilitate the rapid introduction and marketing of new biopesticides that addressed grower needs for safe and effective pest control. AgraQuest identified, developed and manufactured biopesticides made from naturally occurring microbes that had been screened for their commercial viability as pest management tools.

Through its systematic and proprietary discovery process, AgraQuest built a library of more than 13 000 microbes - all pre-screened for their potential as bioinsecticides, biofungicides and bionematicides for farm and home pest management. The company's debut product, the biofungicide Serenade®, was discovered within the first 1000 samples tested and was brought to market in

3 years. Chemical companies typically screen 20 000–200 000 AIs before finding an effective new product. Serenade was first introduced in South America in 1999 and received its US EPA registration for grape, nut and vegetable crops in 2000. Bayer, which acquired AgraQuest in 2012, now sells serenade-Opti in the USA (Pangae 2012; Bayer 2017). Bayer also acquired Athenix in 2012. Indeed, Bayer is not the only mega pesticide corporation to make strategic investments in biopesticides. In 2011, Monsanto bought Divergence and Beeologics (Pangae 2012). In 2012, Syngenta bought Pasteuria and BASF bought private equity-backed Becker. According to Pangae (2012), the principal driver behind these recent biopesticide company acquisitions was the EU implementation of the Plant Protection Products Regulation. This piece of European legislation required that, aside from inherent effectiveness, crop protection products needed to prove that, other than the target organism, they posed a risk to neither human health nor the environment. At the time of implementation, only 26% of existing pesticides (many of them biopesticides) passed the new stringent tests (Pangae 2012).

7.3.2 Creating space for biotechnologies in sustainable agro-food systems

Since the early 1990s, life science corporations have been attempting to create space for emerging biotechnologies in sustainable agro-food systems (see Fig. 7.2). In many respects, the case of biotechnologies, particularly GMOs, is similar to that of pesticides, in as much as they generated substantial public opposition to their use. However, whereas the use of pesticides received a honeymoon phase throughout which little opposition to their use materialised, organised opposition was already in place before the first GMO was even released. In addition, although the pesticide/life science industry's political influence and economic power had grown, they did not expect the level of public resistance to GMOs and, indeed, the speed and effectiveness of anti-GMO campaigns. Whilst public opinion remains predominantly hostile towards GMOs, the latent potential of new biotechnologies to significantly contribute to food security, human and animal nutrition and health care, and environmental enhancement, has endowed emerging biotechnologies with many allies.

Whilst the private sectors, particularly corporations involved in agro-food systems and pharmaceuticals, are the most ardent supporters of biotechnologies, they have found staunch allies in a growing number of academics and, perhaps more importantly, from a growing number of national and international public institutions.

Not surprisingly, the World Bank has been one of the most ardent supporters of GMOs. Indeed, Ismail Serageldin, Vice President of the World Bank, is on record saying 'biotechnology will be crucial part of expanding agricultural productivity in the 21st Century' and 'if safely deployed, could be a tremendous help in meeting the challenge of feeding an additional three billion human beings, 95% of them in the poor developing countries, on the same amount of land and water currently available' (World Bank 2003).

Following suit, the FAO has signalled its support for biotechnology, stating that 'biotechnology provides powerful tools for the sustainable development of agriculture, fisheries and forestry, as well as the food industry. When appropriately integrated with other technologies for the production of food, agricultural products and services, biotechnology can be of significant assistance in meeting the needs of an expanding and increasingly urbanised population in the next millennium' (FAO 2004). In addition, both the Agenda 21 and CSD highlight the potential importance of biotechnologies. In Chapter 16, Agenda 21 recognises that, 'although biotechnology cannot provide solutions to all the fundamental problems of environment and development, it could nevertheless contribute substantially to sustainable development by improvements in food production and feed supply, health care and environmental protection' (UNEP 2003c). This sentiment is echoed in CSD's position, namely that; 'agricultural biotechnologies have a major potential for facilitating and promoting sustainable agriculture and rural development. They could also generate environmental benefits, especially where renewable genetic inputs can be effectively used to substitute for dependency on externally provided agrochemical inputs' (UN 2000).

Finally, if a little apprehensively, UNEP and WHO acknowledge the latent potential of emergent biotechnologies. Indeed, whilst insisting on the need for biosafety institutional capacity building in developing countries, UNEP has acknowledged that, 'biotechnology is a potentially powerful tool for enhancing the productivity of crops, and hence improving both food security and peoples' livelihoods' (UNEP 2003d). In a similar vein, whilst insisting on the establishment of 'sound safety and risk assessments for foods derived from modern biotechnology', the WHO has acknowledged that 'the application of modern biotechnology to food production presents new opportunities for human health' (WHO 2004). Indeed, in 1992, an international conference of experts convened by the World Bank, the UNDP and FAO concluded that a solution to the problem of securing world food supplies while preserving the environment is virtually inconceivable without recombinant genetics and biotechnology (IFPRI 1996). The UN ECOSOC has also espoused the benefits of biotechnologies, including the potential of apomixis in out-crossing crop species; higher nutrient levels; marker-assisted selection; abiotic stress resistance; pest and disease resistance and so forth (UN 2000).

The CGIAR is also pro-biotechnology. In its Agricultural Biotechnology and the Poor Report (2000), it states that 'molecular biology and other tools of modern biotechnology add elegance and precision to the pursuit of solutions to thwart poverty, malnutrition and food insecurity' (BIO 2004d). 'Scholars of today seek to use new discoveries in molecular biology and genetics to understand and protect the natural world and to improve the productivity of agricultural systems' (BIO 2004d). Biotechnology can be used to develop crop plants with reliably high yields at the same or lower cost (IFPRI 1996). The Asian Development Bank stated that 'agricultural biotechnology is expected to contribute significantly towards poverty reduction and food security in Asia through increased productivity, lower production costs and food prices, and improved nutrition' (BIO 2004d).

China has invested considerable amounts of public money into agricultural biotechnology research since the 1980s in order to ensure food security for its 1.3 billion people (Planet Ark 2003d). China has promoted GM in high-yield potential areas and where disease resistance traits were expected to make a considerable impact. Conversely, it has also promoted GM-free areas in order to secure access to profitable exports markets where many consumers are reluctant to accept GM ingredients. Although, as with countries such as Mexico and Brazil, the planting of GM crops, particularly GM cotton, is allegedly widespread even in non-GM areas as Bt cotton farmers apply only 2–4 insecticide sprays compared to up to 40 sprays for conventional cotton (Planet Ark 2003d). According to Planet Ark (2003a), China has developed the largest plant biotechnology capacity outside of the USA. Publicly funded Chinese researchers are working on more than 60 crops (Planet Ark 2003a). In India and South Africa, GM crops have allegedly increased crop yields, while, at the same time, reducing input costs and environmental degradation (Planet ark 2003c).

In addition to the debate over the compatibility of biotechnology and sustainable agriculture, concerns are growing over the appropriation of the world's biodiversity by the powerful life science corporations. This appropriation not only includes the potentially high levels of financial return on patent-protected genetic modification, but also the patenting of genes from unimproved crop varieties and wild relatives and also genetic material from non-agricultural species. In 1997, the pharmaceutical industry alone was estimated to benefit between US$75 billion/year and 150 billion/year from biodiversity. Indeed, ten of the world's top-selling drugs in that year were from natural sources (UNDP, UNEP, WB, WRI 2000). Because of the potential returns on gene patenting, life science and biotechnology corporations engaged in a gene prospecting and patenting spree throughout the 1990s.

Whilst the relative unconditional support for biotechnologies from these institutions creates room for cynicism, particularly as many still support the use of controversial pesticides, there seems to be a genuine belief in the underlying potential of these emergent technologies, and a willingness and openness, for dialogue to address public anxieties over GMOs. Indeed, whilst remaining committed to the release of biotechnologies, many governments have suspended the release of GMOs within their national boundaries and imposed import restrictions on GMF until more is known about the potential adverse effects of GMOs on human health and the environment. Whilst most governments remain potentially pro-GMO, decisions to suspend the release of GMOs within their national boundaries have caused significant political repercussions from life science corporations and their supportive governments. Intergovernmental institutions have also publicly offered support for the development of plant and animal biotechnologies.

The divestment of chemicals and focus on life science allows business to claim the moral high ground of sustainability (Chemical Week 1997). Biotechnology industry publicists promote the idea that the new agro-biotechnologies are being designed and deployed in the service of 'humanity' (McAfee 2003). The US$52-million-dollar media campaign sponsored by the US BIO suggests that genetically engineered crop varieties are more productive and more nutritious than conventional crops and therefore hold the key to feeding the world's burgeoning population – that is, if their proliferation is not impeded by excessive regulation or scare-mongering by technology critics (McAfee 2003). In addition, many food systems analysts are convinced that output trait biotechnologies have the capacity to change consumer attitudes in favour of biotechnologies (Søgaard 2001).

Pesticide industry associations such as CropLife International and BIO constantly push the message that they are in favour of developing crop protection and agricultural biotechnologies that help to increase productivity, whilst minimising impacts on human health and the environment (CropLife International 2002). On the one hand, they stress that the use of GM crops can slow the rate of natural habitat destruction by reducing the need to bring additional land into agricultural production by increasing crop yields and decreasing crop losses and that productivity-enhancing GM crops and other technologies can help to preserve the natural biodiversity. On the other hand, they promulgate that biotechnology, when used as part of ICM and IPM, can provide many benefits compatible with sustainable agriculture. These include the following: (1) Hardier disease and pest-resistant crops that will allow greater conservation of resources by requiring less fuel, labour and water, and the enhancement of crop diversity. (2) Insect-resistant crops reduce the use of broad-spectrum insecticides that would otherwise have direct and indirect effects on natural communities dwelling near agricultural fields. These new technologies also show reduced impact on non-target insects compared to indiscriminate impact of the traditional broad-spectrum insecticides. (3) They can also reduce the cost of agricultural inputs and help ensure the quantity and quality of crops by reducing the impact of P&D. Indeed, crops developed using biotechnology are suggested to make agriculture more sustainable by reducing synthetic chemical inputs and promoting more environmentally preferable agricultural practices, such as no- or low-till farming (BIO 2004a). For example, HT plants allowed farmers to use new generation herbicides with low toxicity and reduced persistence in the environment (BIO 2004a). Pest-resistant Bt crops reduce pesticide use. According to BIO (2004a), cotton and maize Bt varieties reduced insecticide use by 4.5 million pounds in 2001(BIO 2004a). In addition, CropLife International (2004) suggests that biotechnology also enables farmers to escape the poverty treadmill.

Recently, the biotechnology sector has been keen to highlight the development of future second- and third-generation output trait biotechnologies and their direct benefits to consumers (McAfee 2003). CropLife International (2004) cites the development of canola with high beta-carotene and cassava and fruits with more vitamins, and the production of bananas with hepatitis B vaccine. Another key output trait flagged by the biotechnology sector is 'golden' rice containing provitamin A, which will allegedly reduce blindness among the undernourished poor (McAfee 2003).

Individual pesticides corporations have also been actively engaged in promoting their biotechnology activities as being compatible with and advantageous to sustainable agriculture. In 1986, Hoechst initiated research in what the company described as 'green biotechnology' (Heissler 1995). Latterly, AgrEvo (Hoechst and Schering) stated that the development and application of biotechnology 'will result in greater product diversity and new farming methods that conserve resources and help ease the burden on the environment worldwide' (Hoechst 2004). AgrEvo expected that plant biotechnology and plant protection products were expected to complement each other. AgrEvo announced that it intended to broaden its activities to include ICM, of which pesticide products would continue to play an important role (Hoechst 2004). In 2001, Aventis CropScience (AgrEvo and Rhône-Poulenc), alongside BASF, Dow Chemical, DuPont, Novartis and Zeneca launched massive biotechnology advertising campaigns in the USA (Le Monde diplomatique 2001).

While Bayer recognises that concerns over GMOs in Western Europe and Japan are real, it expects current anxieties will subside over time due to breakthroughs in pharmaceutical, which will most likely eliminate widespread resistance to biotechnology and build acceptance of agricultural biotechnologies (Søgaard 2001). In 2001, Bayer's biotechnology business generated US$3.1 billion; this was expected to exceed US$5 billion by 2005 as markets in China, India and Brazil expanded (Bayer 2003g). Based primarily on its acquisition of Aventis CropScience, Bayer's BioScience division has generated a very strong patent portfolio in biotechnology traits as well as a strong position in seeds in certain crop sectors. Bayer also attempted to promote environmental commitment based on an 'inner conviction' rather than legislation. Indeed, Bayer promotes a three-point programme in which it asserts that it is constantly in search of improved AIs and more environmentally friendly methods of application. In addition, Bayer has pledged itself to the principles of IPM, using as little synthetic pesticide as possible (Bayer 2003g). Bayer regards genetic engineering as an essential component of its attempt to generate new products and solve technological and science-based problems, particularly in relation to medicine and agriculture. Ultimately, Bayer states that its 'overriding concern is to ensure that safety and protection of the environment are given the same priority as quality and commercial viability' (Bayer 2003g). Bayer suggests that its genetic engineering is governed by ethical values, particularly its respect for life and human dignity, and is open to productive dialogue with the general public (Bayer 2003g).

Syngenta is also committed to marketing biotechnologies as not only suitable but essential for sustainable agriculture. Indeed, the Syngenta Foundation for Sustainable Agriculture insists that 'genetic engineering and biotechnology help in the fight for food security'. They suggest that the maintenance and stability of high yields and lower costs of production costs could be achieved through enhanced pest and disease resistance and tolerance, and enhanced capacity to cope with abiotic stresses climatic variability, drought, poor soil quality and low soil nitrogen (Syngenta 2004). Many companies, including Syngenta, have attempted to improve their image and that of biotechnologies, by highlighting flagship activities such as announcing the sale of 'golden rice' in the developed world, whilst making the technology freely available to the world's poorest countries (New York Times 2000). Syngenta waived 70 patents in order to make the technology required for golden rice free of charge (Le Monde diplomatique 2001). Syngenta hopes that golden rice and others will be more acceptable to the public, particularly if they serve humanitarian purposes (New York Times 2000). Scientists said they believed more than 124 million people worldwide suffer from a vitamin A deficiency, and that up to 2 million children die each year in early childhood and another 500 000 become afflicted with an irreversible form of blindness each year because they rely on a rice diet that is deficient in vitamin A (New York Times 2000).

Not surprisingly, after its biotechnology gamble, Monsanto was virtually dependent on public acceptance of biotechnologies for their very existence. After the unexpected resistance to GMOs in Western Europe, Monsanto quickly learnt that for its biotechnologies to be widely accepted, they must bring direct and obvious benefits to either the consumer or the environment or both. In January 2001, Monsanto published its new pledge containing five commitments to its customers. These were dialogue, transparency, respect, sharing and benefits (Le Monde diplomatique 2001). Recently, Monsanto began selling its GMOs as 'information-intensive' techno-science that will displace the polluting practices of productivist agriculture. Such arguments draw upon vague notions of a post-industrial 'new economy' which, almost miraculously, creates values from information with relatively little need for mundane material inputs or labour (McAfee 2003). In order to further enhance the role of biotechnology in sustainable food systems, Monsanto has been involved in technology and knowledge transfers, including the development of feathery mottle virus resistance in sweet potatoes in Kenya, and the control of virus resistance in potatoes in Mexico (Groundup 2003m). Monsanto has also been actively involved in the organisation of micro-credit (Groundup 2003m).

However, many critics of Monsanto are not convinced of its 'substantive change of direction' (Washington Post 1999). Indeed, John Elkington, from the London-based SustainAbility Company, suggested that, while Monsanto was keen to involve a wide range of stakeholders in their internal discussions of GMO issues, the company was almost genetically programmed to pioneer GMOs whatever the consensus view happened to be. Internally, problems associated with the acceptance of GMOs were clearly based on uninformed attitudes, which could be rectified by information and education (Washington Post 1999). Education has become a principal mechanism by which Monsanto has attempted to change hearts, and more importantly, minds. Co-founded by Monsanto, the Council for Biotechnology Information bought TV slots for pro-GM advertisements (Le Monde diplomatique 2001). Partly funded by Monsanto, the Biotechnology Challenge 2000 Programme immersed 33% of Ireland's secondary school pupils in the subject of biotechnology and its role in food production. Indeed, David Byrne, the European Commissioner responsible for protecting consumers' health justified the Programme, stating that 'there's no doubt in my mind that there is a link between consumers' reluctance to accept biotechnology and the serious lack of information on the subject'. In 2001, Patrick O'Reilly, director of Monsanto Ireland, hoped for wider participation in 2002, stating that 'these students are tomorrow's discerning consumers and decision-makers' (Le Monde diplomatique 2001). In addition, Monsanto's Vista Project, based on consumers' value systems, aimed to create conformity between the image of GMO products and consumer expectations, and to create conformity of thinking through intensive advertising and information. As a result, Monsanto has developed advertising with high public impact, including the pro-biotech slogan 'less pesticides on your plates' (Le Monde diplomatique 2001).

Many have also criticised technology transfer and development initiatives of Monsanto and other biotechnology companies. Whilst the transfer of certain technologies to developing African countries can be interpreted as a sign of Monsanto's commitment to sustainability, they can also be interpreted as a way of creating new markets for their proprietary products. For example, the granting of patents waivers for the transfer of transgenic sweet potato to South Africa could aid the future introduction of YieldGuard or Roundup Ready Maize (Le Monde diplomatique 2001). Indeed, Monsanto's micro-credit programme in Thailand generated accusations of using alliances with established and well-respected development groups in order to provide money with which to purchase Monsanto's herbicides (Genet 1999).

BASF was relatively a late to entrant into the biotechnology race, joining towards the end of the 1990s. In 1999, one of BASF's first attempts to enter the market materialised in the form of BASF Plant Science, a joint venture with the Swedish seed company Svalöf Weibull, and announcement of the creation of a new North American biotechnology research centre. In order to close the biotechnology gap between itself and leading biotechnology corporations, BASF committed 700 million euros for biotechnology research up to 2010. The BASF biotechnology research in North America focused on traits that would directly benefit Canadian growers, such as increased cold and drought tolerance. The development of crops with improved output traits such as higher protein, oil and carbohydrate levels is also a priority. This would increase the value of these crops, ultimately benefiting growers as well as consumers (Moore 2000).

BASF is proud to be the first company to introduce new state-of-the-art strobilurin chemistry to the Canadian market. Sovran®, a fungicide to control leaf and fruit scab on apples, is the first strobilurin approved for use in Canada. Sovran® is based on a naturally occurring substance produced by small, edible mushrooms that grow on decaying pinecones. In the woods, the mushroom fungus produces 'Strobilurin A' to control the growth of other fungi that would otherwise compete for nutrients. The AI of Sovran®, kresoxim-methyl, was discovered when researchers modified the naturally occurring compound to produce a stable, synthetic strobilurin. Studies have shown that Sovran® does not harm bees, earthworms, predatory mites and other beneficial organisms. It also rapidly decomposes in the soil.

Dow AgroSciences has also attempted to align biotechnologies with sustainable agriculture. It stresses the importance of agricultural biotechnology for improving crops and increasing yields on existing cultivated areas, thus saving 'untold quantities of land as pristine areas for wildlife habitat and forest' (Dow 2003d). Dow is also keen to stress the importance of agricultural biotechnology in the production of non-food products such as the usual pharmaceuticals and nutriceuticals. In addition, Dow stresses the role of biotechnology in sustainable development by linking it to the production of degradable plastics from carbohydrates, instead of hydrocarbons. In its agricultural biotechnology rhetoric, Dow acknowledges that 'bio-technology has deeper philosophical implications, and scientific understanding of the issues alone is not enough'. To support this, Dow developed a range Guiding Principles for Biotechnology in order to convince the public that it is 'participating in the dialogue and working to understand and address people's honest concerns about such topics as labelling and biodiversity' (Dow 2003d). Its Guiding Principles are a commitment to the following: (1) Actively listen to and dialogue with stakeholders to understand their concerns and to help us progress responsibly. (2) Inform the public about relevant benefits, risks and potential implications of our biotechnology products and processes, and encourage others to do the same. (3) Participate in outreach efforts and explore opportunities to make the benefits of biotechnology available to developing countries. The principles also respect the rights of indigenous people to have access to local germplasm. (4) Promote research on the potential benefits and safety of our biotechnology products and services for humans, animals, and the ecosystem. (5) Support the development and implementation of internationally harmonised approaches to biotechnology safety analysis and promote the creation of a predictable and scientifically sound regulatory framework to reduce scientific uncertainty, manage potential risks, and assure public confidence. (6) Apply our established corporate Environment, Health & Safety Risk Review Process, which includes a thorough consideration of the impact on humans, animals, the environment and society, throughout the lifecycle of all our biotechnology products and services and will take appropriate corrective actions. (7) Support the patentability of inventions as determined by the applicable laws of the countries in which we do business and will respect the IPRs of others and not knowingly infringe upon valid patents. (8) Support the conservation of biological diversity and the sustainable use of biological resources. (9) Promote these principles throughout the industry and value chain. Dow is keen to stress implementation of its in-house Environmental, Health and Safety policy and its commitments to Product Stewardship, Responsible Care®, Code of Business Conduct and Sustainable Development Principles (Dow 2003d).

DuPont is also committed to the rhetoric of sustainable growth through the application of science to create value for society and DuPont's shareholders. DuPont stated that biotechnology has an enormous role to play in making our future world more sustainable (DuPont 1999). It states that 'our sustainability is related to our commitment to core values of safety, environmental stewardship, ethics and the fair treatment of people' (DuPont 1999). DuPont intends to continue its past role of introducing innovative technologies in the form of biotechnologies that will deliver 'nutritional and health improvements in foods and pipeline technologies such as polyester from plants and microbes and bioprocesses' (DuPont 1999). DuPont states that it is on a journey to become a sustainable growth company committed to increasing shareholder and societal value while reducing its environmental footprint, a journey which is highly dependent on biotechnology (DuPont 1999). Like all biotechnology corporations, DuPont endorses the development and application of biotechnology as a technology with the potential bring significant benefits to society, particularly in the food, materials, energy generation, polymers, sensors and electronics sectors (DuPont 2004). Biotechnology sustainability credentials include the lower costs of production, higher product quality, reduced reliance on fossil fuels and other environmental benefits. However, alongside other biotechnology corporations, it has responded to consumer concerns by developing Bioethics Guiding Principles to research which include a commitment to the following. (1) Food/feed safety. (2) The environment and biodiversity. (3) Transparency of information. (4) Engage stakeholders. (5) Advocate independent research. (6) Contribute to developing economies. (7) Formalising access to genetic resources (DuPont 2004).

In its commitment to food and feed Safety, DuPont states that it will develop products derived from biotechnology that are at least as safe as their conventionally produced counterparts for both human food and animal feed using the best available scientific knowledge. Its commitment to the environment focuses on the application of biotechnology in production systems to achieve a net gain for the environment, and one in which it will attempt to protect and conserve biodiversity. DuPont also stresses that it is committed to transparency of information, in as much as it believes in the individual's right to clear and accessible information regarding product safety. DuPont suggests that it is committed to routinely engaging stakeholders, including shareholders, customers, society and employees, and to considering their diverse viewpoints in its biotechnology decision-making process. DuPont is also keen to stress that it seeks opportunities to advocate and/or fund biotechnology research important to its business in public institutions, research centres and NGOs that follow accepted science protocols and peer review standards. DuPont also states that it endeavours to be socially and culturally responsible as it shares knowledge and appropriate technology in developing economies to help improve food, nutrition and the quality of life; and that it seeks to utilise IPR in ways that help alleviate hunger. Lastly, DuPont strives to identify the owner(s) of natural biological resources and knowledge selected for research and product development and will develop fair and equitable business arrangements that recognise the contributions of the involved parties.

DuPont uses the traditional argument of how to provide a 'safe and abundant supply of food, feed and fibre for a growing population while using less land, water, chemicals and nutrients'. DuPont is convinced that biotechnology, as part of an integrated science strategy, can redress to continuous battle to feed a growing population while reducing agriculture's impact on the environment. It promotes the idea that biotechnologies can generate more affordable, efficient and environmentally sustainable solutions to food system problems. DuPont also stresses the need for ethical discussions that include the benefits and consequences of applying and not applying biotechnology to the world's problems. DuPont's rhetoric is also committed to comprehensive stewardship of biotechnology that includes caution and care. This is the careful consideration of consumer concerns, reduced environmental impacts and sustainable intensification of agriculture (DuPont 2004).

7.3.2.1 New Breeding Techniques (NBTs)

New Breeding Techniques (NBT) also named New Genetic Engineering Techniques include techniques such as gene editing, via CRISPR/Cas9 and CRISPR/Cpf1, which enables more precisely modification of DNA. This is achieved through turning selected genes on or off or cutting and pasting specific DNA sequences. These new techniques can remove a genetic weakness or insert desired traits already found in the same species. CRISPR is significantly cheaper and easy than conventional genetic modification and, importantly, does not involve the insertion of genetic material from other species (Paul et al. 2017). Whilst only invented in late 2014, the first working gene drive system, which uses the CRISPR/CAS9 gene editing system, generated significant interest from life science corporations. Since 2014, a wide variety of gene drive applications have been developed, from work focused on attacking the vector for malaria to non-browning mushroom developed at Pennsylvania State University (Greenaway 2017).

NBTs, especially CRISPR/Cas9 and CRISPR/Cpf1, have received considerable interest from life science corporations. For example, in the case of Monsanto, Greenaway (2017), quoting Tom Adams, biotechnology lead for Monsanto, stated that, whilst not a replacement for biotechnology, 'genome-editing techniques have great potential to improve and unlock capabilities across our leading germplasm and genome libraries to enable a wide variety of improvements across crop systems'. This interest has led to an avalanche of patent applications for a range of genes and biotechnology processes. According to Abergel (2011), by 2010, BASF, Bayer, DuPont, Dow, Monsanto and Syngenta had filed 1663 patent documents related to 'climate ready' genes able to tolerate abiotic stresses such as drought, heat, cold, floods and saline soils. These companies were clearly attempting to re-brand themselves as climate saviours (ETC 2008). According to the ETC Group (2011), these companies controlled over 77% of the patent families.

The rush to gain access to NBTs, especially CRISPR/Cas9 and CRISPR/Cpf1, has been a major driver behind the current spate of acquisitions and strategic alliances (Epoch Times 2016; Greenaway 2017). In the last 2 years, Monsanto announced a number of licensing agreements, which are targeted towards ensuring access to gene editing technologies, such as CRISPR/Cas9 and CRISPR-Cpf1 (Greenaway 2017). This includes a gaining access to a tool developed by Dow AgroScience called EXZACT™ Precision Technology® Platform and a USD$1.5 billion partnership with BASF to engineer stress tolerant plants (Greenaway 2017). Bayer has also been actively securing access to CRISPR technologies, entering into a joint venture named Casebia Therapeutics with CRISPR Therapeutics (Epoch Times 2016). DuPont has also made similar investments to ensure access to CRISPR technologies, entering into 'research collaborations and intellectual property licenses with Vilnius University and with UC-Berkeley-based Caribou Biosciences', which was founded by Jennifer Doudna – one of the patent holders of CRISPR/Cas9 technology (Abergel 2011). According to Neal Gutterson, vice president of R&D for DuPont Pioneer, 'This is just the beginning' (Epoch Times 2016).

NBTs are currently being used to improve crop performance (such as yield and drought tolerance) of major staple crops (maize, wheat and rice) as well as soybean and canola. In the case of DuPont alone, CRISPR technology is being used to develop high-yielding and drought-resistant maize, wheat, rice, soybeans and canola (McMichael 2016). There is significant excitement related to the potential of this new technology and the speed in which new products can be brought to market. According to Rachel Haurwitz, another cofounder of Caribou Biosciences, using CRISPR technology, products can be taken to market in 5–10 years, compared to the 10–17 years currently required to get conventional biotech products to the market (McMichael 2016).

One of the most appealing characteristics of NBTs such as CRISPR technology is the fact that they do not involve the transfer of genetic material across the species barrier; this has been the Achille's Heel of previous genetic modification work, which led to labels such as Frankenstein Food. This fact has allowed life science companies to successfully argue that products developed using CRISPR technology do not have to follow the same rigorous testing and regulatory processes that earlier GM products were obliged to follow. This was confirmed when the USDA informed Pennsylvania State University scientists that they would not be regulating field-testing of their non-browning mushrooms developed using CRISPR technology (Greenaway 2017). Key members of the scientific community have also endorsed this critical distinction. Indeed, editors of the Journal 'Nature' back in 2010 stated that genome-edited crops and livestock should not be treated differently from the products of traditional breeding (Abergel 2011). More importantly, after lobbying by the USA during TTIP trade talks, the EU has also agreed to undertake a U-turn and support the US position that biotech products produced using CRISPR technology do not constitute GM products, thus avoiding the need to undergo rigorous safety testing and labelling (Tickell 2016).

Not surprisingly, this decision and the lead-up to this decision sparked significant backlash in consumer and environmental NGOs as well as concerned individuals from the scientific community (Nature 2016). Indeed, attempts to slip new biotech products developed by NBTs through the EU's GM regulatory backdoor, prompted key environmental and consumer NGOs (including EcoNexus, GeneWatch, GreenPeace and Friends of the Earth) to submit a joint letter urging the EU to classify NBTs as GMOs (Greenpeace-CEO-GeneWatch UK 2016).

7.4 Key outcomes of pesticide and biotechnology development and application

7.4.1 GEF and civil responsibility

In 1996, the GEF Secretariat recommended that the GEF Council encourage private sector funding of GEF and the involvement of the private sector in GEF's activities (GEF 2003a). In 1998, 36 donors pledged a further US$2.75 billion to fund GEF's work into the new millennium (GEF 2003a). As of March 2000, the GEF had approved 138 biodiversity-based activities and 157 aimed at ameliorating climate change, with total funding of US$123 million. This assistance empowered nations to 'assess biodiversity and climate change challenges from their own perspective, determine the most promising opportunities for project development, and then pursue full-scale projects with support from the international community' (GEF 2003a). By 2000, GEF had been able to secure nearly US$8 billion in additional funds to support its work, in excess of US$2 billion came in the form of matched funding from recipient countries. Additional contributions were derived from GEF's implementing agencies, other development institutions, other governments, project beneficiaries and, increasingly, the private sector. GEF has supported development of conservation trust funds, including funds in Bhutan, Brazil, Mexico, Peru, Uganda and, more recently, Suriname, which allow activities to continue beyond the duration of the funding phase (GEF 2003a).

October 2000 signalled the beginning of GEF's third replenishment. In December 2001, the GEF Council proposed including the designation of land degradation as a focal area. Their rationale was based on recognition of land degradation as a global issue and that, by addressing land degradation, the GEF would be making a significant contribution to poverty alleviation and sustainable development. In addition, GEF was designated as the executive agency for the financial mechanism of the Stockholm POPs Convention 2001, Article 13. In February 2001, in addition to the more familiar approach of eliciting financial support from national governments and intergovernmental organisations, UNEP's Governing Council appealed to the private sector and NGOs, as 'private civil society' stakeholders, to provide the financial resources required for implementation of the POPs Convention. So far, private sector integration in the GEF remained limited compared to the involvement of other actors (GEF 2003a). In May 2002, the GEF Council formalised its changing relationship with the private sector by proposing amendment to the GEF Instrument. In addition to the inclusion of land degradation as a focal area, it also proposed that POPs should be considered as a focal area and eligible for funding (GEF 2003a). The Second GEF Assembly, in Beijing, China, 16–18 October 2002, approved these amendments. Two new subparagraphs were added to paragraph 2 of the GEF Instrument, to include two new focal areas: land degradation, primarily desertification and deforestation and POPs. Paragraph 3 was revised to read as follows: 'The agreed incremental costs of activities to achieve global environmental benefits concerning chemicals management as they relate to the above focal areas shall be eligible for funding. The agreed incremental costs of other relevant activities under Agenda 21 that may be agreed by the Council shall also be eligible for funding insofar as they achieve global environmental benefits by protecting the global environment in the focal areas (GEF 2003a)'. In 2002, 32 donor countries pledged US$3 billion, matched by almost US$8 billion more in co-financing, to fund operations between 2002 and 2006, making GEF the largest single source of funding for the global environment (GEF 2003a). In October 2017, GEF began its seventh replenishment (GEF 2017).

7.4.2 The Global Compact

UN Secretary-General Kofi Annan introduced the concept of the Global Compact in an address to the World Economic Forum on 31 January 1999. During the address, he 'challenged world business leaders to help build the social and environmental pillars required to sustain the new global economy and make globalization work for the entire world's people' (UNEP 2003a). The Global Compact is not a formal regulatory mechanism but a 'value-based platform' aimed at promoting transparent dialogue and institutional learning in order to identify and disseminate ethical business practices based on a system of universal principles (UNEP 2003a). However, many believe that process of globalisation requires the establishment of international legal standards (International Council on Human Rights Policy 2002), and the establishment of voluntary codes of conduct relating to corporate social responsibility (Williams 2000). According to Novartis (2004), the Global Compact was an attempt by the UN to incorporate universal values into the business ethos of global corporations, attain key social goals and secure open markets. The Compact encompasses nine such principles, drawn from the Universal Declaration of Human Rights, the ILO's Fundamental Principles on Rights at Work and the Rio Principles on Environment and Development (UNEP 2003a). According to UNEP (2004a), the Global Compact involves all the relevant social actors. (1) Governments, who defined the principles on which the initiative is based. (2) Companies, whose actions it seeks to inform. (3) Labour, in whose hands the concrete process of global production takes place. (4) Civil society organizations, representing the wider community of stakeholders. (5) The United Nations, the world's only truly global political forum, as an authoritative convener and facilitator (UNEP 2003a). In order to sign up for the Compact, corporate CEOs must first submit a letter to the UN Secretary-General expressing a 'clear commitment to support the Compact and its nine principles, and a willingness to participate in its activities' (UNEP 2003a). Many of the most powerful global corporations have responded with positive and constructive rhetoric, while very few have undertaken concrete actions. To date, 9670 companies have signed up to the Global Compact (UN 2017).

7.4.3 CBD versus the WTO

By June 1994, the CBD had been signed by 167 states and the European Community, and had been ratified by 64. Soon after the establishment of the Convention, it was recognised that there was a need for a protocol to govern issues of biosafety. Specifically, there were growing concerns over the potential impact on biodiversity of transboundary movements of living modified organisms (LMOs) (Clapp 2002). Reflecting this concern, the text of the CBD asks parties to consider the negotiation of a protocol on biosafety. Talks on the biosafety protocol began in 1995 and were expected to be concluded in 1998. In May 1997, negotiations opened the possibility to negotiate a legally binding Biosafety Protocol to assess and minimise risks in the field of transboundary transfer, handling and use of GMOs. Eventually, agreement was reached in January 2000. Three key issues prolonged the negotiations. First, discussions included the management of contentious GMOs. Second, debate raged over the determination of a rational for refusing imports of GMOs. There were two principal options, the precautionary principle and the principle of familiarity. Actors were split over whether or not to apply the precautionary principle or principle of familiarity, which relied on a version of the US principle of substantial equivalence and sound scientific evidence in the process of assessing risk.

Finally, heated discussions took place to determine the relationship of the biosafety protocol with respect to international trade rules under the WTO (Clapp 2002). In negotiations, the so-called 'Miami Group' (USA, Canada, Japan, Australia, Chile, Argentina and Uruguay) postulated freer trade in LMOs (Meyer 1999). They argued that only GM seed for intentional release into the environment should be regulated and not food and feed (Clapp 2002). They also argued for science-based rules, in accordance with WTO, rather than the precautionary principle. WTO guidelines generally treat technologies as safe unless 'sound science' has produced evidence that they have caused harm (McAfee 2003). Conversely, the Like-Minded Group, which included Western European and most developing countries, wanted the regulation of LMOs to cover all GM commodities (Meyer 1999). In addition, they insisted on the right to refuse imports of LMOs based on the precautionary principle, and assurances that the protocol would not be subordinate to the WTO when conflicts arose (Clapp 2002). Both industry and environmental NGOs lobbied respective governments to adopt their positions. At the first talks in Aarhus, Demark 1996, eight industry groups were represented. By the time of the Cartagena meeting in 1999, this had risen to 20 groups (Clapp 2002). Key life science corporations were represented in many of the meetings. These included Monsanto, DuPont and Syngenta as well as industry lobby groups such as BIO, BIOTECanada, Japan Bioindustry Organisation, the International Chamber of Commerce and the International Association of Plant Breeders for the Protection of Plant Varieties (Clapp 2002). Company representatives have been influential consulting observers at CBD sessions and Biosafety Protocol talks (McAfee 2003). Not surprisingly, they were 'strongly opposed to the adoption of strict rules that would impede the production and/or trade of GMOs, seeds and commodities. Regulation of commodities was the most important issue, as regulation would slow trade unnecessarily and damage profits' (Clapp 2002).

Agribusiness interests in the USA and other LMO-exporting countries have seen the CBD and its Biosafety Protocol as threats to their own strong position in agro-food export markets (McAfee 2003). Unchanged, the Protocol would permit countries to implement their own policies, based on socioeconomic criteria as well as a precautionary approach to environmental risk, on whether or not to allow imports of certain biotechnology products (McAfee 2003). US agribusiness firms have been reluctant to take on the costs of segregating genetically altered from conventional crops and preferred not to compete with producers of 'GMO-free' foods (McAfee 2003). Indeed, according to Clapp (2002), in 1999, a representative of the North American Grain Associated insisted that it was almost impossible to segregate GMO commodities, and that labels indicating that 'may contain modified crops' .... 'is about as stringent... as we can live with'. For example, in 1999, if Brazil, the USA's principal competitor in international soybean market, continued to restrict the importing and planting of GM soybean, and if the Biosafety Protocol recognises its right to do so, Brazil could maintain an advantage in markets in Europe or Japan, where there is a great demand for LMO-free soy products (McAfee 2003). In effect, biotechnology interests vetoed US support of the CBD on the grounds that its IPR language was not strong enough and that its provisions for transfer of biotechnology might threaten their commercial interests (McAfee 2003). In accord with Miami Group, the biotechnology industry was keen to ensure that the science-based rules of WTO would be adopted in the Biosafety protocol.

Informal consultations continued over the summer of 1999, and reconvened in Montreal in January 2000. By this time, there was a high level of public concern over GMOs in Japan and the USA, in addition to Western Europe. Pressures increased to resolve the LMO issue, particularly in light of the disastrous WTO meeting in Seattle, where the USA had promoted the rapid establishment of a Biotechnology Working Group to appropriate the Biosafety Protocol's authority over GMO trade issues (Clapp 2002; McAfee 2003). The USA perceived that bringing biotechnology regulation under the aegis of the WTO was required to facilitate profitable trade in GM products (McAfee 2003).

Industry was well coordinated at the Montreal meeting, having formed the Global Industry Coalition in order to promote the sense of coherence and common purpose in the biotechnology sector. The coalition claimed to represent 2200 companies in over 130 countries (Canada Newswire 2000). Two months after collapse of WTO negotiation in Seattle, US negotiators and the biotechnology industry reluctantly accepted the Cartagena Protocol (McAfee 2003). The principal outcome of the Montreal meeting was the Advanced Informed Agreement (AIA), similar to the PIC procedure (Clapp 2002). It aimed to institutionalise a process that empowered importing countries to apply the precautionary principle to reject seed and commodity LMOs in the absence of science-based supporting evidence. However, an important stipulation insisted that the precautionary measures should be cost-effective (Clapp 2002). While this outcome was decidedly unfavourable to the USA and biotechnology industry, the US delegation did secure agreement that shipments of LMO products and seeds destined for processing and animal feed will not have to be labelled as LMOs (McAfee 2003). In addition, while the Miami Groups had conceded inclusion of the precautionary principle, they were shrewd enough to ensure the inclusion in the Protocol text that 'trade and environmental agreements should be mutually supportive, and that the Protocol shall not be interpreted as implying a change in the rights and obligations of a Party under any existing international agreements'. This vagueness basically implies that import restrictions relating to LMOs would not contravene free trade rules of the WTO (Clapp 2002). In the case of the Biosafety Protocol, industry actors strongly resisted strict regulation over the trade of GMOs, including GM seed, and in the end advanced rules for the trade in GMOs that are weak and vague (Clapp 2002). These events revealed a growing coalescence of interests between many developing countries and Western European nations against the US-led food trade liberalisation and biotechnology deregulation (McAfee 2003).

Another important field for negotiation has been the so-called terminator technology issue. Terminator technology relates to the use of biotechnology to facilitate the expression of genetically enhanced crop traits in the first season, but which renders the resulting seed sterile or unable to complete its lifecycle. This enables biotechnology corporations to control their privately held crop genetic resources in places where their IPR claims are not recognised or are defied, as some farmers' organisations have pledged to do (McAfee 2003). The debate over terminator technology illustrated how economic and molecular-genetic reductionism can reinforce each other (McAfee 2003). In June 1999, an alliance of developing countries asked the CBD's scientific advisory body to call for a worldwide moratorium on the field-testing and commercialisation of 'terminator' technologies (McAfee 2003). Developing country delegates opposed to 'terminator' were in the majority when the 1999 CBD advisory body session began, but the unofficial US delegation and a few OECD states adamantly resisted their call for a moratorium (McAfee 2003).

Developed by the USDA and commercial biotech companies, gene protection systems or GURTs followed quickly on the heels of Terminator Technology (McAfee 2003). A scientific expert commissioned by the CBD Secretariat linked two arguments in support of 'Genetic-Use Restriction Technologies' (McAfee 2003). First, GURTS could be re-engineered so that the seeds will still germinate and grow, but will not express their premium traits unless treated with a chemical inducer, which farmers would have to purchase. Second, farmers in developing countries, being 'good businessmen', will then make a rational choice to purchase or not to purchase the superior seed, based on whether its benefits to them outweigh the added cost (McAfee 2003). In other words, release the technology and let the market decide its fate. Both forms of reductionism are pivotal in this reasoning. First, the notion that small-scale farmers in hungry countries have unencumbered market-choice options to adopt or reject new biotechnologies is an economic-reductionist idea (McAfee 2003). It ignores the multiple ways that farmers' real-world choices are constrained by economic asymmetries and non-economic factors: lack of cash, credit and information, the limited availability of seed varieties as farming becomes more commercialised and globally integrated, contracts that require the use of particular commercial varieties, and the heavy promotion of 'miracle' hybrid and transgenic varieties (McAfee 2003). Second, the idea that 'terminator' crops can be neatly engineered so that traits can reliably be switched on and off is a genetic-reductionist claim (McAfee 2003). However, while the economic purpose of terminator technology is to expand seed markets by preventing farmers from saving proprietary seed is in perfect accordance with the WTO, proponents of a moratorium on terminator technology maintain that such technologies are at odds with CDB commitments to conserving crop biodiversity and the relevant practices of local communities (McAfee 2003).

In 2000, Conference of Parties 5 of the CBD called for a ban on the field-testing of GURT crops and a moratorium on their development until their impact has been fully tested (Corporate Watch UK 2003d). However, instead of banning GURT or calling for a moratorium, the CBD's scientific body (SBSTTA) gave a green light to their commercialisation. The SBSTTA decision even restricts the rights of countries to impose national bans on terminator technology by linking moratoria to trade sanctions. The CBD is accused of bowing to corporate pressure and US bullying (RAFI 1999).

7.5 Summary

Chapter 7 focuses on the impact of the Sustainability Paradigm on the strategies employed by pesticide corporations from the mid-1970s to 2017. Based on the 'limits to growth' concept of 1972, calls for sustainable agriculture have been, and continue to be, orchestrated through environmentalists and public sector institutions. Within the context of the Sustainability Paradigm, pesticide companies responded to several key external drivers, namely the following: (1) Growing concerns over the loss of biodiversity and natural habitats. (2) Concerns over the productivity, stability and resilience of industrial agroecosystems. (3) Emerging concept of sustainable development. (4) Increased activity of environmental NGOs, human welfare groups and participatory social movements. (5) Emergence of new national and intergovernmental institutions with an overt environmental agenda, and subsequent emergence of environmental policymaking. (6) Emergence of an environmental agenda in agricultural policymaking, and the subsequent emergence of agri-environmental policies. (7) Promotion of 'sustainable agriculture'.

The pesticide industry responded to these drivers in six distinct ways. First, it attempted to create space for pesticides in sustainable agro-food systems by down-playing environmentalists' ecocentric characterisations of sustainable agriculture in favour of more technocentric alternatives, based on the emergent principles of corporate environmentalism, which are underpinned by the legitimate use of pesticides. Pesticide corporations continually emphasised the centrality of synthetic pesticides in sustainable agriculture by highlighting their pivotal role in PA, IPM, ICM, IFM, conservation agriculture and zero-till agriculture. Second, the pesticide industry has committed increased resources to the development of biopesticides. Third, the pesticide industry is attempting to sell new input and output trait biotechnologies as sustainable technologies. Fourth, through the creation of foundations and institutes, devoted to the promotion of sustainable development, the pesticide industry has attempted to define the way towards a more agriculturally and environmentally sustainable future. Fifth, the pesticide industry launched a well-funded and vehement defence of pesticide use, in legal, mass media and political arenas. Lastly, pesticide corporations have stepped up their offensive by channelling more resources into pro-pesticide education, marketing and public relations.

To date, the pesticide industry and other corporate agro-food systems actors have managed to secure the operationalisation of technocentric characterisations of sustainable agriculture. Much of this has been achieved through constant and affective lobbying of national and intergovernmental policymaking apparatus, and public relations based on civil responsibility and corporate ethics through work with the GEF and the UN's Global Compact. While the establishment of the Global IPM Facility, and its notable successes based on the ecological and human capital aspects of IPM, is noteworthy, adoption of integrated forms of agriculture has tended to be based on more technocentric interpretations, based on the use of pesticides.

Lastly, designed to protect and promote the sustainable use of global biodiversity, the CBD has, so far, proved no match for corporate considerations orchestrated through the WTO.

Ultimately, the evolution of the pesticide/life science industry, and the food systems of which it is a key constituent part, has been both a dynamic and complex process. Once almost totally independent of external off-farm inputs, food systems, across much of the world, evolved to increasingly depend on a wide range of non-farm-based inputs and processes, including pesticides and biotechnologies. Importantly, and in congruence with past civilisations, the political, economic and social stability of nation states still remains dependent on the durability of contemporary food systems of which they, in concert with private sector interests, have constructed. However, food systems in the contemporary period display considerable heterogeneity, which, in turn, has prompted a range of divergent pesticide/life science industry strategies. Over the past 40 years, four major food system paradigms (Residual Productivist, Neo-Productivist, Post-Productivist and Sustainability) have emerged as a potential successor to the Fordist-based Productivist Paradigm and the Second Food Regime. So far, there has been no distinct paradigm shift and ascendancy of a coherent Third Food Regime. In turn, administrations of nation states will continue to pursue food system strategies that continue to maintain or strengthen their political legitimacy, authority, and stability, and, ultimately, economic performance. Likewise, intergovernmental institutions, NGOs and citizen groups will continue to pursue strategies that, while addressing their specific mandate, will enhance their sphere of influence and assure their financial and political solvency. Ultimately, public sector food system strategies will continue to be driven by changing global and local political, economic, socio-cultural and environmental imperatives. Success of the pesticide/life science industry will depend on their ability to shape the evolution of contemporary and future food systems that maximise opportunity, profitability and investor returns, and, in concert with other private sector interests, on their ability to evolve with contemporary and future political, economic, socio-cultural and bio-physical systems.

Chapter 8

The future of pesticides and biotechnologies

8.1 Key characteristics of agricultural development

Global population is expected to increase from nearly 7 billion in 2017 to 8 billion by 2030, and almost 10 billion by 2050, boosting agricultural demand by some 50% compared to 2013 (FAO 2017). In some regions, such as SSA, the population is likely to double by 2050 (PRB 2013). Rising income levels of an increasing growing urbanised population are also expected to dramatically increase, and change, the demand for cereals (Ignaciuk and Mason-D'Croz 2014). This growth will increasingly challenge the ability of the global agri-food system to supply healthy and nutritious food, feed, fuel and other outputs whilst conserving key elements of the globe's natural resources.

The production of maize, wheat and rice combined will need to increase by 60–70% by 2050 to meet this growing demand (Cairns et al. 2013; Godfray et al. 2010; Ignaciuk and Mason-D'Croz 2014; Watson 2017). In line with shifting diets towards increased animal protein and milk consumption, global meat production is expected to increase by more than 58 million metric tonnes (MMT) during the next decade. Poultry continues to dominate the meat sector, as reflected in production growth of 27% by 2023 relative to 2014. This represents almost half of the additional meat that will be produced globally by 2023 (OECD-FAO 2014). It is suggested that, over the next decade, 75% of additional agricultural output will be produced in developing regions (OECD-FAO 2014). Much of this demand is linked to the increased production of pork and poultry, which relies on the use of maize feed (OECD-FAO 2014). Indeed, Timsina et al. (2011) suggest that, by 2020, demand for maize in Asia may increase by as much as 87% due to increased demand for animal, especially poultry, feed. Demand for wheat is expected to increase between 34% and 60% by 2050 (Valin et al. 2014), mostly linked to projected increased consumption in developing countries. Global demand for rice is expected to increase from 479 million tons of milled rice in 2014 to 536–551 million tons in 2030 (RICE 2016). Today, 54% of the world's population lives in urban areas, a proportion that is expected to increase to 66% by 2050. Projections show that urbanisation combined with the overall growth of the world's population could add another 2.5 billion people to urban populations by 2050, with close to 90% of the increase concentrated in Asia and Africa; especially India, China and Nigeria (FAO 2017).

Feeding the global population will still predominantly rely on the production of the three major cereal crops, maize, wheat and rice (Shiferaw et al. 2011; Ignaciuk and Mason-D'Croz 2014). Maize is the most important cereal in terms of production, with 1016 MMT produced on 184 million ha (M ha) (FAOSTAT 2013) in both temperate and tropical zones around the globe. Maize is the major source of food security in SSA and Latin America and the Caribbean (LA), and is among the top three crops in Asia (Watson 2017). It is used directly for human consumption, but it is also extremely important as a livestock feed and industrial uses, including use as a biofuel. Maize alone contributes over 20% of total calories in human diets in 21 low-income countries, and over 30% in 12 countries that are home to a total of more than 310 million people. Of the 22 countries in the world where maize forms the highest percentage of calorie intake in the national diet, 16 are in Africa. Maize consumption rates are highest in eastern and southern Africa, accounting for almost half of the calories and protein consumed, and one-fifth of the calories and protein consumed in West Africa (Shiferaw et al. 2011). Wheat is grown in temperate and subtropical zones across the globe. Like maize, wheat is an industrial crop produced on large farms in developed countries. In contrast, smallholder systems dominate wheat production across 75 developing countries (Nelson et al. 2010). Wheat provides approximately 20% of protein and calories consumed worldwide (WHEAT 2016). Conversely, rice is grown predominantly in the tropics and subtropics, with only a small proportion of rice area (3%) being located in developed countries (Fischer et al. 2014). Rice is a staple food for some 4 billion people worldwide, and it provides 27% of the calories in low- and middle-income countries (RICE 2016).

The advent of climate change, especially the greater frequency of temperature extremes and both erratic and extreme precipitation, will increasingly challenge the ability of maize, wheat and rice agri-food systems to meet growing global demand for food and feed. The challenge to agri-food systems is threefold. First, increasing temperature (especially extreme temperature events) and reduced or erratic rainfall limit the ability of these crops to produce a harvestable product, especially in rain-fed and underdeveloped agri-food systems. Second, industrial agri-food systems are extremely energy intensive and, both directly and indirectly, produce a considerable amount of greenhouse gases, which further exacerbates climate change. Third, changes in temperature and rainfall will require adjustment to cropping geographies and integration of more drought- and heat-tolerant crops, especially in the predicted climate change hotspots. The good news is that there is significant genetic variation for heat and drought/submergence tolerance in the global maize, wheat and rice genebanks.

To meet the growing needs of the global food system, whilst at the same time mitigating the effects of climate change and other production limiting factors and reducing negative environmental externalities, the global agri-food system will be required to sustainably intensify production. In the irrigated systems of developing countries, significant scope exists for increasing water use efficiency through new soil, water and fertiliser management approaches and crop diversification. In addition to the development of weather agro-advisory services and weather index insurance, the options available in rain-fed systems differ markedly from those available in irrigated systems, namely, to optimise every drop of available rainfall or to avoid drought stress situations. Whilst breeding for heat tolerance and diversified cropping systems are likely to be the principal short-term responses to increased mean global temperatures and extreme heat events, changes over the longer term are predicted to be quite dramatic, especially the expansion of temperate crops towards the poles. Whilst significant opportunities exist to ameliorate the effects of climate change, these opportunities generally involve risk. In developed countries, the private sectors (seed, fertiliser, pesticide, irrigation, credit and insurance suppliers) are generally at hand to advise farmers on how to address the challenges posed by climate change. Conversely, in most developing countries, there are few sources of advice and support for smallholder farmers. This situation leaves many smallholder farmers in developing countries without the advice and support that they desperately need. Ultimately, the most vulnerable farmers and communities, namely, smallholder subsistence and market-oriented farmers in marginal environments, are those who face the most extreme climate-change-related challenges at the same time as being the least able to adapt. Whilst international agricultural research centres (CGIAR), advanced R&D-focused centres of developed countries and both local and international NGOs strive to develop and translate evolving crop management approaches, the dissemination of climate-smart agricultural practices is extremely slow.

8.2 Key drivers of pesticide industry evolution

Set against the tendencies outlined above, what is likely to drive the evolution of pesticides, biotechnologies and the global agri-food system within the next 10–15 years? Well, given there is no clear hegemon, it is likely that all paradigms will continue vying for position. Having said this, depending on the nature of influence wielded by consumers and food retailers, it is likely that the neo-productivist model will continue to prevail over the largest land area and have highest level of support from government, intergovernmental institutions (such as the WTO, etc.) and the private sector (especially those related to upstream sections of agricultural value chains, such as agricultural input suppliers, large-scale producers and processers).

8.2.1 Political support for agriculture

Considering the existing and predicted challenges associated with climate change, the limited land area suited for agriculture, and the strategic importance of current, and potential future, high-value outputs from agriculture, most national governments are expected to continue prioritising agriculture. In the case of developed economies, it is likely that the focus on self-reliance will continue or, depending on the relative cost of domestic agricultural support, may even strengthen. It is likely that food production (which is the only subsidised component of the food system) will become increasingly more market oriented, hence a strengthening of the neo-productivist paradigm. For example, the EU's CAP is criticised for its high cost and production and trade distorting nature. Many critics would prefer to rely on a more competitive and seemingly cost-effective free world market in agricultural goods and services. Globalisation of food systems is likely to continue. This said, the argument for agricultural protectionism and national and regional self-sufficiency across developed economies remains. In part, this relates to two fundamental premises upon which food self-reliance stands. These are that food consumers have enough money in their pockets to meet their dietary needs and that the global food system can produce the amounts of food demanded. In many developing countries, especially those that were rocked by the food riots, support for national food self-sufficiency - dependent on neo-productivist agricultural production systems - is unlikely to disappear. In addition, farmers are still likely to receive at least some support for the protection of valued semi-natural habitats, landscapes and tourist/recreational spaces, or as a vehicle to maintain the rural economy or promote local food systems. Financial support may also increase for organic or less industrial local food systems - although, depending on the level of expressed demand, these could further expand without the need for public financial support. Post-Brexit Britain will be an interesting test case in this regard.

The food retailer sector remains a bit of a wild card when it comes to predicting what food systems will look like in the next 10–15 years. Ultimately, retailer power is likely to continue growing and increasingly shape agri-food systems policies. Indeed, there are signs that TNCs, especially retailer TNCs proposing to act in the interest of consumers, will continue usurping the power of national governments. The big question is how the retailer sector will use this power. In many respects, food retailers have been happy to sell the agricultural products originating from productivist, neo-productivist and post-productivist production systems. They have been able to role with the punches and provide consumers with choice, low-cost standardised processed foods through to high-end food products endowed with both improved health and environmentally sensitive credentials. The difficulty lies in predicting consumer behaviour. If their behaviour in relation to the purchase of food could be reduced to simple economics, namely, they would buy seemingly healthier and environmentally friendly foods if they earned enough money to do so, it would be easier to determine consumer and therefore retailer behaviour. However, in many cases, this does not happen. For example, many wealthy consumers, despite purporting to be in favour of healthy eating and conservation of the environment, continue to purchase food at the lowest cost. Conversely, some poor consumers continue to spend a relatively high proportion of their income on seemingly healthier and more environmentally friendly foods. Indeed, in many developed countries, traditional food retailers face increasing competition from new entrants into the food retail market that predominantly trade on bargain basement prices.

8.2.2 Retailer and consumer influences

Whilst multinational, regional and national food retailers will continue to strengthen their grip on food sales globally, we should take a moment to acknowledge the expected growth in short-chain food networks based on growing support for alternative food delivery such as local food networks, 'food from somewhere', organic and ethical movements. Indeed, it is likely that the volume of food produced by, and the area of land devoted to, these alternative approaches will increase. Indeed, there are signs that a growing number of consumers are willing to accept misshaped fruits and vegetables with cosmetic blemishes. However, it is difficult to predict if this will become the norm or will remain a minority of customers. Most of these alternative approaches reject the use of synthetic pesticides and first-generation biotechnologies. Ultimately, unless proponents of these approaches can provide clear-cut evidence that their adoption does not come with lower productivity and higher risk than industrial agriculture, it is unlikely that they will find the political support they perhaps deserve. The extent to which these networks will replace industrial food chains is difficult to predict.

So, how are multinational food retailers and food consumers likely to affect the context of food system development, and, in turn, the evolution of pesticides and biotechnologies?

It appears that there will be a hard road ahead for pesticides and first-generation biotechnologies. Food quality and safety concerns regarding GMOs and human health are unlikely to dissipate. It is likely that long retailer supply chains (especially for fruit and vegetables) linked to consumers in developed economies, and increasingly developing economies, will demand safer foods and display reduced tolerance to pesticide residues. Indeed, it is likely that a large number of consumers will continue insisting on pesticide and GMO-free foods. Retailers are likely to explore ways to reduce pesticides and first-generation GMOs in their food supply chains. Retailers will increasingly invest in methods/approaches to reduce the use of pesticides – including research into pesticide reduction methods, and increasingly deploy rigorous and exacting quality control policies and procedures and elaborate systems of food traceability (Barrett et al. 1999). Indeed, it is likely that retailers will invest directly, or encourage investment, in plant breeding for pest and disease resistance and possibly breeding to enhance competition against weed species, thus drastically reducing the need for pesticide applications. Indeed, it is hoped that retailers will partner much more with the CGIAR and associated research institutes, who have been producing pest- and disease-resistant crops for over 50 years. Currently, this is a missed opportunity. Breeding for P&D resistance will be increasingly pushed for by processors and retailers but will also make sense from an IPM, ICM and IFM standpoint.

So, does this spell the end for pesticides and biotechnologies? The answer is, unlikely. For example, retailers will increasingly invest, and encourage investment, in the development of biological pesticides. Indeed, retailers are also likely to support the development of increasingly more selective, less toxic and lower dose pesticides. The case of second- and third-generation biotechnologies is interesting. Whilst a vocal portion of food consumers are fervently against GMOs in general, it is highly likely that retailers will be more discerning, especially regarding second- and third-generation biotechnologies. Given their expected health and nutrition benefits, retailers and food processors foresee greater acceptance of second- and third-generation biotechnologies.

8.2.3 Summary

Unless game-changing global events occur within the next 10–15 years, we can expect slow and incremental development of the life science industry. Whilst alternative food system models may expand a little, especially in specific geographies such as Western Europe, it is likely that the Neo-Productivist model will continue to dominate global agri-food systems. If game-changing events, such as climate-change-related crop failures, and associated food shortages and food price hikes occur, this will most likely serve to strengthen the Neo-Productivist paradigm. Indeed, this may even garner acceptance of climate-smart crop varieties developed using gene editing or other NBTs. Conversely, if food scares occur during this period and second and third generation are not widely accepted, it is likely that local food networks, IFM, organic and perhaps other agricultural approaches may make further headway, especially in developed economies.

Whilst an important movement, the promotion of post-productivist agriculture has somewhat stalled. It has not been established as the dominant paradigm. And, whilst second- and third-generation output trait biotechnologies will increasingly address both general and specific consumer needs, these needs are likely to be met in a much more neo-productivist manner than a post-productivist manner. Indeed, the result could be a hybridisation between the Neo-Productivist and Post-Productivist paradigms but one where the Neo-Productivist paradigm dominates. For example, in this hybridisation, diverse crop and livestock outputs may more closely resemble the Post-Productivist paradigm but with no price premium and less emphasis on the protection of biodiversity, landscapes, traditions and place. In other words, post-productivism could be viewed as a successful pilot experiment but one where many aspects are not seen as appropriate for going to scale. Put another way, Neo-productivism could easily tackle higher levels of product differentiation in globalised food markets but would not carry the rest of the post-productivist baggage, unless its inclusion was subsidised or legislated.

8.3 Key characteristics of life science corporation development

Ultimately, life science corporations and other key corporate food system actors are likely to remain entrenched in the policymaking instruments of both national, regional and international government institutions.

8.3.1 Food system concentration

It is also likely that concentration in agri-food system will continue. However, whilst horizontal mergers will lead to increased concentration and power within agri-food systems, it is likely that mergers between the pesticide and seed giants have probably reached their zenith. Most of the concentration activity is expected to take place between corporations along the food chain (i.e. vertical concentration). The degree of concentration is likely to depend on how competitive agri-food systems become and on the level of control required within increasingly 'contractualised' food systems. Indeed, it is likely that some of the current strategic alliances and experimental food production and processing arrangements may indeed become more formalised through future vertical M&A. As the position of farmers weakens, and profit becomes increasingly linked to control, vertical mergers/acquisitions are likely to increasingly involve agri-food cartels directly controlling the means of production (i.e. land), like the Cooperative Wholesale Society (CWS) example. The growing demand for food, feed, fibre and specialised biotech outputs will be met by either farmers (in both developed and developing countries) or agribusiness corporations. Indeed, if the IPR surrounding new biotechnologies is contestable and difficult to enforce, retailers and processors may opt to form cartels with supplier cartels to secure profits. If, however, IPR becomes less contested and easily enforceable, more 'cooperative capitalism' may prevail.

8.3.2 High-tech and 'contractualised' food systems

During the next 10–15 years, it is likely that food production will become increasingly more precise. PA is likely to expand exponentially. This precision will be aided through the increased usage of drone and satellite technologies, especially in association with the production of higher value agricultural products. PA will be applied to control as much biological variability as possible; in effect, more closely mimicking the factory floor–production conveyor belt.

The world is also likely to witness a greater 'contractualisation' of agricultural production. This contractualisation will increasingly be coordinated by cartels of input suppliers – pesticides, seed, machinery, fertiliser, credit/insurance, processors and retailers, especially as quality, traceability and output trait biotechnologies become mainstreamed.

How would this work? One approach would be to develop closed loop arrangements between farmers and input suppliers, such as the mega life science companies, who would act as a general contractor for retailers or food processors. Life science companies (or other powerful actors) would have two choices. First, depending on the level of control needed, companies may decide to buy land outright, especially where land is cheap and in large blocks. Termed as 'land grab', this trend is already evident and is being practised by both private enterprises and governments. Second, life science companies would subcontract farmers to grow food crops or rear animals offering farmers risk insurance to mitigate the effects of biological and climatic systems that lay outside the control of the companies. Farmers would be expected to follow detailed production protocols, which would stipulate almost every decision in the production process. This would be like the contracts currently operated by supermarkets to produce high-value fruit and vegetables. Ultimately, farmers are likely to be increasingly eliminated from farm-level management and decision-making as they become renters of proprietary germplasm.

If farmers cannot unite, they will become progressively weaker. In response, the next 10–15 years may see the establishment of more farmer cooperatives. Their success will rely on the direct relationship they have with the mega processors and retailers and their access to new technologies. Farmer groups will increasingly need to assert their critical role in food production, namely access to land (either owned or rented), their central role in the hands-on production of marketable outputs such as cereals, pulses, meat and milk. And, in their stewardship of multifunctional landscapes (conservation of land, biodiversity, landscape and traditions). Indeed, depending on their success in negotiating with bigger more powerful food system actors, individual farmers and farmer groups may increasingly explore the development of much shorter, less industrialised, food chains (local or regional food systems), where they have significantly more control and can extract higher levels of value added.

8.3.3 New pesticides

The big four pesticide producers are likely to voluntarily undertake pesticide reduction to improve their public image. Any new pesticide products in the pipeline will become increasingly selective, use extremely low doses and nanotechnology, and are likely to be more natural, such as biopesticides, rather than synthetic. The big four are likely to distance themselves from unpopular (off-patent) pesticides. And, whilst generic producers will still find a market for their products in developing countries, it is likely that they will come under increasing pressure to phase out or restrict use of the most controversial pesticides. In this light, generic producers are likely to continue reformulate/reinvent older pesticides, making them less dangerous to consumers, more selective and less impactful on the environment.

It is increasingly likely that the big four pesticide producers will openly acknowledge that pesticides no longer offer a panacea for crop protection against pests, diseases and weeds. As part of the process of food system concentration and contractualisation, it is likely that most pesticide producers will increasingly recognise and recommend application of economic damage thresholds, and IPM, ICM and IFM approaches. This will both ensure the quantity, quality and cost of agricultural production as well as flashing their environmental credentials.

Within the next 10–15 years, governments of both developed and developing countries will increasingly regulate the production and application of pesticides. It is likely that virtually all organochlorine and organophosphate pesticides, even carbamates will have disappeared, although some may still hang on in reformulated pesticides. Anti-pesticide NGOs will have achieved their aims.

8.3.4 New biotechnologies

Advances in science and technologies, especially those linked to crop breeding and precision crop production, will continue to provide new opportunities for those involved in agri-food systems. Advances in both sectors within the next decade or so are likely to be much more significant than advances made during the past 30–40 years. New applications for science and technology will appear relatively faster, and be more specific/targeted and cheaper than at any time in history.

The NBTs, linked primarily to the development of second- and third-generation biotechnologies and precision food production approaches could turn the tide for life science corporations. Indeed, NBTs could offer a world unknown and, given the rate of development in this area (alongside nanotech), who knows what NBTs will emerge in the next 10–15 years. Whilst there is strong support for second- and third-generation biotechnologies from western governments and EU, public acceptance of these new GMOs is by no means guaranteed. Even though these new GMOs offer tangible health and nutrition benefits to the public, life science companies will still need to deploy hard sell tactics, such as explicit support from prominent social figures and health experts. Either way, life science companies can expect to be blasted by anti-GMO campaigns. Ultimately, new products and approaches are likely to be produced considerably cheaper and have a higher net worth than pesticides and first-generation biotechnologies. There could be new halcyon days for the life science corporations.

8.3.5 Sustainable agriculture

The life science industry will continue to define sustainable agriculture to serve their own ends. Conversely, whilst acknowledging the negative environmental impacts of certain technologies, those caused by organochlorine, organophosphate pesticides and so forth, the life science industry is likely to continue adopting an extreme technocentric position, insisting that innovative new technologies are essential if the global agri-food system is to both feed a growing world population and conserve valuable natural resources. Indeed, the life science industry will define sustainable agriculture through their active engagement with, and promotion of, IPM/ICM/IFM, incorporating second- and third-generation biotechnologies, climate-resilient germplasm, biopesticides, nanotechnologies, ultra-low dose and highly selective pesticides. Life science companies are also more likely to favour land sparing as opposed to land sharing approaches. Namely, they will favour the intensification of production on existing land rather than advocating the expansion of agricultural land into natural and semi-natural habitats. They are likely to dispel concerns about the impact of rapidly declining biodiversity on the productivity, stability and resilience of industrialised agroecosystems as most of the intensively farmed land globally is still highly productive. This undermines the legitimacy of those purporting a collapse of intensive industrial agriculture.

References

AAP (1999). *Australian Associated Press*, 26 July 1999.

Abergel, E. A. (2011). Climate-ready crops and bio-capitalism: Towards a new food regime? *International Journal of Sociology of Agriculture and Food*, vol. 3, pp. 260–74.

Ackerly, R. L. (1981). DDT: A re-evaluation, part II. Chemical times and trends. *Chemical Times and Trends* (October), p. 55.

Advisory Committee on Toxic Chemicals (1969). *Review of Organochlorine Pesticides in Britain*. Department of Education and Science, London, UK.

AFR (1999). *Australian Finance Review*, 17 May 1999.

Agence France Presse (1999). *Agence France Presse*, 9 September 1999.

Agranova (1999). Agrochemicals in perspective: Analysis of the worldwide demand for agrochemical active ingredients. *Agranova*. Agranova.co.uk/agtalk99.htm (accessed on 19 August 2003).

Agranova (2001). Global herbicide directory, 3rd ed. *Agranova*. agranova.co.uk/herbhist.htm (accessed on 19 August 2003).

Agriculture Outlook (2000). U.S. farm policy: The first 200 years. *Agriculture Outlook*, March 2000. USDA Economic Research Service, Washington DC.

Agrios, G. N. (1988). *Plant Pathology*, 3rd ed. Academic Press, San Diego, CA, 803p.

Agrow (1993). Pan-uk.org/pestnews/pn22/pn22p11 (accessed on 15 November 2003).

Agrow (1995a). *World Crop Protection News*, vol. 222, p. 7.

Agrow (1995b). *World Crop Protection News*, vol. 224, p. 3.

Agrow (2000). GM seed sales rise. *World Crop Protection News*, vol. 352, 19 May 2000.

Agrow (2017). Dow/DuPont merger to close on August 31st. https://agrow.agribusiness-intelligence.informa.com/AG028197/DowDuPont-merger-to-close-on-August-3 (accessed on 19 August 2017).

AHRP (2016). How Monsanto rigged the system: Through politics & propaganda. *Alliance for Human Research Protection (AHRP)*, 15 November 2016. ahrp.org/a-profit-enhancing-policy-shift-trumps-public-safety-scientific-integrity

Aigner, E. J., Leone, A. D. and Falconer, R. L. (1998). Concentrations and enatiomeric ratios of organochlorine pesticides in soils from the U.S. corn belt. *Environmental Science & Technology*, vol. 32, pp. 1162–68.

Albright, R. K., Kram, B. W. and White, R. P. (1983). Malathion exposure associated with acute renal failure. *Journal of the American Medical Association, vol.* 250, no. 18, p. 2469. doi:10.1001/jama.1983.03340180031010.

Aldridge, W. N., Miles, J. W., Mount, D. L. and Verschoyle, R. D. (1979). The toxicological properties of impurities in malathion. *Archives in Toxicology*, vol. 42, pp. 95–106.

Allen, R. (1980). *How to Save the World: Strategy for World Conservation*. Kogan Page, London, UK.

Altieri, M. A. (1995). *Agroecology: The Science of Sustainable Agriculture*, 2nd ed. Westview Press, Boulder, CO.

AMA (2003). *American Marketing Association*. http://www.marketingpower.com/live/mg-dictionary-view3957.php (accessed on 15 July 2003).

Amin, A. (2000). Post-Fordism: Model, fantasies and phantoms of transition. In: Amin, A. (Ed.), *Post-Fordism: A Reader*. Blackwell Publishers, Oxford, UK, pp. 1-39.

Amin, A. and Thrift, N. (1994). *Globalization, Institutions, and Regional Development in Europe*. Oxford University Press, Oxford and New York.

Anderson, S. and Cavanagh, J. (2000). *Top 200: The Rise of Corporate Global Power*. Institute for Policy Studies, Washington DC.

Anderson, F. R., Glicksman, R. L., Mandelker, D. R. and Tarlock, A. D. (1999). *Environmental Protection: Law and Policy*, 3rd ed. Aspen Law and Business, New York, NY.

Andrew, D. A. (1991). Vegetational diversity and arthropod population response. *Annual Review of Entomology*, vol. 36, pp. 561-86.

Angus, I. (2008). Food crisis: 'The greatest demonstration of the historical failure of the capitalist model'. *Global Research* (28 April 2008).

ANOHS (1997). African newsletter on occupational health and safety. Pest management notes no. 5.

Anon. (1935). In: University of Illinois (1990). http://www.fec.cropsci.uiuc.edu/1990/report4/ (accessed on 16 October 2003).

AON (2001). *Ag Online News*, 8 March 2001.

ARWPPR (1992). *Annual Report of Working Party on Pesticide Residues*. HMSO, London.

Associated Press (2002). Monsanto seeks to profit from new biotech products. *Associated Press*, 1 April 2002.

Assouline, G. and Pierre-Benoit, J. (2001). Rhone-Poulenc agrochimie: An uncertain future. *AgBioForum*, vol. 4, no. 1, pp. 26-33.

Atkins, P. and Bowler, I. (2001). *Food in Society. Economy, Culture, Geography*. Arnold, London, UK.

ATSDR (1997). https://www.atsdr.cdc.gov/hs/hsees/Cum1993-1997.html

ATT (2003). Att.net/tangents/issue/globalpop.ht (accessed on 16 October 2003).

Attaran, A., Roberts, D. R., Curtis, C. F. and Wenceslaus, L. K. (2000). Balancing risks on the backs of the poor. *Nature Medicine*, vol. 6, no. 7, pp. 729-31.

Aventis (2001). Bayer agrees to acquire Aventis CropScience. *Aventis*, press release, 2 October 2001.

Aventis (2003). http://www.aventis.com/main/print.asp?pageid=633221490251445656 4&folderid=&1 (accessed on 10 July 2003).

Avery, N., Drake, M. and Lang, T. (1993). *Cracking the Codex - An Analysis of Who Sets World Food Standards*. National Food Alliance, London.

Bailey, L. H. (1909). *Cyclopedia of American Agriculture: A Popular Survey of Agricultural Conditions, Practices and Ideals in the United States and Canada*, vol. 2. Macmillan Company, New York, NY.

Baker, E. (1978). Epidemic malathion poisoning in Pakistan malaria workers. *The Lancet*, vol. 1, pp. 31-4.

Baldock, D. and Bennett, G. (1991). *Agriculture and the Polluter Pays Principle: A Study of Six EC Countries*. Institute for European Environment Policy, London, UK.

Ball, P. (1996). Living factories. *New Scientist*, 3 February 1996.

Barker, R., Herdt, R. W. and Rose, B. (1985). *The Rice Economy of Asia*. Resources for the Future, Washington DC. In cooperation with the International Rice Research Institute, Philippines.

Barrett, H. R. and Browne, A. (1996). Export horticulture production in sub-Saharan Africa: The incorporation of The Gambia. *Geography*, vol. 81, pp. 47–56.

Barrett, H. R., Ilberry, B. W., Browne, A. W. and Binns, T. (1999). Globalization and the changing networks of food supply: The importation of fresh horticultural produce from Kenya into the UK. *Transactions of the Institute of British Geographers*, vol. 24, pp. 159–74.

BASF (1998). Basf.de/en/corporate/overview/geschichte/1998.htm (accessed on 10 July 2003).

BASF (2000). *BASF*, press release, Ludwigshafen, Germany, 13 March 2000.

BASF (2003). BASF AG: Milestones in history: 1865–1901. *BASF*. Basf.de/en/corporate/overview/geschichte/1901.htm (accessed on 10 July 2003).

Baumol, W. J., Panzar, J. C. and Willig, R. D. (1992). *Contestable Markets and the Theory of Industry Structure*. Harcourt Brace Jovanovich, New York, NY.

Bayer (1988). Bayer.com/en/bayer/history/printview.php?page1=ug_1988.php (accessed on 01 September 2003).

Bayer (2003a). Bayer.com/en/bayer/history/printview.php?page1=ug_1863.php (accessed on 01 September 2003).

Bayer (2003b). Bayer.com/en/bayer/history/printview.php?page1=ug_1881.php (accessed on 01 September 2003).

Bayer (2003c). Bayer.com/en/bayer/history/printview.php?page1=ug_1925.php (accessed on 01 September 2003).

Bayer (2003d). Bayer.com/en/bayer/history/printview.php?page1=ug_1951.php (accessed on 01 September 2003).

Bayer (2003e). Bayer.com/en/bayer/history/printview.php?page1=ug_1974.php (accessed on 01 September 2003).

Bayer (2003f). Bayer.com/en/bayer/printview.php?page1=orga.php (accessed on 10 July 2003).

Bayer (2003g). http://www.bayer.com/en/bayer/ub_up_gentechnik.php (accessed on 10 July 2003).

Bayer (2003h). http://www.bayerus.com/msms/about/programmes/index.htm (accessed on 10 July 2003).

Bayer (2017). https://www.cropscience.bayer.us/products/fungicides/serenade-opti

BCERF (1998). Pesticides and breast cancer risk, an evaluation of Heptachlor. Cornell University program on breast cancer and environmental risk factors in New York State. Fact Sheet 12, March 1998.

Belsie, L. (2001). *Christian Science Monitor*, 15 June 2001.

Berenbaum, M. R. (1995). *Bugs in the System: Insects and Their Impact on Human Affairs*. Addison-Wesley, Reading, MA.

Beyond Pesticides (2003). Beyond pesticides daily news archive. *Beyond Pesticides*, daily news blog, 10 January 2003. http://www.beyondpesticides.org/main.html (accessed on 11 July 2003).

Beyond Pesticides (2017). The ever-revolving door: Industry and the EPA. *Beyond Pesticides*, daily news blog. http://beyondpesticides.org/dailynewsblog/2017/10/ever-revolving-door-industry-epa/ (accessed on 03 June 2017).

Bijman, J. (2001a). Advanta: Worldwide challenges. *AgBioForum*, vol. 4, pp. 34-9.

Bijman, J. (2001b). AgrEvo: From crop protection to crop production. *AgBioForum*, vol. 4, pp. 20–5.

Bijman, J. and Joly, P. (2001a). Introduction to the pita project. *AgBioForum*, vol. 4, no. 1, pp. 1-3.

Bijman, J. and Joly, P. (2001b). Innovation challenges for the European agbiotech industry. *AgBioForum*, vol. 4, no. 1, pp. 4–13.

BIO (2004a). Biotechnology delivers clear benefits to the farmer and the environment. *Biotechnology Industry Organization*. www.bio.org (accessed on 20 February 2004).

BIO (2004b). Agricultural biotech products on the market. *Biotechnology Industry Organization*. www.bio.org (accessed on 20 February 2004).

BIO (2004c). Codex alimentarius. *Biotechnology Industry Organization*. www.bio.org (accessed on 20 February 2004).

BIO (2004d). Agricultural biotechnology can help developing countries gain food security. *Biotechnology Industry Organization*. www.bio.org (accessed on 20 February 2004).

Blandford, D. and Fulponi, L. 1999. Emerging public concerns in agriculture. *European Review of Agricultural Economics*, vol. 26, no. 3, pp. 409–24.

Bloom, D. E. (1995). International public opinion the environment. *Science*, vol. 269, pp. 354–8.

Boardman, R. (1986). *Pesticides in World Agriculture: The Politics of International Regulation*. St Martin's Press, New York, NY.

Boltanski, L. and Thévenot, L. (1991). *De la justification. Les économies de la grandeur*. Gallimard, Paris, France.

Bosso, C. J. (1987). *Pesticides and Politics: The Life Cycle of a Public Issue*. University of Pittsburgh Press, Pittsburgh, 312pp.

Bourdieu, P. (1984). *Questions de Sociologie*. Minuit, París, France, pp. 222-50.

Bowler, I. R. (1988). Agricultural geography. *Progress in Human Geography*, vol. 12, pp. 538–48.

Bowler, I. R. (1990). Agricultural geography. *Progress in Human Geography*, vol. 14, pp. 569–78.

Bowler, I. R. (1992). *The Geography of Agriculture in Developed Market Economies*. Longman Scientific and Technical, New York, NY.

Britton, D. (1990). *Agriculture in Britain: Changing Pressures and Policies*. CAB International, Wallingford, UK.

Bull, D. (1982). *A Growing Problem: Pesticides and the Third World Poor*. Oxfam, Oxford, UK.

Bullock, W. O. (2000). Institute for Biotechnology Information Research, Triangle Park, NC. ISB-News Report.

Burgett, M. (2003). *Oregon State University*, Department of Entomology. oregonstate.edu/dept/entomology/burgett/X0019_ENT30022.txt.html (accessed on 23 September 2003).

Burkhardt, J. (1992). On the ethics of technical change: The case of BST. *Technology in Society*, vol. 14, pp. 221–43.

Busch, L. and Juska, A. (1997). Beyond political economy: Actor networks and the globalization of agriculture. *Review of International Political Economy*, vol. 4, pp. 688–708.

Buttel, F. H. (1992). Environmentalism: Origins, processes, and implications for rural social change. *Rural Sociology*, vol. 57, no. 1, pp. 1-27.

Buzby, J. and Skees, J. (1994). Consumers want reduced exposure to pesticides on food. *Food Review*, vol. 17, pp. 19–26.

BWTND (2003). *Bridges Weekly Trade News Digest*, vol. 3, no. 42.

Byé, P. and Font, M. (1991). Technical change in agriculture and new functions for rural spaces in Europe. Paper presented at the American Sociological Association, Cincinnati, OH.

CABI (2003). http://www.cabi-publishing.org/Bookshop/ReadingRoom/0851993400/3400ch1.pdf (accessed on 23 September 2003).

Cairns, J. E., Hellin, J., Sonder, K., Araus, J. L., MacRobert, J. F., Thierfelder, C. and Prasanna, B. M. (2013). Adapting maize production to climate change in sub-Saharan Africa. *Food Security*, vol. 5, pp. 345–60. doi:10.1007/s12571-013-0256-x.

Callon, M. (1986). Some elements of a sociology of translation: Domestication of the scallops and the fishermen of St Brieuc bay. In: Law, J. (Ed.), *Power, Action and Belief: A New Sociology of Knowledge*. Routledge & Kegan Paul, London, UK, pp. 196–233.

Campbell, H. (1978). Estimating the marginal productivity of agricultural pesticides: The case of tree fruit farms in the Okanagan Valley. *Canadian Journal of Agricultural Economics*, vol. 24, no. 2, pp. 23–30.

Canada Newswire (2000). Biosafety protocol: A major step forward to sustain development of biotechnology. *Canada Newswire*, 29 January 2000.

Cecchine, G., Golomb, B. A., Hilborne, L. H., Spektor, D. M. and Anthony, C. R. (2000). *A Review of the Scientific Literature As It pertains to Gulf War Illness, Volume 8: Pesticides*. National Defense Research Institute, RAND, Santa Monica, CA.

Census (2003). Census.gov/ipc/www/worldhis.html (accessed on 12 July 2003).

Chataway, J. (2001). Novartis: New agribusiness strategy. *AgBioForum*, vol. 4, no. 1, pp. pp.14–19.

Chemical and Engineering News (1998). Divesting conoco and other crown jewels. *Chemical Engineering News* (June 1998), p. 46.

Chemical and Engineering News (2001). Rohm and Haas also streamlines operations. *Chemical and Engineering News*, vol. 79, no. 19, pp. 11. ISSN: 0009-2347. Pubs.acs.org/cen/topstory/7919/7919notw2.html

Chemical Week (1997). Hurdles for Dow-Sentrachem. *Chemical Week*, 21 May, p. 3.

Chemical Week (1998). *Chemical Week*, 25 March.

Cheminova (2002). Cheminova.com/en/cheminova/news_views/20021129_04.htm (accessed on 01 September 2003).

Chemische Industrie (1989). *Chemische Industrie*, vol. 8, no. 89, p. 28.

Chemische Industrie (1992). *Chemische Industrie*, vol. 1, no. 92, p. 38.

Chemische Industrie (1994). *Chemische Industrie*, vol. 3, no. 94, pp. 28–32.

Christian Aid (1996). *The Global Supermarket*. Christian Aid, London.

Ciba-Geigy (1993). *Sustainable Agriculture: We Know It's Important. What Does It Mean?* Ciba broadcast. Ciba-Geigy Corp, Greensboro, NC.

Ciba-Geigy (1996). FTC accord in Ciba-Geigy/Sandoz merger to prevent slowdown in gene therapy development and preserve competition in corn herbicides, flea control markets. *Federal Trade Commission*, press release. ftc.gov/opa/1996/12/ciba.htm (accessed on 22 January 2004).

Clapp, J. (2002). *Transnational Corporate Interests and Global Environmental Governance: Negotiating Rules for Agricultural Biotechnology and Chemicals*. Trent University, Ontario, Cananda.

Cloke, P. and Goodwin, M. (1992). Conceptualizing countryside change: From post-fordism to rural structural coherence. *Transactions of the Institute of British Geographers*, vol. 17, pp. 321–36.

Clunies-Ross, T. and Hildyard, N. (1992). *The Politics of Industrial Agriculture*. Earthscan, London, UK.

Cobb, D., Feber, R. and Hopkins, A. (1999). Integrating the environmental and economic consequences of converting to organic agriculture: Evidence from a case study. *Land Use Policy*, vol. 16, no. 4, pp. 207–21.

CODEX (1989). *Procedural Manual*, 7th ed. FAO, Rome, Italy, p. 31.

CODEX (2003). http://www.codexalimentarius.net/ (accessed on 12 May 2003).

Cohen, R. and Kennedy, P. (2000). *Global Sociology*. MacMillan, London, UK, p. 377.

Collins, H. B. (2000). Terminator technology not terminated. *Agra/Industrial Biotechnology Legal Letter*, vol. 1, no. 1, p. 4.

Copplestone, J. (1977). A global view of pesticide safety. In: Watson, D. and Brown, W. (Eds), *Pesticide Management and Insect Resistance*. Elsevier Academic Press, New York, NY, pp. 147–55.

Corporate Watch UK (2002a). Syngenta: A company profile. Corporatewatch.org.uk/profiles/Syngenta/syngenta2.htm (accessed on 25 November 2003).

Corporate Watch UK (2002b). Advanta B.V. Corporatewatch.org.uk/profiles/biotech/Advanta/advanta1.htm (accessed on 25 November 2003).

Corporate Watch UK (2002c). Advanta B.V. Corporatewatch.org.uk/profiles/biotech/Advanta/advanta4.htm (accessed on 25 November 2003).

Corporate Watch UK (2003a). Corporatewatch.org.uk/profiles/Syngenta/syngenta4.htm (accessed on 25 November 2003).

Corporate Watch UK (2003b). Corporatewatch.org.uk/profiles/dupont/dupont.htm (accessed on 25 November 2003).

Corporate Watch UK (2003c). Corporatewatch.org.uk/profiles/dupont/dupont2.htm (accessed on 25 November 2003).

Corporate Watch UK (2003d). Corporatewatch.org.uk/profiles/Syngenta/syngenta5.htm (accessed on 25 November 2003).

Costanza, R., d'Arge, R., de Groote, R., Farber, S., Grasso, M., Hannon, B., Limburg, K., Sutton, P. and van der Belt, M. (1997). The value if the world's ecosystem services and natural capital. *Nature*, vol. 387, pp. 253–60.

Cowan, R. and Gunby, P. (1996). Sprayed to death: Path dependencies, lock-in and pest control strategies. *The Economic Journal*, vol. 106, pp. 521–42.

Cronin, D. (2013). Pesticides and the revolving door. *Counterpunch*. https://www.counterpunch.org/2013/06/11/pesticides-and-the-revolving-door/

CropLife International (2002). http://www.croplife.org/website/pages/background.aspx (accessed on 23 August 2003).

CropLife International (2004). http://www.croplife.org (accessed on 23 February 2004).

CropLife International (2014). Global snapshot of no-till farming. https://croplife.org/news/global-snapshot-of-no-till-farming/ (accessed on 25 October 2017).

CSE (2003). *Center for Science & Environment*. http://www.cseindia.org/html/eyou/geg/factsheet/fact11.pdf (accessed on 23 December 2003).

CSE (2011). Endosulfan industry's dirty war to save its toxic product: Summary of recent events by CSE. *Center for Science & Environment*. http://www.cseindia.org/content/endosulfan-industrys-dirty-war-save-its-toxic-product-summary-recent-events-cse-0 (accessed on 24 November 2017).

Daly, H. E. (1999). Globalization versus internationalization – some implications. *Ecological Economics*, vol. 31, pp. 31–7.

Davignon, L. F., St-Pierre, J., Charest, G. and Tourangeau, F. J. (1965). A study of the chronic effects of insecticides in man. *Canadian Medical Association Journal*, vol. 92, pp. 597–602.

DEFRA (2000). Design of a tax or charge scheme for pesticides. *Department for Environment, Food & Rural Affairs*, 29 April 2000.

DEFRA (2001). Chemical stakeholder forum: Meetings - Intergovernmental Fora on Chemical Safety (IFCS). defra.gov.uk/environment/chemicals/csf/03082001/07.htm (accessed on 09 February 2004).

Demmig-Adams and Adams (2002). Planteark.org/avantgo/dailynewsstory.cfm?newsid=19069 (accessed on 29 April 2003).

Dent, D. R. (2000). *Insect Pest Management*, 2nd ed. CABI Publishing, Wallingford, UK.

Dent, D. R. (2003). History of pest management. http://www.pestmanagement.co.uk/lib/history.shtml (accessed on 01 February 2004).

Detrick (2003). http://www.detrick.army.mil/cutting_edge/chapter5.cfm (accessed on 24 November 2003).

Deutsche Bank (1999). GMOs are dead. 12 July. http://www.biotech-info.net/Deutsche.html (accessed on 20 September 2003).

Dinham, B. (1994). Pesticide production in the South. Linking production and trade. *Pesticides News*. agris.fao.org.

Dinham, B. (1996). Pesticide production in the south. In: Dinham, B. (Ed.), *Growing Food Security: Challenging the Link Between Pesticides and Access to Food*. The Pesticide Trust, London, UK, pp. 12–16.

Dittus, K. M. and Hillers, V. N. (1996). Attitudes about the nutritional benefits and pesticide exposure risk from fruit and vegetable consumption. *Family and Consumer Sciences Research Journal*, vol. 4, no. 2, pp. 406–21.

Domain-B (2003). Tough times ahead for Aventis CropScience. *domain-b.com*, 10 March 2000. Domain-b.com/companies/companies_a/Aventis_cropscience_india/200003 10aventiscropscience (accessed on 18 July 2003).

Donga (2003). History of pesticides. Home.donga.ac.kr/pesticide/history.htm

Dow (2003a). This is dow about dow history. Dow.com/about/aboutdow/history/1890s.htm (accessed on 18 August 2003).

Dow (2003b). This is dow about dow history. Dow.com/about/aboutdow/history/1940s.htm (accessed on 18 August 2003).

Dow (2003c). This is dow about dow history. Dow.com/about/aboutdow/history/1950s.htm (accessed on 18 August 2003).

Dow (2003d). This is dow about dow history. Dow.com/about/aboutdow/history/2000s.htm (accessed on 18 August 2003).

Dow Jones (1999a). *Dow Jones News*, 30 June 1999.

Dow Jones (1999b). *Dow Jones News*, 25 August 1999.

Downtoearth (2011). Pesticides lobby's coup. *Downtoearth*, 15 January 2011. http://www.downtoearth.org.in/news/pesticides-lobbys-coup-2763 (accessed on 25 November 2017).

Duffy, M. (2001). Who benefits from biotechnology? *American Seed Trade Association Meeting*, 5-7 December 2001. http://www.leopold.iastate.edu/pubinfo/paperspeeches/biotech.html

Duke, S. O. (1990). Natural pesticides from plants. In: Janick, J. and Simon, J. E. (Eds), *Advances in New Crops*. Timber Press, Portland, OR.

Dunlap, T. R. (1981). *DDT: Scientists, Citizens, and Public Policy*. Princeton University Press, Princeton, NJ.

DuPont (1999). *DuPont*, 22 September 1999. Dupont.com/corp/overview/history (accessed on 20 February 2004).

DuPont (2004). *DuPont*. dupont.com/biotech/difference/principles.html (accessed on 20 February 2004).

Dutton, G. (1998). Ag biotech ripe for investment. *Chemical Market Reporter*, 9 November 1998.

EC (1998). Council Regulation 1139/98 concerning the compulsory indication the labelling of certain foodstuffs produced from genetically modified organisms. *Official Journal of the European Communities*, vol. L159, p. 4.

Ecologist (1980). *Ecologist*, March 1980.

E-council (2003). ecouncil.ac.cr/rio/focus/report/English/subsidies/chap6.htm (accessed on 05 September 2003).

EDF v Ruckelshaus (1971). *EDF v. Ruckelshaus*, 439 F.2d 584 (D.C. Cir. 1971).

Edwards, J. G. (1978). *Testimony and Affidavit*. California Department of Food and Agriculture, Los Angeles, CA.

Edwards, D. (1995). *Free to be Human: Intellectual Self-Defense in an Age of Illusion*. Resurgence, Green Books Ltd, Dartington, UK, p. 188.

Efron, E. (1985). *The Apocalyptics*. Touchstone/Simon & Schuster, New York, NY, p. 268.

Ekboir, J. (Ed.). (2002). *World Wheat Overview and Outlook*. CIMMYT, Mexico City, Mexico, pp. 1–38.

Enderson, J. H. and Berger, D. D. (1968). Chlorinated hydrocarbons in peregrines from northern Canada. *Condor*, vol. 70, pp. 149–53.

Engels, J. M. M. (1995). *In-situ* conservation and use of plant genetic resources for food and agriculture in developing countries. Report of a *DSE/ATSAF/IPGRI Workshop*, 2–4 May 1995, Bonn-RÖttgen, Germany.

Environmental Defense Fund (1999). Environmental defense fund collection. http://www.sunysb.edu/library/edfhist.htm (accessed on 25 February 2004).

Environmental Liaison Centre (1987). Monitoring and reporting the implementation of the international code of conduct on the use and distribution of pesticides. Final report. ELC, Nairobi, p. 12.

EPA (1972a). DDT ban takes effect. *EPA*, press release, 31 December 1972. epa.gov/history/topics/ddt/01.htm.

EPA (1972b). The challenge of the environment: A primer on EPA's statutory authority. *EPA*, December 1972. epa.gov/history/topics/fwpca/05.htm (accessed on 12 January 2004).

EPA (1972c). EPA to ask for comments on new pesticide law. *EPA*, press release, 8 November 1972. epa.gov/history/topics/fifra/03.htm (accessed on 12 January 2004).

EPA (1975a). Federal register. *EPA*, 8 April 1974, p. 15, 962.

EPA (1975b). Train stops manufacture of heptachlor/chlordane. Cites imminent cancer risk. *EPA*, press release, July 1975. epa.gov/history/topics/legal/01.htm (accessed on 12 January 2004).

EPA (1983). *EPA*, press release, 30 September 1983. epa.gov/history/topics/legal/02.htm.

EPA (1988). FIFRA amendments of 1988. *EPA*, press release, 26 October 1988. epa.gov/history/topics/fifra/01.htm (accessed on 12 January 2004).

EPA (1990). The spirit of the first Earth Day. *EPA*, January/February 1990. epa.gov/history/topics/earthday/01.htm (accessed on 12 January 2004).

EPA (1992a). *Environmental Protection Agency*, March 1992.

EPA (1992b). *EPA*, press release, 19 May 1992. epa.gov/history/topics/sdwa/02.htm

EPA (1993). Environmental protection and toxic substances. *EPA*, September 1993.

EPA (1997). epa.gov/oppbead1/pestsales/97pestsales/tables_charts1997_5.html (accessed on 23 May 2003).

EPA (1999). Methyl parathion risk management decision. Office of pesticide programs, August 1999.

EPA (2000a). Environmental protection and toxic substances. *EPA*, September 2000.

EPA (2000b). Chlorpyrifos revised risk assessment and agreement with registrants. Office of pesticide programs, June 2000.

EPA (2001). Biotechnology corn approved for continued use. *EPA*, press release, 16 October 2001. http://www.usepa.gov (accessed on 28 November 2003).

EPA (2003a). DDT regulatory history: A brief survey (to 1975). *EPA*. epa.gov/history/topics/ddt/02.htm (accessed on 12 January 2004).

EPA (2003b). epa.gov/safewater/dwh/t-soc/endrin.html (accessed on 22 August 2003).

EPA (2003c). epa.gov/safewater/dwh/t-soc/toxaphene.html (accessed on 22 August 2003).

EPA (2003d). epa.gov/safewater/dwh/t-soc/lindane.html (accessed on 22 August 2003).

EPA (2003e). epa.gov/docs/RED/factsheets/endosulfan_fs.htm (accessed on 22 August 2003).

EPA (2003f). epa.gov/docs/RED/factsheets/lindane_fs.htm (accessed on 22 August 2003).

EPA (2003g). epa.gov/safewater/dwh/t-soc/simazine.html (accessed on 22 August 2003).

EPA (2003h). epa.gov/safewater/dwh/t-soc/atrazine.html (accessed on 22 August 2003).

EPA (2003i). epa.gov/safewater/dwh/t-soc/edb.html (accessed on 22 August 2003).

EPA (2003j). http://www.epa.gov/REDs/carbaryl_ired.pdf (accessed on 22 August 2003).

EPA (2003k). http://www.epa.gov/oppsrrd1/REDs/0028red.pdf (accessed on 22 August 2003).

EPA (2004a). Pesticides and public health. epa.gov/history/publications/formative6.htm (accessed on 12 January 2004).

EPA (2004b). The Guardian: EPA's formative years 1970–1973. epa.gov/history/publications/print/formative.htm (accessed on 12 January 2004).

EPA (2004c). The Guardian: Origins of the EPA. epa.gov/history/publications/print/origins.htm (accessed on 12 January 2004).

Epoch Times (2016). New technology spurs consolidation in seed industry. *Epoch Times*, 27 September 2016.

Erlich, P. R. and Erlich, A. H. (1996). *Betrayal of Science and Reason: How Anti-Environmental Rhetoric Threatens Our Future*. Island Press and Shearwater Books, Washington DC.

Erlichman, J. (1991). The deadly cocktails. *The Guardian*, 2 March 1990.

Escalada, M. M. and Heong, K. L. (1993). Communication and implementation of change in crop protection. In: Ciba Foundation (Ed.), *Crop Protection and Sustainable Agriculture*. John Wiley and Sons, Chicester, UK, pp. 191–202.

Estienne, C. and Liebault, J. (1600). *Maison Rustique or the Countrie Farme*, translated by R. Surflet. E. Bollifant, London.

ETC (1996). The life industry. *ETC Communique*, 30 September 1996. Etcgroup.org/article.asp?newsid=198

ETC (1998). Seed industry consolidation: Who owns whom? *ETC Communique*, 30 July 1998. Etcgroup.org/article.asp?newsid=186

ETC (1999). The gene giants: Masters of the universe? *ETC Communique*, 30 March 1999. Etcgroup.org/article.asp?newsid=180

ETC (2001a). Globalization, Inc. concentration in corporate power: The unmentioned agenda. *ETC Communique*, issue 71, July/August 2001.

ETC (2001b). Seeds, biotech, agrochemicals D.A.M.N.D. *ETC Group*, 27 August 1997. http://www.etcgroup.org/article.asp?newsid=71 (accessed on 14 November 2003).

ETC (2003). Oligopoly, Inc. concentration in corporate power: 2003. *ETC Communique*, issue 80, November/December 2003.

ETC (2005). Global seed industry concentration - 2005. *ETC Communique*, issue 90, September/October 2005.

ETC (2008). Patenting the 'Climate Genes'... And capturing the climate agenda. *ETC Communique*, issue 99, May-June.

ETC (2011). Who will control the Green economy? *ETC Communiqué*, issue 107, November 2011.

ETC (2015). Breaking Bad: Big ag mega-mergers in play dow + duPont in the pocket? Next: Demonsanto? *ETC Communique*, issue 116, December 2015.

ETC (2016). Software vs. Hardware vs. Nowhere: Deere & Co. is becoming 'Monsanto in a Box'. *ETC Group*, December 2016.

EUREP (1999). Good Agricultural Practice [GAP] protocol. *Euro-Retailer Produce Working Group*. http://www.eurep.org

European Union (2017). http://ec.europa.eu/environment/pubs/pdf/factsheets/7eap/en.pdf (accessed on 20 July 2017).

Evans, L. T. (1998). *Feeding the Ten Billion: Plants and Population Growth*. Cambridge University Press, Cambridge, UK.

Evans, D. A. (2000). New era, new challenges, new solutions. In: *The BCPC Conference – Pests and Diseases 2000*. British Crop Protection Council, Farnham, Surrey, pp. 3-18.

Facts versus Fear (1998). Facts versus Fear: DDT, 3 June 1998.

Fairly, P. (1996). NAFTA commission eyes DDT ban. *Chemical Week*, vol. 158, no. 5, p. 41.

Falck-Zepeda, J. B., Traxler, G. and Nelson, R. G. (2000). Surplus distribution from the introduction of a biotechnology innovation. *American Journal of Agricultural Economics*, vol. 82, pp. 360–9.

FAO (United Nations-Food and Agriculture Organization) (1997). *State of the World's Forests*. FAO, Rome, Italy.

FAO (1998). IPM-trained farmers in Indonesia escape pest outbreaks. http://www.fao.org/news/1998/981104-ehtm (accessed on 05 November 2003).

FAO (2001). Baseline study on the problem of obsolete pesticide stocks. FAO, Rome, Italy.

FAO (2002). http://www.fao.org/ag/agp/agpp/pesticid/code/Download/Revised%20Version_ICOC.pdf

FAO (2003a). http://www.fao.org/WAICENT/FAOINFO/AGRICULT/AGP/AGPP/Pesticid/Disposal/index_en.htm (accessed on 17 May 2005).

FAO (2003b). http://www.fao.org/DOCREP/005/AC806E/ac806e06.htm (accessed on 17 May 2003).

FAO (2003c). http://www.fao.org/globalipmfacility/home.htm (accessed on 17 May 2003).

FAO (2004). http://www.foa.org/biotech/stat.asp (accessed on 17 January 2004).

FAO (2017a). FAO cereal supply and demand brief. http://www.fao.org/worldfoodsituation/csdb/en/ (accessed on 27 September 2017).

FAO (2017b). Implementation of Article 27.3(b): Drafting and Enacting National Legislation (Sui Generis Systems). http://www.fao.org/docrep/003/x7355e/x7355e07.htm (accessed on 04 October 2017).

Farm Chemicals (1996). *Farm Chemicals*, March 1996, p. 78.

Farmers Weekly (1999). *Farmers Weekly*, 3 December 1999.

FAZ (1995). *Frankfurter Allgemeine Zeitung*, 28 March 1995, p. 21.

FDA (2003a). http://www.fda.gov/opacom/backgrounders/miles.html (accessed on 10 July 2003).

FDA (2003b). http://www.fda.gov/oc/historyoffda/section3.html (accessed on 10 July 2003).

Feedstuffs (2002). *Feedstuffs*, issue 35.

Fernandez-Cornejo and McBride (2002). *Feedstuffs*, issue 35.

Ferrell, J. A. and Witt, W. W. (2002). Comparison of glyphosate with other herbicides for weed control in corn (*Zea mays*): Efficacy and economics. *Weed Technology*, vol. 16, no. 4, pp. 701–6.

Financial Daily (2000). *Financial Daily*, 2 April 2000.

Financial Times (1998a). From corn to cancer. *Financial Times*, 2 June 1998.

Financial Times (1998b). The facts of life: Chemical and pharmaceutical companies see their future in biological innovation. *Financial Times*, 9 December 1998.

Finch, J. (2000). Mr Del Monte says yes to cholesterol lowering juice. *The Guardian*, 20 April.

Finkel, A. M. (1995). Toward less misleading comparisons of uncertain risks: The example of aflatoxin and alar. *Environmental Health Perspectives*, vol. 103, no. 4, pp. 376–85.

Fischer, R. A., Byerlee, D. and Edmeades, G. O. (2014). *Crop Yields and Global Food Security: Will Yield Increase Continue to Feed the World?* Australian Centre for International Agricultural Research, Canberra, Australia. Retrieved from http://aciar.gov.au/publication/mn158.

FMC (2003). fmc.co./corporate/v2/generaldetail/0,1478,16,00.html (accessed on 19 August 2003).

FOE (2003). *Friends of the Earth*. FOE.co.uk/resource/briefings/scimac.html (accessed on 15 November 2003).

Ford Foundation (2003). http://www.fordfound.org/publications/recent_articles/docs/celebrating_indonesia/02_english.pdf (accessed on 03 October 2003).

Freudenburg, W. R. and Pastor, S. K. (1992). Public responses to technological risks: Towards a sociological perspective. *The Sociological Quarterly*, vol. 33, pp. 389–412.

Friedland, W. (1984). Commodity systems analysis: An approach to the sociology of agriculture. *Research in Rural Sociology and Development*, vol. 1, pp. 221–35.

Friedland, W. (1994). The new globalization: The case of fresh produce. In: Bonnano, A. (Ed.), *From Columbus to Conagra*. University of Kansas Press, Lawrence, KS, pp. 210–31.

Friedman, M. (1962). *Capitalism and Freedom*. University of Chicago Press, Chicago, IL. ISBN:13:9780226264004.

Friedmann, H. (1993). The political economy of food. *New Left Review*, vol. 197, pp. 29–57.

Friedmann, H. (2005). From colonialism to green capitalism: Social movements and the emergence of food regimes. In: Buttel, F. H. and McMichael, P. D. (Eds), *New*

Directions in the Sociology of International Development: Research in Rural Sociology and Developments, vol. 11. Elsevier, Amsterdam, the Netherlands, pp. 227-64.

Fulton, M. (1997). The economics of intellectual property rights: Discussion. *American Journal of Agricultural Economics*, vol. 79, pp. 1592-4.

Fulton, M. and Giannakas, K. (2001). Agricultural biotechnology and industry structure. *AgBioForum*, vol. 4, no. 2, pp. 137-51.

FWS (2003). U.S. Fish and Wildlife Service. laws.fws.gov/laswsdigest/fednvp.html (accessed on 15 March 2003).

GAIA Foundation (1999). *GAIA Foundation*, UK, 12 May 1999, accessed on 19 October 2003.

Gasson, R. and Potter, C. (1988). Conservation through land diversification: A survey of farmers' attitudes. *Journal of Agricultural Economics*, vol. 39, pp. 340-51.

Gaugler, R. (1997). Rethinking approaches to commercializing bio-pesticides. *The Israel Journal of Plant Protection Sciences*, vol. 25, no. 3.

GCPF (1997). The benefits of using crop protection products with specific reference to habitat preservation and biodiversity. *GCPF*. http://www.gcpf.org/positions/bio-diver.html (accessed on 13 July 2003).

GCPF (2000). *Global Climate Partnership Fund*. www.gcpf.org/ (accessed on 13 July 2003).

GEF (2002). Africa Stockpile Programme. *Global Environment Facility*. In: World Bank, World Development Report 2003 (accessed on 13 July 2003).

GEF (2003a). http://www.gefweb.org/outreach/Outreach-PUblications/Agenda_21_.pdf (accessed on 13 July 2003).

GEF (2003b). http://www.gefweb.org/index.html (accessed on 13 July 2003).

GEF (2017). https://www.thegef.org/events/gef-7-replenishment (accessed on 26 October 2017).

Genentech (2003). http://www.gene.com/gene/about/corporate/history/ (accessed on 19 September 2003).

Genet (1999). Monsanto & IRRI push pesticides on Thai farmers through Thai development. *NGO*, 6 April 1999. http://www.gene.ch/genet/1999/Apr/msg00005.html (accessed on 09 October 2017).

Genewatch (2003a). The Monsanto Company. Genewatch.org/GeneSrch/Companies/Monsanto.htm (accessed on 13 November 2003).

Genewatch (2003b). Dupont, E. I. De Nemours & Company (INC). Genewatch.org/GeneSrch/Companies/DuPont.htm (accessed on 13 November 2003).

Genewatch (2003c). Syngenta. Genewatch.org/GeneSrch/Companies/Syngenta.htm (accessed on 13 November 2003).

Geo-PIE (2004). Public issues education project. Geo-pie.cornell.edu/regulation/reg.html (accessed on 27 January2004).

Georghiou, G. P. (1986). The magnitude of the problem. In: Glass, E. H. (Ed.), *Pesticide Resistance: Strategies and Tactics for Management*. National Academy Press, Washington DC.

Georghiou, G. and Taylor, C. (1976). Pesticide resistance as an evolutionary phenomenon. In: *Proceedings of the XVth International Congress of Entomology*, 19-27 August 1976, Washington DC, pp. 759-85.

Gerberg, E. J. and Wilcox III, H. (1977). Environmental assessment of malaria and control projects - Sri Lanka. *Agency for International Development*, vol. 20, pp. 32-3.

Giannakas, K. and Fulton, M. (2002). *Consumption Effects of Genetic Modification: What if Consumers are Right?* International Food Policy Research Institute, Washington DC.

Giller, K. E., Beare, M. H., Izac, A. M. and Swift, M. J. (1997). Agricultural intensification, soil biodiversity and agroecosystem function. *Applied Soil Ecology*, vol. 6, pp. 3–16.

Glicksman, R. L., Markell, D. L., Mandelker, D. R., Tarlock, A. D. and Anderson, F. R. (2003). *Environmental Protection: Law and Policy*, 4th ed. Aspen Publishers, New York, NY.

Godfray, H. C. J., Beddington, J. R., Crute, I. R., Haddad, L., Lawrence, D., Muir, J. F., Pretty, J., Robinson, S., Thomas, S. M. and Toulmin, C. (2010). Food security: The challenge of feeding 9 billion people. *Science*, vol. 327, no. 5967, pp. 812–18.

Goklany, I. M. (1999). Meeting global food needs: The environmental tradeoffs between increasing land conversion and land productivity. In: Morris, J. and Bate, R. (Eds), *Fearing Food: Risk, Health and Environment*. Butterworth-Heinemann, Oxford, UK, pp. 256–89.

Goklany, I. M. (2000). Applying the precautionary principle to DDT. 2 December 2000. Available at: http://www.fightingmalaria.org (cited 6 March 2001).

González Esteban, A. L. (2017). Patterns of world wheat trade 1945–2010: The long hangover from the second food regime. *Journal of Agrarian Change* (forthcoming) (accepted 26 January 2017). doi:10.1111/joac.12219.

Goodland, R. (1991). The case that the world has reached its limits: More precisely that the current throughput growth in the global economy cannot be sustained. In: Goodland, R., Daly, H., El Serafy, S. and von Droste, B. (Eds), *Environmentally Sustainable Economic Development: Building on Brundtland*. UNESCO, Paris, France.

Goodman, D. and Redclift, M. (1991). *Refashioning Nature. Food, Ecology and Culture*. Routledge, New York, NY.

Goodman, D. and Watts, M. (1994). Reconfiguring the rural or fording the divide? Capitalist restructuring and the global agro-food system. *The Journal of Peasant Studies*, vol. 22, no. 1, pp. 1–49.

Goodman, D. and Watts, M. J. (Eds) (1997). *Globalising Food: Agrarian Questions and Global Restructuring*. Routledge, London and New York.

Gosovic, B. (2002). Global intellectual hegemony and the international development agenda. *International Social Science Journal*, vol. 52, no. 166, pp. 447-56.

GPC (2000). The WTO and pesticide reform. *Global Pesticide Campaigner*, vol. 10, no. 1.

Green, M. (1976). *Pesticides, Boon or Bane?* Electra Books, London, UK.

Green, B. (1985). *Countryside Conservation: The Protection and Management of Amenity Ecosystems*, The resource management series. Chapman and Hall, London, UK.

Greenaway, T. (2017). http://civileats.com/author/tgreenaway/

Greenland, D. J. and Szabolcs, I. (Eds) (1994). *Soil Resilience and Sustainable Land Use*. CAB International, Wallingford, UK.

Green Peace (1999). *Green Peace*, Germany, 26 August 1999.

Green Peace (2002). *Green Peace*. http://archive.greenpeace.org/~toxics/tbg/tbg2.html (accessed on 20 August 2003).

Greenpeace-CEO-GeneWatch UK (2016). Commission fails to regulate new GMOs after intense US lobbying. *Media Briefing*, 21 April 2016.

Greenyearbook (2003). http://www.greenyearbook.org/igo/csd.htm (accessed on 20 August 2003).

Greer, J. and Singh, K. (2000). A brief history of transnational corporations. Globalpolicy.org/socecon/tnc/historytncs.htm (accessed on 20 November 2003).

Grooms, L. (1999). With merger completed, Harris Moran focuses on future. *Seed and Crops Digest*, January 1999.

Groundup (2003a). Groundup.org/dupont/inputs.htm (accessed on 25 November 2003).

Groundup (2003b). Groundup.org/dupont/lobby.htm (accessed on 25 November 2003).

Groundup (2003c). Groundup.org/dupont/merger.htm (accessed on 25 November 2003).

Groundup (2003d). Groundup.org/dupont/subsid.htm (accessed on 25 November 2003).

Groundup (2003e). Groundup.org/dupont/oqg.htm (accessed on 25 November 2003).

Groundup (2003f). Groundup.org/dupont/fields.htm (accessed on 25 November 2003).

Groundup (2003g). Groundup.org/dupont/ae.htm (accessed on 25 November 2003).

Groundup (2003h). Groundup.org/dupont/strat.htm (accessed on 25 November 2003).

Groundup (2003i). Groundup.org/dupont/pp.htm (accessed on 25 November 2003).

Groundup (2003j). Groundup.org/Monsanto/empire.htm (accessed on 25 November 2003).

Groundup (2003k). Groundup.org/Monsanto/pipeline.htm (accessed on 25 November 2003).

Groundup (2003l). Groundup.org/Novartis/research.htm (accessed on 25 November 2003).

Groundup (2003m). Groundup.org/fcorp.htm (accessed on 25 November 2003).

Gupta, A. S. (2003). MNCs: Some less known facts. delhiscienceforum.org/wto5.html (accessed on 24 November 2003).

Hamburger, J. (2001). IPM that works: The UN FAO IPM Programme and the Global IPM Facility. *Global Pesticide Campaigner*, vol. 11, no. 1.

Handler, P. (1971). The federal government and the scientific community. *Science*, vol. 171, no. 3967, pp. 144-51.

Hansen, J. W. (1996). Is agricultural sustainability a useful concept? *Agricultural Systems*, vol. 50, pp. 117–43.

Hansen-Kuhn, K. and Suppan, S. (2013). *Promises and Perils of the TTIP. Negotiating a Transatlantic Agricultural Market*. Institute for Agriculture and Trade Policy, Minneapolis and Heinrich Böll Foundation, Berlin.

Hartnell, G. (1996). The innovation of agrochemicals: Regulation and patent protection. *Research Policy*, vol. 25, no. 3.

Harvard University (2000). Realizing the promise of green biotechnology for the poor. cid.harvard.edu/cidbiotech/pp/policypaper1.htm (accessed on 17 February 2004).

Harvey, D. (1996). *Justice, Nature and the Geography of Difference*. Blackwell Publishers, Oxford, UK.

Hay, A. (1982). *The Chemical Scythe - Lessons of 2,4,5,T and Dioxin*. Plenum Press, New York, NY.

Hayenga, M. (1998). Structural change in the biotech seed and chemical industrial complex. *AgBioForum*, vol. 1, no. 2.

Hayes (1991). http://feql.wsu.edu/entom558/110303ppt.pdf (accessed on 15 September 2003).

Headley, J. C. (1968). Estimating the productivity of agricultural pesticides. *American Journal of Agricultural Economics*, vol. 50, pp. 13-23.

Heissler, M. (1995). New style merging: AgrEvo. *Biotechnology and Development Monitor*, no. 23, pp. 21-2.

Henson, S. and Northern, J. (1998). Economic determinants of food safety controls in supply of retailer own branded products in the United Kingdom. *Agribusiness*, vol. 14, no. 2, pp. 113-26.

Hicks, J. (1992). DDT - friend or foe? *Pesticide News, Journal of the Pesticide Trust*, 17 September 1992.

Higgot, R. and Reich, S. (1998). Globalization and sites of conflict: Towards definition and taxonomy. CSGR working paper no. 01/98. University of Warwick and CSGR and University of Pittsburgh.

Hodgson, G. (1999). *Economics and Utopia: Why the Learning Economy is Not the End of History*. Routledge, London, UK.

Hoechst (1999). Hoechst wants to speed up merger with Rhone-Poulenc. *Hoechst*, press release. Archive.hoechst.com/English/news/99/pm180399.html (accessed on 22 January 2004).

Hoechst (2003a). Archive.hoechst.com/English/index.html (accessed on 21 November 2003).

Hoechst (2003b). Archive.hoechst.com/English/life_sciences/ernaehrung/lw_zahlenfakten.html. (accessed on 24 November 2003).

Hoechst (2004). Research for nutrition in the Hoechst Group. http://www.archive.hoechst.com/english_3er/publikationen/future/ernaehr/art7.html (accessed on 25 March 2004).

Holmes, M. (2001). *Melancholy Consequences: Britain's Long Relationship with Agricultural Chemicals since the mid-Eighteenth Century*. White Horse Press, Cambridge, UK.

Horstmeier, G. D. (1996a). Cataloguing the seed industry. *Farm Journal*, September 1996.

Horstmeier, G. D. (1996b). Strategic bedfellows. *Farm Journal*, October 1996.

Horstmeier, G. D. (1997). What's next for designer seeds? *Farm Journal*, January 1997.

Hough, P. (1998). *The Global Politics of Pesticides. Forging Consensus from Conflicting Interests*. Earthscan Publications Ltd, London, UK.

Howard, P. H. (2009). Visualizing consolidation in the global seed industry: 1996–2008. *Sustainability*, vol. 1, pp. 1266–87. doi:10.3390/su104126.

IFAS (2003). *Institute of Food and Agricultural Science*, University of Florida. flnem.ifas.ufl.edu/history/nematicide his.htmin (accessed on 13 September 2003).

IFPRI (1996). International Food Policy Research Institute. 2020 vision brief 35.

Ignaciuk, A. and Mason-D'Croz, D. (2014). Modelling adaptation to climate change in agriculture. OECD Food, Agriculture and Fisheries Papers, No. 70. OECD Publishing.

Ikerd, J. (1996). Ag industrialization: Who should fund it? *Agweek*, 29 April 1996, p. 5.

Ilbery, B. W. (1992). From Scott to ALURE- and back again. *Land Use Policy*, April 1992.

Ilbery, B. W. and Bowler, I. R. (1998). From agricultural productivism to post-productivism. In: Ilbery, B. (Ed.), *The Geography of Rural Change*. Longman, Harlow, UK, pp. 57–84.

Ilbery, B. W. and Kneafsey, M. (1998). Product and place: Promoting quality products and services in the lagging rural regions of the European Union. *European Urban and Regional Studies*, vol. 5, pp. 329–41.

Ilbery, B. W. and Kneafsey, M. (2000). Producer constructions of quality in regional speciality foods production: A case study from southwest England. *Journal of Rural Studies*, vol. 16, pp. 217–30.

ILO (1999). The ILO programme on occupational safety and health in agriculture. *International Labour Organization*. ilo.org/public/English/protection/safework/agriculture/agrivf01.htm (accessed on 16 December 2003).

Independent (1999). *Independent*, UK, 28 May 1999.

International Council on Human Rights Policy (2002). Beyond voluntarism: Human rights and the developing international legal obligations to companies. Geneva, 2002. www.ichrp.org.

IPCS (2003). Environmental health criteria 104, Principles for the toxicological assessment of pesticide residues in food. *International Programme on Chemical Safety*. http://www.inchem.org/documents/ehc/ehc/ehc104.htm#PartNumber:2 (accessed on 15 April 2003).

Irwin, A., Rothstein, H., Yearley, S. and McCarthy, E. (1997). Regulatory science-Towards a sociological framework. *Futures*, vol. 29, no. 1, pp. 17–31.

ISAAA. (2016). Global status of commercialized biotech/GM crops: 2016. ISAAA brief no. 52. ISAAA, Ithaca, NY.

Ishii-Eiteman, M. (2000). Countering corporate influence at the World Bank: The mobilization of transnational coalitions to promote sustainable community-based development. Paper presented at the workshop *Human Rights and Globalization: When Transnational Civil society Networks Hit the Ground*, University of Santa Cruz, 1–2 December 2000.

Islam, N. (Ed.) (1995). *Population and Food in the Early Twenty-First Century: Meeting Future Food Demand of an Increasing Population*. IFPRI, Washington DC.

Ismael, Y., Beyers, L., Lin, L. and Thirtle, C. (2001). Smallholder adoption and economic impacts of Bt cotton in the Makhathini Flats, South Africa. Paper presented at the *5th ICABR International Conference on Biotechnology, Science and Modern Agriculture: A New Industry at the Dawn of the Century*, Ravello, Italy.

IUCN (1957). *Proclamation of the 5th General Assembly*. IUCN, Gland, Switzerland, pp. 31–85.

IUCN (1967). *Proclamation of the 9th General Assembly*. IUCN, Gland, Switzerland, pp. 199–200.

IUCN, UNEP, WWF (1980). *World Conservation Strategy: Living Resource Conservation for Sustainable Development*, Sections 2–5. IUCN, Gland, Switzerland.

IUCN (The World Conservation Union), UNEP (United Nations Environment Program) and WWF (World Wildlife Fund) (1991). *Caring for the Earth: A Strategy for Sustainable Living*. IUCN, UNEP and WWF, Gland, Switzerland. http://www.iucn.org/about/index.htm

James, C. (1998). Global status of transgenic crops. ISAAA brief no. 15. ISAAA, Ithaca, NY.

James, C. (2000). Global status of commercialised transgenic crops: 2000. ISAAA brief no. 21. ISAAA, Ithaca, NY.

James, C. (2001). Global review of commercialised transgenic crops: 2001. ISAAA brief no. 24. ISAAA, Ithaca, NY. http://www.isaaa.org/publications/briefs/brief_24.htm.

James, C. (2015). Special issue on agri-biotech studies from policy and regulatory perspectives: Preface. *AgBioForum*, vol. 18, no. 1, pp. 1–2.

JEN (1999). *Japan Economic Newswire*, 27 September 1999.

Jeyaratnam, J. (1987). Survey of acute pesticide poisoning among agricultural workers in four Asian countries. *Toxicology Letters*, vol. 33, pp. 195–201.

Jeyaratnam, J., de Alwis Seneviratne, R. S. and Copplestone, J. F. (1982). Survey of pesticide poisoning in Sri Lanka. *Bulletin of WHO*, vol. 60, no. 4, pp. 615–9.

Jones, M. and MacLeod, G. (1999). Towards a regional renaissance? Reconfiguring and rescaling England's economic governance. *Transactions of the Institute of British Geographers*, vol. 24, pp. 295–313.

Kalaitzandonakes, N. and Hayenga, M. (2000). Structural change in the biotechnology and seed industry complex: Theory and evidence. In: Lesser, W. H. (Ed.), *Transitions in Agbiotech: Economics of Strategy and Policy*. Food Marketing Policy Center and University of Connecticut, Storrs, CT.

Kempf (1999). Catunesco.upc.es/revista/numero2/season.htm (accessed on 24 October 2003).

Keohane, R. (1984). *After Hegemony. Cooperation and Discord in the World Political Economy*. Princeton University Press, Princeton, NJ.

King, J. L., Wilson, N. and Naseem, A. (2002). A Tale of two mergers: What we can learn from agricultural biotechnology event studies. *AgBioForum*, vol. 5, no. 1, pp. 14–19.

Kloppenburg, J. (1984). The social impacts of biogenetic technology in agriculture: Past and future. In: Berardi, G. M. and Geisler, C. C. (Eds), *The Social Consequences and Challenges of New Agricultural Technologies*. Westview Press, Boulder, CO.

Knight, D. (1997). Who benefits, who suffers? The global politics of pesticide use in Brazil. *Zmagazine*. http://zena.secureforum.com/znet/zmag/aricles/jan97knight.htm.

Kolata, G. (1997). Study discounts DDT role in breast cancer. *The New York Times*, 10 October 1997, p. A26.

Krauer, A. (1992). Sustainable development – a window of opportunity for progressive companies. *Ciba-Geigy Journal*, vol. 2, pp. 6–11.

Krimsky, S. and Plough, A. (1988). *Environmental Hazards: Communicating Risk as a Social Process*. Auburn House, Dover, MA.

Kuyek, D. (2000). Lords of poison: The pesticide cartel, June 2000. http://www.grain.org/publications/jun00/jun003.htm (accessed on 28 November 2003).

Lahteenmaki, R. and Fletcher, L. (2001). Public biotech 2000 – the numbers. *Nature Biotechnology*, vol. 19, pp. 407–12.

LA Times (1999). *LA Times*, 31 December 1999.

Latour, B. (1987). *Science in Action: How to Follow Scientists and Engineers through Society*. Harvard University Press, Cambridge, MA. ISBN:9780674792913.

Lawrence, F. (2003). Growers' market. *Guardian UK*, 17 May 2003.

Le Monde diplomatique (2001). New Monsanto and GMO Propaganda. http://www.bio-tech_info.net/monsanto_propaganda1.html

Le Monde diplomatique (2004). *Le Monde diplomatique*, December 2004, pp. 13–15. www.bhopal.net

Lee, K. (1989). Food neophobia: Major causes and treatments. *Food Technology* (December), pp. 62-73.

Lerner, I. (2001). Fungicides are key to Dow, Rohm and Haas Ag chem merger. *Chemicals Market Reporter*, 19 March 2001.

Levidow, L. and Bijman, J. (2002). Farm inputs under pressure from the European food industry. *Food Policy*, vol. 27, no. 1, pp. 31–45.

Lipietz, A. (1987). *Mirages and Miracles*. Verso Press, London, UK.

Lipin, S., Kilman, S. and Warren, S. (1999). DuPont agrees to purchase seed firm for $7.7 billion. *Wall Street Journal*, 15 March 1999

Lipschutz, R. (1995). *Global Civil Society and Global Environmental Governance*. Sunny Press, New York, NY.

Longfield, J. (1993). In: Avery, N. (1993). Fears over food quality standards – Pesticide residue levels harmonised to promote trade. *Pesticide News – Journal of The Pesticide Trust*, vol. 20, p. 5.

Lovering, J. (2010). The coming regional crisis (and how to avoid it). *Regional Studies*, vol. 35, no. 4, pp. 349–54. doi:10.1080/00343400124009.

Lowe, P., Murdoch, J., Marsden, T., Munton, R. and Flynn, A. (1993). Regulating the new rural spaces the uneven development of land. *Journal of Rural Studies*, vol. 9, pp. 205–22.

MacEachern, C. (1992). Farmers and conservation: Conflict and accommodation in farming politics. *Journal of Rural Studies*, vol. 8., pp. 159-71.

MacFarlane, R. (2002). Integrating the consumer interest in food safety: The role of science and other factors. *Food Policy*, vol. 27, pp. 65-80.

MAFF (1989). *Environmentally Sensitive Areas*. HMSO, London, UK.

MAFF (1994). *Circular: Nitrate Vulnerable Zones*. HMSO, London, UK.

Magdoff, F., Foster, J. B. and Buttel, F. (1998). Introduction. *Monthly Review*, July/August 1998.

Maintoba Co-operator (1999). *Maintoba Co-operator*, Canada, 10 June 1999.

MarketandMarkets (2016). Biopesticides Market by Type (Bioinsecticides, Biofungicides, Bioherbicides, and Bionematicides), Origin (Beneficial Insects, Microbials, Plant-incorporated Protectants, and Biochemicals), Mode of Application, Formulation, & Crop Type - Global Forecast to 2022. http://www.marketsandmarkets.com/Market-Reports/biopesticides-267.html?gclid=EAIaIQobChMIyc7Y9fvj1gIVDld-Ch3M7AHgEAAYASAAEgJXsvD_BwE

Marsden, T. (1996). Rural geography trend report: The social and political bases of rural restructuring. *Progress in Human Geography*, vol. 20, pp. 246-58.

Marsden, T. K. (1997). Creating space for food: the distinctiveness of recent agrarian development. In: Goodman, D. and Watts, M. (Eds), *Globalizing Food*. Routledge, London, UK, pp. 169-91.

Marsden, T. K. (1998). Agriculture beyond the treadmill? Issues for policy, theory and research practice. *Progress in Human Geography*, vol. 22, no. 2, pp. 265-75.

Marsden, T. K. (1999). Rural futures: The consumption countryside and its regulation. *Sociologia Ruralis*, vol. 39, no. 4, pp. 501-20.

Marsden, T. K. (2000). Food matter and the matter of food: Towards a new food governance? *Sociologia Ruralis*, vol. 40, no. 1, pp. 20-9.

Marsden, T., Murdoch, J., Lowe, P., Munton, R. and Flynn, A. (1993). *Constructing the Countryside*. UCL Press, London, UK.

McAfee, K. (2003). Neoliberalism on the molecular scale. Economic and genetic reductionism in biotechnology battles. *Geoforum*, 34, pp. 203-19.

McBratney, A., Whelan, B. and Ancev, T. (2005). Future directions of precision agriculture. *Precision Agriculture*, vol. 6, pp. 7-23.

McCalla, A. (1993). Agricultural trade liberalization: The ever allusive grail. *American Journal of Agricultural Economics*, vol. 75, 5, pp. 1102-12.

McEwen, F. and Stephenson, G. (1979). *The Use and Significance of Pesticides in the Environment*. John Wiley and Sons, New York, Chichester, Brisbane and Toronto.

McMichael, P. (2005). New directions in the sociology of global development research. *Rural Sociology and Development*, vol. 11, pp. 269-303.

McMichael, P. (2009a). A food regime analysis of the 'world food crisis'. *Agric Hum Values*, vol. 26, pp. 281-95. doi:10.1007/s10460-009-9218-5.

McMichael, P. (2009b). A food regime genealogy. *Journal of Peasant Studies*, vol. 36, no. 1, pp. 139-69.

McMichael, P. (2013). Land grabbing as security mercantilism in international relations. *Globalizations*, vol. 10, no. 1, pp. 47-64. doi:10.1080/14747731.2013.760925.

McMichael, P. (2016). Commentary: Food regime for thought. *The Journal of Peasant Studies*, vol. 43, no. 3, pp. 648-70. doi:10.1080/03066150.2016.1143816

Meadows, D. H., Meadows, D. L., Randers, J. and Behrens W. W. (1972). *Limits to Growth: A Report for the Club of Rome's Project on the Predicament of Mankind*. Earth Island, London, UK.

Meiners, R. E and Morriss, A. P. (2001). Pesticides and property rights. PERC policy series. http://www.perc.org/pdf/ps22.pdf (accessed on 13 October 2003).

Metcalfe, M., McWilliams, B., Hueth, B, Van Steenwyk, R., Sunding, D., Swoboda, A. and Zilberman, D. (2002). The economic importance of organophosphates in California agriculture. University of California Cooperative Extension. California Department of Food and Agriculture.

Meyer, H. (1999). Precise precaution versus sloppy science. *Bulletin of Science, Technology and Society*, vol. 19, no. 2, pp. 91–5.

Milius, S. (1988). Birds' eggs started to thin long before DDT. *Science News*, vol. 153, no. 17, p. 261.

Mitsui (2003). mitsuiup.com/cgi-bin/webobjects/fif2s5/f0f.woa/wa/read/f20885d3e3+++/ (accessed on 31 October 2003).

Monbiot (1999). *The Guardian*, 16 December 1999.

Monsanto (2001). www.monsanto.com (accessed on 11 December 2003).

Monsanto (2002). Dow agrosciences and Monsanto reach agreement enabling broader worldwide licensing of biotechnology events. *Monsanto*, press release, 2 October 2002 (accessed on 11 December 2003).

Monsanto (2003a). http://www.monsanto.com/Monsanto/layout/pf.asp?pf (accessed on 10 July 2003).

Monsanto (2003b). Monsanto.com/Monsanto/layout/about_us/timeline/popups/first_moves_bio.asp (accessed on 11 December 2003).

Moore, S. K. (2000). BASF bags Cynamid for $3.8 Billion. *Chemical Week*, 29 March 2000.

Morris, J. and Bate, R. (1999). *Fearing Food: Risk, Health and Environment*. Butterworth-Heinemann, Oxford, UK.

Morris, C. and Young, C. (2000). 'Seed to shelf', 'teat to table', 'barley to beer' and 'womb to tomb': Discourses of food quality and quality assurance schemes in the UK. *Journal of Rural Studies*, vol. 16, pp. 103–15.

Morriss, A. P. (1997). Pesticides and environmental federalism: An empirical and qualitative analysis of § 24(c) registrations. In: Anderson, T. L. and Hill, P. J. (Eds), *Environmental Federalism*. Rowman & Littlefield, Lanham, MD, pp. 133–74.

Mortimer, J. (1721). *The Whole Art of Husbandry*, vol. 1, 5th ed. George Grierson, Dublin, Ireland.

Moukheiber, Z. (1998). A hail of silver bullets. *Forbes*, 26 January 1998.

Murray, D. L. and Taylor, P. L. (2000). Claim no easy victories: Evaluating the pesticide industry's global safe use campaign. *World Development*, vol. 28, no. 10, pp. 1735–49.

NAS (1970). *The Life Sciences*. National Academy of Science Press, Washington DC.

NAS (1980). *Regulating Pesticides*. National Academy of Science Press, Washington DC, 192p.

Nature (2016). Fast-spreading genetic mutations pose ecological risk. *Nature*, 08 June 2016.

Nelson, G., Rosengrant, M., Palazzo, A., Gray, I., Ingersoll, C., Robertson, R., Tokgoz, S., Zhu, T., Sulser, T., Ringler, C., Msangi, S. and You, L. (2010). *Food Security, Farming, and Climate Change to 2050: Scenarios, Results, Policy Options*. International Food Policy Research Institute, Washington DC.

New Scientist (2002). Much ado about nothing, 18 May. http://www.biotech-info.net/much_ado.html

New York Times (1969). Special report of united nations food and agriculture organization. *New York Times*, 29 November 1969.

New York Times (1998). Hoechst is planning to spin off Celanese. *New York Times*, 18 November 1998.

New York Times (1999a). *New York Times*, 25 September 1999.

New York Times (1999b). *New York Times*, 12 November 1999.

New York Times (2000). *New York Times*, 16 May 2000.

New York Times (2003).

NGIN (2002). Bayer/Monsanto/DuPont. *Norfolk Genetic Information Network (NGIN)*. Ngin.tripod.com/070402a.htm (accessed on 22 January 2004).

NNFA (1999). *National Nutritional Foods Association*, 20 July 1999.

Norris, R. (Ed.) (1982). *Pills, Pesticides and Profits - The International Trade in Toxic Substance*. North River Press, Croton-On-Hudson, NY.

Novartis (2003a). Novartis.com/about_novartis/en/3companies.html (accessed on 14 July 2003).

Novartis (2003b). Novartis.com/about_novartis/en/2companies.shtml (accessed on 14 July 2003).

Novartis (2004). foundation.novartis.com/Novartis_un_global_compact_globalisation_print.htm (accessed on 25 March 2004).

NRC (2000). The future role of pesticides in US agriculture. Committee on *The Future Role of Pesticides in US Agriculture*, National Research Council.

NTTC (2003). *National Technology Transfer Center*. Wheeling College Inc. Nttc.edu/products/guide/seca02.html (accessed on 15 January 2004).

NUS (2003). *National University of Singapore*. http://scholars.nus.edu.sg/landow/victorian/history/hist5.html (accessed on 15 June 2003).

NWS (2003). *National Wildlife Federation*. http://www.nwf.org/about (accessed on 13 July 2003).

OECD (1993). *Agricultural Policies, Markets and Trade: Monitoring and Outlook 1993*. OECD, Paris, France.

OECD (1994). *Agricultural Policies, Markets and Trade: Monitoring and Outlook 1994*. OECD, Paris, France.

OECD-FAO (2014). *OECD-FAO Agricultural Outlook 2014*. OECD Publishing, Paris, France.

Oerke, E. C. (2005). Crop losses to pests. *Journal of Agricultural Science*, vol. 144, pp. 31–43.

Ollinger, M. and Fernandez-Cornejo, J. (1998). Sunk costs and regulation in the U.S. pesticide industry. *International Journal of Industrial Organization*, vol. 16, no. 2, pp. 139–68.

O'Mahoney, P. (Ed.) (1999). *Nature, Risk and Responsibility: Discourses of Biotechnology*. Springer, New York, NY. ISBN:978-0333665886.

Ordish, G. (1976). *The Constant Pest. A Short History of Pests and their Control*. Peter Davies Ltd, London, UK.

O'Riordan, T. (1981). *Environmentalism*. Pion, London, UK.

OTA (1994). *Multinationals and the US Technology Base*. Office of Technology Assessment, Washington DC.

OTA (1995). Biologically based technologies for pest control. Report OTA-ENV-636. US Congress, Office of Technology Assessment, Washington DC.

Otero, G. (2004). Global economy, local politics: Indigenous struggles, citizenship and democracy. *Canadian Journal of Political Science*, vol. 37, no. 2, pp. 325–46.

Ott, S. (1990). Supermarket shoppers' pesticide concerns and willingness to purchase certified pesticide residue-free produce. *Agribusiness: An International Journal*, vol. 6, no. 6, pp. 593–602.

Paarlberg, S. (1993). Managing pesticide use in developing countries. In: Haas, P. M., Keohane, R. O. and Levy, M. A. (Eds), *Institutions for the Earth: Sources of Effective Environmental Protection*. MIT Press, Cambridge, MA.

Padel, S., Lampkin, N. and Foster, C. (1999). Influence of policy support on the development of organic farming in the European Union. *International Planning Studies*, vol. 4, no. 3, pp. 303–16.

Pampana, E. and Russell, P. (1955). *Malaria. A World Problem, Chronicle of the WHO 9*. WHO, Geneva.

PAN (1987). *Summary Position Paper for the 24th FAO Conference*, Rome, 7–26 November 1987.

PAN (1994). Newsletter. http://www.panna.org (accessed on 11 November 2003).

PAN (1998a). Global enterprises dominate commercial agriculture. *PAN*, 9 February 1998. http://www.panna.org (accessed on 11 November 2003).

PAN (1998b). Prior informed consent. Pesticide management notes no. 5. November 1998.

PAN (1999). PAN Asia and Pacific, 21 June 1999. http://www.panna.org (accessed on 11 November 2003).

PAN (2002). *PAN*, 28 October 2002. http://www.panna.org (accessed on 11 November 2003).

PAN (2003a). Pesticide Action Network. PAN-fact sheet-toxic-trail. http://www.panna.org (accessed on 11 November 2003).

PAN (2003b). *PAN*, 7 July 2003. http://www.panna.org (accessed on 11 November 2003).

Pangae (2012). Biopesticides: The next crop of cleantech home runs. Wednesday, 24 October 2012. http://www.pangaeaventures.com/blog/biopesticides-the-next-crop-of-cleantech-home-runs

PANNA (1995). http://www.panna.org/resources/pestis/pestis.1995.39.html (accessed on 25 November 2003).

PANNA (2003). Leaping ahead or running in place: Pesticide regulation in the European Union and United States. *Global Pesticide Campaigner*, vol. 13, no. 1.

PAN-UK (2000). http://www.panna.org (accessed on 11 November 2003).

PAN-UK (2001). News release, 12 April 2001. http://www.panna.org (accessed 11 November 2003).

PAN-UK (2002). http://www.panna.org (accessed on 11 November 2003).

PAN-UK (2003a). Review of the global pesticide market. Pan-uk.org/pestnews/pn22/pn22p11.htm (accessed on 19 August 2003).

PAN-UK (2003b). The perils of paraquat. Sales in developing countries. Pan-uk.org/pestnews/pn60/pn60p4.htm (accessed on 25 November 2003).

PANUPS (2003). *PANUPS*, 13 June 2003. http://www.panna.org (accessed on 11 November 2003).

Paul, H., Bucking, E. and Steinbrecher, R. A. (2017). 'New Breeding Techniques' and synthetic biology – genetic engineering by another name. *Ecologist*, 4 April 2017.

Pease, W. S. and California Policy Seminar, University of California, Berkeley and Center for Occupational and Environmental Health (1993). *Preventing Pesticide-related illness in California Agriculture: Strategies and Priorities*. School of Public Health and University of California, Berkeley, CA.

Pechlaner, G. and Otero, G. (2010). The neoliberal food regime: Neoregulation and the new division of labor in North America. *Rural Sociology*, vol. 75, no. 2, pp. 179–208.

Peck, J. and Tickell, A. (2000). Searching for a new institutional fix. In: Amin, A. (Ed.), *Post-Fordism: A Reader*. Blackwell Publishers, Oxford, UK.

Peet, R. (1969). The spatial expansion of commercial agriculture in the 19th Century: A von Thunen interpretation. *Economic Geography*, vol. 45, pp. 283–301.

Perkins, J. H. (1997). *Geopolitics and the Green Revolution: Wheat, Genes, and the Cold War*. Oxford University Press, Oxford, UK.

Perraton, J., Goldblatt, D., Held, D. and McGrew, A. (1997). The globalisation of economic activity. *New Political Economy*, vol. 2, no. 2, pp. 251–78.

Pesticide News Briefs (2003). http://www.pan-uk.org/pesticide-news/

Pesticides and You (2014). Weed and insect resistance caused by genetically engineered crop failure treadmill. *Pesticides and You*, vol. 34, no. 1.

Pesticides and You (2015). Glyphosate causes cancer: Is monsanto ready for the consumer response? *Pesticides and You*, vol. 35, no. 2.

Pesticides News (1998a). Off the treadmill: Cotton IPM in Pakistan. Pesticides news no. 40, June 1998, pp. 12–13. http://www.pan-uk.org/pesticide-news/

Pesticides News (1998b). Pesticides news no. 41, September 1998, pp. 10–11.

Pesticides News (1998c). Prior Informed Consent-a long haul to gain accountability. Pesticide news no. 42, p. 10. http://www.pan-uk.org/pestnews/Issue/pn42/pn42p10.htm

Pesticides News (2000). Merger mania in world agrochemical market. Pesticide news no. 49, September 2000.

Pesticides News (2001). Pesticide news no. 54, December 2001.

Peterson, E. and Harris, R. (1981). *Background Report on the Executive Order on Federal Policy Regarding the Export of Banned or Severely Restricted Substances*. White House, Washington DC.

Pimentel, D. (1983). Environmental aspects of pest management. In: Schemilt, L. (Ed.), *Chemistry and World Food Supplies: The New Frontiers*. CHEMRAWN II, Pergamon Press, Oxford, UK.

Pimentel, D. (1986). Agroecology and economics. In: Kogan, M. (Ed.), *Ecological Theory and Integrated Pest Management Practice*. John Wiley and Sons, New York, NY.

Pimentel, D. and Pimentel, M. (1979). *Food, Energy and Society*. Edward Arnold, London, UK.

Pimentel, D., Acquay, H., Biltonen, M., Rice, p., Silva, M., Nelson, J., Lipner, V., Giodano, S., Horowitz, A. and D'Amore, M. (1993). Assessment of environmental and economic impacts of pesticide use. In: Pimental, D. and Lehman, H. (Eds), *The Pesticide Question: Environment, Economics, and Ethics*. Chapman & Hall, New York, NY, pp. 47–84.

Pimentel, D., Harvey, C., Resosudarmo, P., Sinclair, K., Kurz, D., McNair, M., Crist, S., Shpritz, L., Fitton, L., Saffouri, R. and Blair, R. (1995). Environmental and economic costs of soil erosion and conservation benefits. *Science*, vol. 267, pp. 1117–23.

Pimm, S. I., Russell, G. J., Gittelman, J. L. and Brooks, T. M. (1995). The future of biodiversity. *Science*, vol. 269, pp. 347–50.

Pingali, P. L. and Rogers, P. A. (Eds) (1995). *Impacts of Pesticides on Farmer Health and the Rice Environment*. International Rice Research Institute (IRRI), Philippines.

Pingali, P. L., Hossain, M. and Gerpacio, R. V. (1997). *Asian Rice bowls: The Returning Crisis?* International Rice Research Institute (IRRI), Los Baños, Laguna, Philippines and CAB International, Wallingford, UK.

Piore, M. J. and Sabel, C. F. 1984. *The Second Industrial Divide*. Basic books, New York, NY, 354pp.

Planet Ark (2003a). Planetark.org/avantgo/dailynewsstory.cfm?newsid=19151 (accessed on 07 November 2003).

Planet Ark (2003b). Planetark.org/avantgo/dailynewsstory.cfm?newsid=17959 (accessed on 07 November 2003).

Planet Ark (2003c). Planetark.org/avantgo/dailynewsstory.cfm?newsid=19070

Planet Ark (2003d). Planetark.org/avantgo/dailynewsstory.cfm?newsid=19159

Plunkett, D. L. (1991). Saving lives through agricultural research. Issues in agriculture no. 1. Consultative Group on International Agricultural Research, Washington DC.

Popp, J. (2011). Cost-benefit analysis of crop protection measures. *Journal of Consumer Protection and Food Safety*, vol. 6, no. 1, pp. 105–12. doi:10.1007/s00003-011-0677-4.

Potter, C. (1988). Environmentally sensitive areas in England and Wales. *Land Use Policy*, vol. 5, pp. 301–13.

Pray, C., Paarlberg, R. and Unnevehr, L. (2007). Patterns of political response to biofortified varieties of crops produced with different breeding techniques and agronomic traits. *AgBioForum*, vol. 10, no. 3, pp. 135–43.

PRB (2013). 2013 world population data sheet. Population reference bureau. Available at: www. prb.org/Publications/Datasheets/2013/2013-world-population-data-sheet. aspx (accessed on 20 September 2013).

Pretty, J. N. (1995). *Regenerating Agriculture: Policies and Practice for Sustainability and Self-Reliance*. Earthscan, London, UK.

Pritchard, B. (2009). The long hangover from the second food regime: A world-historical interpretation of the collapse of the WTO Doha Round. *Agricultural and Human Values*, vol. 26, no. 4, pp. 297–307.

PR Newswire (2002). *PR Newswire*, 30 May 2002.

PR Newswire (2016) *PR Newswire*, 5 September 2016.

Pruzin, D. (2000). Global treaty to reduce, eliminate persistent organic pollutants finalized. *Chemical Regulation Reporter*, vol. 24, no. 49.

Purcell, J. P. and Perlak, F. J. (2004). Global impact of insect-resistant (Bt) cotton. *AgBioForum*, vol. 7, no. 1 and 2, pp. 27–30.

RAFI (1998a). *RAFI Communique*, 9 February 1998. http://www.rafi.org (accessed on 11 November 2003).

RAFI (1998b). Out of control. *RAFI Communique*, 28 October 1998. http://www.rafi.org (accessed on 11 November 2003).

RAFI (1999). *RAFI Communique*, 28 June 1999. http://www.rafi.org (accessed 11 November 2003).

RAFI (2000a). *RAFI Communique*, February-March 2000. http://www.rafi.org (accessed on 11 November 2003).

RAFI (2000b). Decoding the Clinton/Blair announcement, 26 March 2000. www.rafi.org (accessed on 11 November 2003).

Raghavan, C. (1996). TNCs control two-thirds of world economy. *Third World Network Features*, 29 January 1996. http://hartford-hwp.com/archives/25/007.html (accessed on 13 August 2017).

Redclift, M. (1989). *Sustainable Development: Exploring the Contradictions*. Routledge, London, UK.

Repetto, R. (1985). *Paying the Price: Pesticide Subsidies in the Developing Countries*. World Resources Institute, Washington DC.

Reuters (1999a). *Reuters*, 8 March 1999. https://www.reuters.com

Reuters (1999b). *Reuters*, 17 June 1999. https://www.reuters.com

Reuters (1999c). *Reuters*, 27 July 1999. https://www.reuters.com

Reuters (1999d). *Reuters*, 13 September 1999. https://www.reuters.com

Reuters (1999e). *Reuters*, 7 October 1999. https://www.reuters.com

Reuters (1999f). *Reuters*, 28 October 1999. https://www.reuters.com

Reuters (1999g). *Reuters*, 16 November 1999. https://www.reuters.com

Reuters (1999h). *Reuters*, 21 December 1999. https://www.reuters.com

Reuters (2001). Money matters to U.S. farmers as they embrace GMOs. *Reuters*, 6 November 2001. http://www.biotech-info.net/money_matters.html.

RICE (2016). *RICE CRP Proposal*. IRRI, Las Banjos, Philippines.

Roberts, M. and Begley, R. (1994). Pesticide producers innovate under recessionary pressure. *Chemical Week*, Vol. 155, no. 9.

Rogers, W. (1970). The persistent problem of the persistent pesticides: A lesson on environmental law. *Columbia Law Review*, vol. 70, p. 567.

Rogers, W. (1994). *Environmental Law*, 2nd ed. West, St Paul, MN.

Rohmhaas (2003). Rohmhaas.com/company/history.html (accessed on 15 May 2003).

Rola, A. C. and Pingali, P. L. 1993. *Pesticides, Rice Productivity and farmers' Health: An Economic Assessment.* International Rice Research Institute, Los Baños, Philippines and The World Resources Institute, Washington DC.

Romer, P. M. (1990). Endogenous technological change. *Journal of Political Economy*, vol. 98, no. 5, pp. 71-102.

Rosenblum, G. (1994). *On the Way to Market: Roadblocks to Reducing Pesticide Use on Produce*. Public Voice for Food and Health Policy, Washington DC.

Rosenthal, E. (2002). Free trade and pesticides in Central America. Panna.org/resources/gpc/gpc_200212.12.3.03.dv.html (accessed on 17 February 2004).

Rothamsted (2003). http://www.rothamsted.bbsrc.ac.uk/res/resorigins.htm (accessed on 17 October 2003).

Rotterdam Convention (2004). http://www.pic.int/en/ViewPage.asp?id=345 (accessed on 13 February 2004).

Rotterdam Convention (2017). http://www.pic.int/TheConvention/Chemicals/AnnexIIIChemicals/

Salais, R. and Storper, M. (1992). The four worlds of contemporary industry. *Cambridge Journal of Economics*, vol. 16, pp. 169-9.

Sansom, C. (1998). Unravelling the human genome. *Scrip Magazine*, September 1998.

Sarre, P. and Blunden, J. (Eds). (1995). *An Overcrowded World? Population, Resources and the Environment*. Oxford University Press, Oxford, UK.

Sauer, M. (1990). Fordist modernization of German agriculture and the future of family farms. *Sociologia Ruralis*, vol. 30, pp. 260-79.

Schierow, L. (2004). *Federal Insecticide, Fungicide, and Rodenticide Act*. National Council for Science and the Environment, Washington DC. ncseonline.org/nle/crsreports/briefingbooks/laws/1.cfm (accessed on 15 January 2004).

Schonfield, A., Anderson, W. and Moore, M. (1995). PAN's dirty dozen campaign. *Global Pesticide Campaigner*, vol. 5, no.3, September 1995. panna.org/resources/pestis/pestis.1995.36.html

Schweiger, T. (1999a). After Unilever: Nestle GE -free! Gene.ch/genet/1999/Apr/msg00013.html (accessed on 05 April 2003).

Schweiger, T. (1999b). GP press release: Tesco goes GMO-free! Gene.ch/genet/1999/Apr/msg00015.html (accessed on 05 April 2003).

Scott, M. L., Zimmerman, J. R., Marinsky, S. and Mullenhoff, P. A. (1975). Effects of PCBs, DDT and mercury compounds upon egg production, hatchability and shell quality in chickens and Japanese quail. *Poultry Science*, vol. 50, pp. 656–9.

Sehgal, S. (1996). IPR driven restructuring of the seed industry. Biotechnology and development monitor no. 29, December 1996, p. 1821.

Selman, P. (2000). *Environmental Planning*, 2nd ed. Sage publications, London, UK.

Senate committee on agriculture, nutrition, and forestry (1988). Report on the FIFRA Reform Act of 1988, S. Rep. No. 100-346, at 7 (1988) in lawstudy. law.ukans.edu/glicks/envprot/7-FIFRA.htm

Sexton, J. (2000). History of the fruit industry in Mesa County. http://www.colostate.edu/programs/wcrc/rc_werefruit.html (accessed on 20 August 2003).

Sheingate, A. D. (2001). The Rise of the Agricultural Welfare State. Princeton, NJ: Princeton University Press (2001), xii, p. 279.

Shiferaw, B., Prasanna, B. M., Hellin, J. and Bänziger, M. (2011). Crops that feed the world 6. Past successes and future challenges to the role played by maize in global food security. *Food Security*, vol. 3, pp. 307–27.

Shucksmith, M. (1993). Farm household behaviour and the transition to post-productivism. *Journal of Agricultural Economics*, vol. 44, pp. 466–78.

Shy, O. (1996). *Industrial Organization: Theory and applications*. The MIT Press, Cambridge, MA.

Siedenburg, K. (1991). Demise of the drins. *GPC*, January 1991.

Smith, C. (2001). Pesticide exports from U.S. ports, 1997–2000. *Global Pesticide Campaigner*, vol. 11, no. 3.

Smith, C. and Root, D. (1999). The export of pesticides: Shipments from U.S. ports, 1995–1996. *International Journal of Occupational and Environmental Health*, vol. 5, no. 2.

Søgaard, V. (2001). Bayer AG – Chemicals and life sciences. *AgBioForum*, vol. 4, no. 1, pp. 68–73.

South West Farm Press (2003). Southwestfarmpress.com/magazinearticle.asp?magazinearticleid=114596&magazineid=19&sitel (accessed on 23 July 2003).

Speed, A. (1659). *Adam out of Eden*. London, UK, p. 445.

Steinfeld, J. L. (1971). Surgeon general testimony before EPA. 9 September 1971.

Steinrucken, H. C. M. and Hermann, D. (2000). Speeding the search for crops chemicals. *Chemical & Industry*, no. 7, 246–9.

Stipp, D. (2000). *Fortune*, vol. 141, no. 4, 21 February 2000.

Stockholm Convention (2003). http://www.pops.int/documents/press/pr2-04SC.pdf

Stockholm Convention (2017). http://chm.pops.int/TheConvention/ThePOPs/AllPOPs/

Storper, M. (1992). The limits of the globalization: Technology districts and international trade. *Economic Geography*, vol. 68, pp. 60–93.

Storper, M. (1997). *The Regional World: Territorial Development in a Global Economy*. Guilford Press, London, UK.

Storper, M. and Salais, R. (1997). *Worlds of Production: The Action Frameworks of the Economy*. Harvard University Press, Cambridge, MA.

Strube, J. F. (2000). BASF banks on plant biotechnology. http://www.basf.de/news/presse.

Successful Farming (2000). BASF will buy Cynamid's crop protection business. *Successful Farming*, 20 April 2000.

Sutton, J. (1998). *Technology and Market Structure*. The MIT Press, Cambridge, MA.

Sweeney, E. M. (1972). EPA Hearing Examiner's recommendations and findings concerning DDT hearings. 25 April 1972 (40 CFR 164.32).

Swezey, S. L., Murray, D. L. and Daxl, R. G. (1986). Nicaragua's revolution in pesticide policy. *Environment*, vol. 28, no.1, pp. 6-36.

Swinbank, A. (1992). A surplus of farm land? *Land Use Policy*, vol. 9, pp. 3-7.

Symes, D. and Marsden, T. K. (1985). Industrialisation of agriculture: Intensive livestock farming in Humberside. In: Healey, M. J. and Ilberry, B. W. (Eds), *The Industrialization of the Countryside*. Geo Books, Norwich, UK, pp. 99-120.

Syngenta (1997). Novartis acquires Mercks crop protection business deal valued at USD 910 million. Syngenta.com/en/media/article.aspx?article_id=34 (accessed on 22 January 2004).

Syngenta (1998). Novartis to become a crop protection leader in South Korea by acquiring business from Oriental Chemical Industries. Syngenta.com/en/media/ (accessed on 21 January 2004).

Syngenta (2003). http://www.syngenta.com/en/about_syngenta/time_prin.aspx (acce ssed 18 August 2003).

Syngenta (2004). syngentafoundation.com/article_biotechnology_food_security.htm (accessed on 17 February 2004).

Tait, J. (2001). Zeneca agrochemicals. *AgBioForum*, vol. 4, no. 1, pp. 63-7.

Tait, J. and Chataway, J. (2003). Risk and uncertainty in genetically modified crop development: The industry perspective. INNOGEN working paper 1, June 2003.

Tansey, G. and Worsely, T. (1995). *The Food System: A Guide*. Earthscan, London, UK.

Tarjan, R. and Kemeny, T. (1969). Multi-generation studies on DDT in mice. *Food Cosmet and Toxicol*, vol. 7, pp. 214-22.

Tebtebba Foundation (1999). *Tebtebba Foundation*, Philippines, 13 September 1999.

Teng, P. S. (1990). Integrated pest management in rice: An analysis of the status quo with recommendations for action. Report submitted to the Multi-Agency Integrated Pest Management Task Force.

The Evening News (1999). *The Evening News (New Glasgow)*, Canada, 1 October 1999.

The Gazette (1999). *The Gazette (Canada)*, 29 December 1999.

The Halifax Daily News (1999). *The Halifax Daily News*, Canada, 28 September 1999.

The Hindustan Times (2000).

The Lancet (1981). *The Lancet*, 8 August 1981.

The Ottawa Citizen (2003). *The Ottawa Citizen*, 3 July 2003.

Thevenot, L. (1998). Innovating in 'qualified' markets: Quality, norms and conventions. Paper presented at *Systems and Trajectories of Innovation Conference*, Institute of International Studies, University of California, Berkley, CA, April 1998.

Thompson, E. P. (1971). The moral economy of the English crowd in the 18th century. *Past & Present*, vol. 50, pp. 76-136.

Thompson Financial Services (2001). M&A in 2000: Fast start, fading finale. 3 January 2001. www.tfsd.com.

Thrift, N. and Olds, K. (1996). Refiguring the economic in economic geography. *Progress in Human Geography*, vol. 20, pp. 311-37.

Thrupp, L. A. (1995). *Bittersweet Harvest for Global Supermarkets*. World Resources Institute, Washington DC.

Tickell, O. (2016). New GMOs are 'not GM' – EU folds under US pressure. *Ecologist*, 21 April 2016.

Timsina, J., Buresh, R. J., Dobermann, A. and Dixon, J. (2011). *Rice-maize Systems in Asia: Current Situation and Potential*. International Rice Research Institute and International Maize and Wheat Improvement Center, Los Banōs (Philippines), 232p.

Tombs, S. (1993). The chemical industry and environmental issues. In: Smith, D. (Ed.), *Business and the Environment: Implications of the New Environmentalism*. Paul Chapman, London, UK, pp. 131–49.

Turner, K. (1993). *The Planner*, May 1993.

Turner, R. K., Pearce, D. and Bateman, I. (1994). *Environmental Economics: An Elementary Introduction*. Harvester Wheatsheaf, Hemel Hempstead, UK, p. 31.

UN (1990). Commodity trade statistics. United Nations statistical paper series D, vol. 38, no. 1-19.

UN (1998). *World Investment Report 1998. Trends and Determinants*. UN, New York and Geneva.

UN (1999). World urbanization prospects: The 1999 revision. http://www.un.org/esa/population/pubsarchive/urbanization/urbanization.pdf (accessed on 16 August 2017).

UN (2000). Sustainable agriculture and rural development. Report of the Secretary General - Addendum - Biotechnology for sustainable agriculture. un.org/documents/ecosoc/cn17/2000/ecn172000-7add2.htm (accessed on 09 December 2003).

UN (2003a). The world at six billion. http://www.un.org/esa/population/publications/sixbillion/sixbilpart1.pdf (accessed on 16 August 2017).

UN (2003b). Sustainable development platform. http://.www.un.org/esa/sustdev/csd/about_csd.htm (accessed on 09 December 2003).

UN (2003c). Multi-stakeholder review of voluntary initiatives and agreements for industry. United Nations Department of Economic and Social Affairs, Division for Sustainable Development. un.org/esa/sustdev/mgroups/viaprofiles_FAO_Pesticides.html (accessed on 09 February 2004).

UN (2017). https://www.unglobalcompact.org/

UNCTAD (2003). Continued upswing of global FDI in 1996 with an increasing participation of developing countries. *UNCTAD*, press release. Unctad.org/templates/webflyer.asp?docID=3305 (accessed on 03 December 2003).

UNDP (2002). http://hdr.undp.org/reports/global/2002/en/indicator/pdf/hdr_2002_table_5.pdf (accessed on 07 August 2003).

UNDP (United Nations Development Program), UNEP (United Nations Environment Program), WB (World Bank) and WRI (World Resources Institute) (2000). *World Resources 2000-2001*. World Resources Institute, Washington DC.

UNEP (United Nations Environment Program) (1992). *Saving Our Planet, Challenges and Hopes, the State of the Environment (1972-92)*. UNEP, Nairobi, Kenya.

UNEP (2002). The hazardous chemicals and wastes conventions. *UNEP*, July 2002 (accessed on 03 August 2003).

UNEP (2003a). http://www.unep.org/Documents/Default.asp?DocumentID=43&ArticleID=3301 (accessed on 03 August 2003).

UNEP (2003b). http://www.unep.org/newdraft/unep/eia/ein/grid/history.htm (accessed on 03 August 2003).

UNEP (2003c). http://www.unep.org/unep/programme/natres/biodiv/irb/unepgds.htm#unepgds01 (accessed on 03 August 2003).

UNEP (2003d). http://www.unep.org/Evaluation/UNEP_Annual_Report_2002.pdf (accessed on 03 August 2003).

UNESCO (2003). Catunesco.upc.es/revista/numero2/season.htm (accessed on 09 January 2004).

UN Habitat (2004). Global trends. www.unhabitat.org/habrdd/global.html (accessed on 05 January 2004).

UNICEF (1999). *Rolling Back Malaria*. UNICEF, New York, NY.

Unilever (2000). Genetically modified organisms. http://www.unilever.com

Unilever (2001). How we manage our impact on the environment. http://www.unilever.com

Union of Concerned Scientists (1999). Union of concerned scientists USA. 11 August 1999.

University of Barcelona (2003). http://bus-web.ad.uab.edu/studentdrive/danthony/Sept%2030/Calgene.pdf (accessed on 05 September 2003).

University of North Colorado (2003). http://geography.unco.edu/department/faculty/DUNN/GIGI%202003/Urban%20Growth%20Les%202%20Sample%20ST.pdf (accessed on 09 October 2003).

USDA (2003). *A History of American Agriculture 1776–1990*. ars.usda.gov/history2text11.htm (accessed on 28 August 2003).

USDA-ERS (2004). Chronological history by decade from the creation of USDA in 1862 until 2000. ers-usda.gov/Briefing/biotechnology/chapter1.htm (accessed on 12 January 2004).

USDA-ERS (2017). Recent trends in GE adoption. https://www.ers.usda.gov/data-products/adoption-of-genetically-engineered-crops-in-the-us/recent-trends-in-ge-adoption.aspx

USDA-ARS (2003). Ars.usda.gov/is/timeline/1940chron.htm?pf=1 (accessed on 12 January 2004).

USDA-ARS (2004a). Chronological history of USDA 1970s. *U.S. Department of Agriculture and Agricultural Research Service*. ars.usda.gov/is/timeline/1970chron.htm?pf=1 (accessed on 12 January 2004).

USDA-ARS (2004b). Chronological history of USDA 1980s. *U.S. Department of Agriculture and Agricultural Research Service*. ars.usda.gov/is/timeline/1980chron.htm?pf=1 (accessed on 12 January 2004).

Valin, H., Sands, R. D., van der Mensbrugghe, D., Nelson, G. C., Ahammad, H., Blanc, E., Bodirsky, B., Fujimori, S., Hasegawa, T., Havlik, P., Heyhoe, E., Kyle, P., Mason-D'Croz, D., Paltsev, S., Rolinski, S., Tabeau, A., van Meijl, H., von Lampe, M. and Willenbockel, D. (2014). The future of food demand: Understanding differences in global economic models. *Agricultural Economics*, vol. 45, pp. 51–67.

Viederman, S. (1997). Multinational corporations are incompatible with sustainable development. Remarks at United Nations Correspondents Association Press Briefing, 23 June 1997.

Vorley, B. (2002). *Sustaining Agriculture: Policy, Governance, and the Future of Family Based Farming*. International Institute for Environment and Development, London, UK.

Vorley, W. and Keeney, D. (Eds) (1998). *Bugs in the System: Redesigning the Pesticide Industry for Sustainable Agriculture*. Earthscan Publications Ltd, London, UK.

Wahlström, B. (2003). *Managing POPs in Sweden*.

Waibel, H. and Fleischer, G. (1998). *Kosten und Nutzen des Chemischen Pflanzenshutzes in der deutschen landwirtschaft aus gesamtwirtschaftilcher*. Sicht Wissenschaftsverlag Vauk, Kiel, Germany.

Wallace, I. (1985). Towards a geography of agribusiness. *Progress in Human Geography*, vol. 9, pp. 401–514.

Wallach, L. M. (1999). International harmonization of social, economic and environmental standards. *Public Citizen*, 1999.

Walth, B. and Pulaski, A. (2003). The history behind new pesticide law. http://www.epa.gov/opppsps1/fqpa/fqpahigh.htm (accessed on 28 August 2003).

Ward, N. (1993). The agricultural treadmill and the rural environment in the post-productivist era. *Sociologia Ruralis*, vol. 33, pp. 348–64.

Ward, N., Marsden, T. K. and Munton, R. (1990). Farm landscape change. Trends in upland and lowland England. *Land Use Policy*, vol. 7, no. 4, pp. 291–302.

Washington Post (1999).

Watson, D. J. (1998). GATT and CAP reform: The redress of European agrarian development. Unpublished PhD, University of Hull, Hull, UK.

Watson, D. J. (2003). Politics and policies. In: *Global Trends Task Force Report*, Background material for CIMMYT Strategic Planning meeting, CIMMYT headquarters, 17–19 February 2003. Internal document.

Watson, D. J. (Ed.) (2017). *Achieving Sustainable Cultivation of Maize*. Burleigh Dodds Science Publishing, Cambridge, UK.

Watts, M. (1995). The agrarian question in the late twentieth century development (or Kautsky reduxe). Paper presented at the *Spatial Dimensions of Development Interaction between Local, National and Global Conference*. University of Copenhagen, 5–10 November.

Watts, M. (1996). Development III: The global agrofood system and late twentieth-century development (or Kautsky *Redux*). *Progress in Human Geography*, vol. 20, pp. 230–45.

Watts, M. and Goodman, D. (1997). Agrarian questions, global appetite, local metabolism: Nature, culture and industry in fin-de-siecle agro-food systems. In: Goodman, D. and Watts, M. (Eds), *Globalizing Food*. Routledge, London, UK, pp. 287-304.

WCED (1987). *World Commission on Environment and Development - Our Common Future*. Oxford University Press, London, UK, p. 383.

Weaver, R. D., Evans, J and Luloff, E. (1992). Pesticide use in tomato production: Consumer concerns and willingness to pay. *Agribusiness: An International Journal*, vol. 8, pp. 131–42.

Webref (2003). http://www.webref.org/agriculture/p/plant_variety_protection_act_of.htm (accessed on 15 April 2003).

Weir, D. (1987). *The Bhopal Syndrome - Pesticides*. Environment and Health and Earthscan, London, UK.

Weir, D. and Shapiro, M. (1981). *Circle of Poison: Pesticides and People in a Hungry World*. Third World Publications for the Institute for Food and Development Policy, San Francisco, CA.

Werden, G. J. and Froeb, L. M. (1998). The entry-inducing effects of horizontal mergers: An exploratory analysis. *Journal of Industrial Economics*, vol. 46, pp. 525–43.

Werner, K. (1999). Variations between U.S., Canadian prices linked to pesticide market size, patents. *Chemical Regulation Reporter*, vol. 23, no. 28, p. 1135.

West, J. (1998). E.I. du Pont de Nemours and Company. Harvard Business School Case Study 699-037, p. 8, 19 November 1998.

Whatmore, S. and L. Thorne (1997). Nourishing networks: Alternative geographies of food. In: Goodman, D. J. and Watts, M. J. (Eds), *Globalizing Food: Agrarian Questions and Global Restructuring*. Routledge, London, UK, pp. 287–304.

WHEAT (2016). *WHEAT CRP Proposal*. CIMMYT, Mexico DF, Mexico.

Wheelwright, E. (1984). Consumers, transnational corporations and the developing world in the 1980s. II: The Pesticide Industry, Australian Consumers Association. Paper presented at *11th World Congress of the International Organisation of Consumer Unions*, Bangkok, Thailand.

WHO (1971). The place of DDT in operations against malaria and other vector-borne disease. Official records no. 190. WHO, Geneva, April 1971, pp. 176–82.

WHO (1973). Safe use of pesticides, 20th report of the WHO Expert Committee on insecticides. Technical report series no. 513. WHO, Geneva.

WHO (1986). *Summary Position Paper for the 24th FAO Conference*, Rome, 7-26 November 1987.

WHO (1990). *Public Health Impact of Pesticides Used in Agriculture*. WHO, Geneva.

WHO (2003a). http://www.who.int/pcs/docs/Classif_Pestic_2000-02.pdf (accessed on 20 November 2003).

WHO (2003b). http://www.who.int/pcs/pcs_about.html (accessed on 20 November 2003).

WHO (2003c). http://www.who.int/pcs/jmpr/what_is_jmpr.html (accessed on 20 November 2003).

WHO (2004). General Information about biotechnology (GM foods). www.who.int/foodsafety/biotech/general/en/print.html (accessed on 20 February 2004).

Whorton, J. (1974). *Before Silent Spring: Pesticides and Public Health in Pre-DDT America*. Princeton University Press, Princeton, NJ, p. 74.

Wield, D. (2001). BASF: AgBio fast follower. *AgBioForum*, vol. 4, no.1.

Wield, D., Chataway, J. and Bolo, M. (2010). Issues in the political economy of agricultural biotechnology. *Journal of Agrarian Change*, vol. 10, no. 3, pp. 342–66.

William, et al. (1993). panna.org/resources/pestis/pestis.1995.36.html

Williams, O. F. (Ed.) (2000). *Global Codes of Conduct: An Idea Whose Time Has Come*. University of Notre Dame Press, Notre Dame, IN.

Williamson, J. G. (1996). Globalization, convergence and history. *Journal of Economic History*, vol. 56, no. 2, pp. 277–306.

Winter, M. (1996). *Rural politics: Policies for Agriculture, Forestry and the Environment*. Routledge, London and New York.

Woomer, P. L. and Swift, M. J. (1992). The biological management of tropical soil fertility. In: Greenland, D. J. and Szabolcs, I. (Eds), *Soil Resilience and Sustainable Land use*. CAB International, Wallingford, UK.

World Bank (1999). *World Development Report 1998/1999*. Oxford University Press, New York, NY.

World Bank (2000). *Entering the 21st Century: World Development Report 1999/2000*. Oxford University Press, New York, NY.

World Bank (2003). http://www.agbioworld.org/biotech_info/topics/agbiotech/worldbank.html (accessed on 11 May 2003).

Wrathall, J. E. (1988). Recent changes in arable crop production in England and Wales. *Land Use Policy*, vol. 5, pp. 219–31.

WRI (1994). *World Resources 1994–95*. World Resources Institute, Washington DC and Oxford University Press, Oxford, UK.

WRI (1998). *World Resources 1998–99*. World Resources Institute, Washington DC and Oxford University Press, Oxford, UK.

WRI (2002). Intensification of agriculture: Chemical inputs. World Resources Institute, Washington DC. Wri.org/wr-98-9/agrichem.htm

WRI (2003). Decisions for the Earth. Wri.org.

WSU (2003a). Washington State University – Food and Environmental Quality Lab. http://feql.wsu.edu/entom558/110303ppt.pdf (accessed on 23 October 2003).

WSU (2003b). *Washington State University*. http://www.wsu.edu/~susdev/WCED87.html (accessed on 23 October 2003).

Zanatta, M. (1999). Brazilian authorities burnt down AgrEvo's HR rice. Gene.ch/genet/1999/Apr/msg00021.html (accessed on 23 July 2003).

Zilberman, D., Schmitz, A., Casterline, G., Lichtenberg, E. and Siebert, J. B. (1991) The economics of pesticide use and regulation. *Science*, vol. 253, pp. 518–22.

ZIPR (1997). Zimbabwe Institute of Permaculture Research survey. Pest management notes no. 5.

Index

CPSIA information can be obtained
at www.ICGtesting.com
Printed in the USA
BVHW05*2130030818
523478BV00003B/8/P

9 781786 762764